PREDICTIVE TOXICOLOGY

PREDICTIVE TOXICOLOGY

edited by
Christoph Helma
University of Freiburg, Germany

Taylor & Francis
Taylor & Francis Group
Boca Raton London New York Singapore

A CRC title, part of the Taylor & Francis imprint, a member of the Taylor & Francis Group, the academic division of T&F Informa plc.

Published in 2005 by
Taylor & Francis Group
6000 Broken Sound Parkway NW
Boca Raton, FL 33487–2742

Printed in the United States of America on acid-free paper
10 9 8 7 6 5 4 3 2 1

International Standard Book Number-10: 0–8247–2397–X (Hardcover)

Library of Congress Cataloging-in-Publication Data

Catalog record is available from the Library of Congress

T&F informa

Taylor & Francis Group
is the Academic Division of T&F Informa plc.

Visit the Taylor & Francis Web site at
http://www.taylorandfrancis.com

Contents

Contributors

Suman Chakravarti Case Western Reserve University, Cleveland, Ohio, U.S.A.

Mark T. D. Cronin School of Pharmacy and Chemistry, John Moores University, Liverpool, U.K.

Bart De Moor ESAT-SCD, K.U. Leuven, Leuven, Belgium

Frank De Smet ESAT-SCD, K.U. Leuven, Leuven, Belgium

Kristof Engelen ESAT-SCD, K.U. Leuven, Leuven, Belgium

Lennart Eriksson Umetrics AB, Umeå, Sweden

Dmitri Filimonov Institute of Biomedical Chemistry of Russian Academy of Medical Sciences, Moscow, Russia

Paolo Frasconi Dipartimento di Sistemi e Informatica, Università degli Studi di Firenze, Firenze, Italy

Wolfgang Guba F. Hoffmann-La Roche Ltd, Pharmaceuticals Division, Basel, Switzerland

Christoph Helma Institute for Computer Science, Universität Freiburg, Georges Köhler Allee, Freiburg, Germany

Julian Ivanov MULTICASE Inc., Beachwood, Ohio, U.S.A.

Erik Johansson Umetrics AB, Umeå, Sweden

Gilles Klopman MULTICASE Inc., Beachwood, Ohio, and Department of Chemistry, Case Western Reserve University, Cleveland, Ohio, U.S.A.

Stefan Kramer Institut für Informatik, Technische Universität München, Garching, München, Germany

David Y. Lai Risk Assessment Division, Office of Pollution Prevention and Toxics, U.S. Environmental Protection Agency, Washington, D.C., U.S.A.

Torbjörn Lundstedt Acurepharma AB and BMC, Uppsala, Sweden

Peter McBurney Department of Computer Science, University of Liverpool, Liverpool, U.K.

Kathleen Marchal ESAT-SCD, K.U. BMC, Leuven, Leuven, Belgium

Simon Parsons Department of Computer and Information Science, Brooklyn College, City University of New York, Brooklyn, New York, U.S.A.

Vladimir Poroikov Institute of Biomedical Chemistry of Russian Academy of Medical Sciences, Moscow, Russia

Herbert S. Rosenkranz Department of Biomedical Sciences, Florida Atlantic University, Boca Raton, Florida, U.S.A.

Roustem Saiakhov MULTICASE Inc., Beachwood, Ohio, U.S.A.

Aleksandr Sedykh Department of Chemistry, Case Western Reserve University, Cleveland, Ohio, U.S.A.

Bhavani P. Thampatty Department of Environmental and Occupational Health, Graduate School of Public Health, University of Pittsburgh, Pittsburgh, Pennsylvania, U.S.A.

Yin-Tak Woo Risk Assessment Division, Office of Pollution Prevention and Toxics, U.S. Environmental Protection Agency, Washington, D.C., U.S.A.

1

A Brief Introduction to Predictive Toxicology

CHRISTOPH HELMA

Institute for Computer Science, Universität Freiburg, Georges Köhler Allee, Freiburg, Germany

1. WHAT IS PREDICTIVE TOXICOLOGY?

The public demand for the protection of human and environmental health has led to the establishment of toxicology as the science of the action of chemicals on biological systems. Toxicological research is focused presently very much on the elucidation of the cellular and molecular mechanisms of toxicity and the application of this knowledge in safety evaluation and risk assessment. This is essentially a *predictive* strategy (Fig. 1): Toxicologists study the action of chemicals in simplified biological systems (e.g., cell cultures, laboratory animals) and try to use these results to predict the potential impact on human or environmental health.

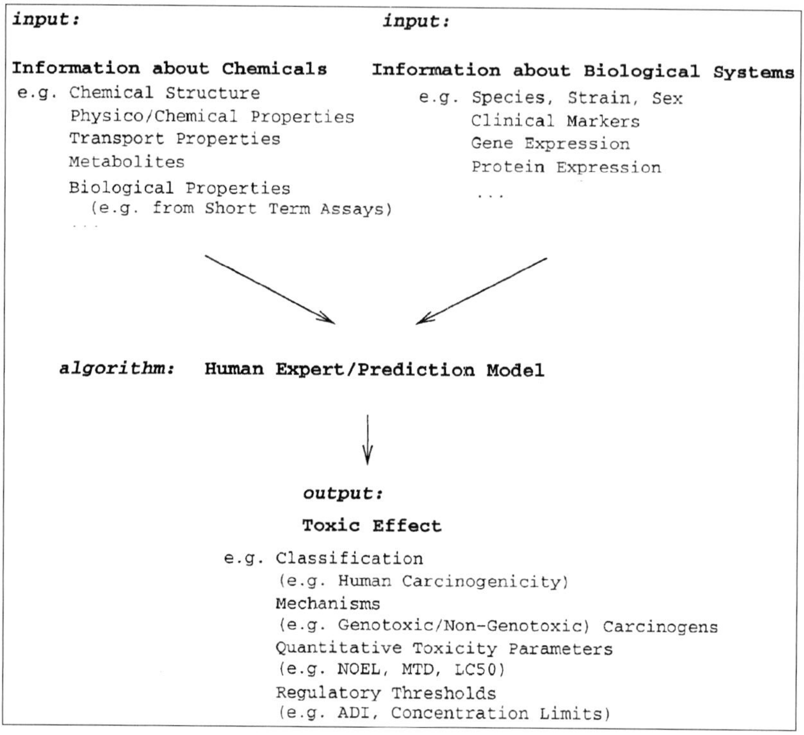

Figure 1 Abstraction of the predictive toxicology process.

Predictive toxicology, as we understand it in this book, does something very similar (Fig. 1): In predictive toxicology, we try to develop procedures (*algorithms* in computer science terms) that are capable to predict toxic effects (the *output)* from chemical and biological information (the *input).*

Figure 1 summarizes also the key ingredients of a predictive toxicology system. First, we need a description of chemicals and biological systems as *input* for predictions. This information is processed by the prediction *algorithm,* to generate a toxicity estimation as *output.* We can also distinguish between *data (input* and *output)* and *algorithms.*

2. INGREDIENTS OF A PREDICTIVE TOXICOLOGY SYSTEM

2.1. Chemical, Biological, and Toxicological Data

Most of the research in predictive toxicology has been devoted to the development of algorithms, but for a good performance, the data aspect is at least equally important. It is in principle possible to use many different types of information to describe chemical and biological systems. The key problem in predictive toxicology is to identify the parameters that are relevant for a particular toxic effect. The situation is relatively easy, if the underlying biochemical mechanisms are well known. In this case, we can determine a rather limited set of parameters, that might be relevant for our purpose. In practice, however, biochemical mechanisms are frequently unknown and/or too complex, to determine a suitable set of parameters a priori. Methods for parameter selection are therefore an important research topic in predictive toxicology.

Toxicity data are needed for two purposes: First of all, we need to *validate* prediction methods, and this can be done by comparing the predictions with realworld measurements. But we can use toxicity data also as *input* to one of the data driven approaches that are capable of generating prediction models automatically from empirical data (Fig. 2). In this case, the quality of the prediction model is largely determined by the quality of the input data.

Despite many possibilities, practical applications have focused on a relatively small set of chemical and biological features. The most popular chemical features are closely related to the chemical structure (e.g., presence/absence of certain substructures) or to properties, that can be calculated from the chemical structure (e.g., physico-chemical properties). As no experimental work is needed to obtain this type of data, the rationale for their choice is obvious, but other substance-related information (e.g., biological activities in screening assays, IR-spectra) can be used as well.

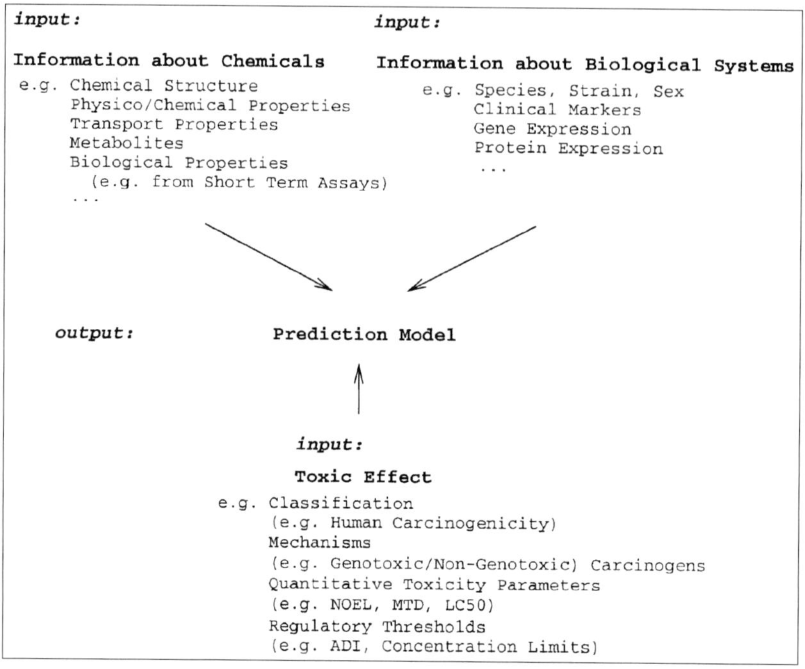

Figure 2 Abstraction of a data driven approach in predictive toxicology.

Up to now information about biological systems has been rarely considered in predictive toxicology. Biological systems have been treated as ensembles of uniform members (e.g., equal individuals), without any biological variance. The explicit consideration of the biological part of the equation will be an interesting research topic of the next years.[a]

Chemical, biological and toxicological data and their representation are the topics of the first section of this book. It contains the chapters *Description and Representation of Chemicals* by Guba (1), *Computational Biology and Toxicogenomics* by Marchal et al. (2), and *Toxicological Information for Use in Predictive Modeling: Quality, Sources, and Databases* by Cronin (3).

[a] The chapter from Marchal et al. (2) provides some examples how to use biological information for predictive purposes.

2.2. Prediction Algorithms

For the prediction algorithm, we have the choice between two strategies. We can try to mimic a human expert by building an *expert system,* or we can try to deduce a prediction model from empirical data by a data-driven approach as in Fig. 2.

The basics of expert systems and some exemplary applications are the topic of Parson and McBurney's chapter, *The Use of Expert Systems for Toxicology Risk Prediction* (4). Two of the programs [*META* (5) and *OncoLogic* (6)] discussed in the section Implementations of Predictive Toxicology Systems are also expert systems.

If we intend to generate a prediction model from experimentally determined toxicity data as in Fig. 2, we have the choice between many different methods. Statistical methods, for example, have been successfully applied in quantitative structureactivity relationships (QSAR) for decades. Eriksson et al.(7) describe statistical techniques in the chapter entitled *Regression- and Projection-Based Approaches in Predictive Toxicology.*

More recently, techniques originating from artificial intelligence research have been used in predictive toxicology. These computer-science oriented developments are summarized in two chapters: *Machine Learning and Data Mining* by Kramer and Helma (8) and *Neural Networks and Kernel Machines for Vector and Structured Data* by Frasconi (9). Three programs of the section Implementations of Predictive Toxicology Systems [*MC4PC* (10), *PASS* (11), *lazar* (12)] use such a data-driven approach.

I want to stress the point that similar predictions can be obtained with a variety of methods. The choice of the method for a particular purpose will depend largely on the scope of the application, present research trends and the personal preferences of the individual researcher.

2.3. Application Areas

The primary aim of predictive toxicology is, of course, the prediction of toxic activities of untested compounds. This enables chemical and pharmaceutical companies, for example, to evaluate potential side effects of candidate structures even without synthesizing them. The same feature is also attractive for governmental authorities that have to deal with compounds with incomplete toxicity information. Present predictive toxicology systems are efficient enough to process thousands if not millions of compounds in a reasonable amount of time. The prioritization of compounds for physical toxicity assays is another important application of predictive toxicology.

Many predictive toxicology systems are capable to provide a rationale for their predictions. If the prediction is based on chemical substructures and/or chemical properties, it is straightforward to use this information for the design of less dangerous, but equally efficient compounds *(computer aided drug design).*

A frequently overlooked but, in my opinion, very important feature of many predictive toxicology systems is their applicability as a tool for the generation and verification of scientific hypotheses. The chapter *Applications of Substructure-Based SAR in Toxicology* by Rosenkranz and Thampatty (13) provides some examples for the creative use of a predictive toxicology system for scientific purposes as well as some mainstream applications.

2.4. Implementations

We can use any statistics and/or data mining package (8) [maybe in conjunction with a computational chemistry package for calculating chemical descriptors (1)] for the prediction of toxic properties, but it is also possible to use a program that has been developed specifically for an application in predictive toxicology.

In the section Implementations of Predictive Toxicology Systems, we describe two expert systems: *OncoLogic: A Mechanism-Based Expert System for Predicting the Carcinogenic Potential of Chemicals* by Woo and Lai (6) and *META: An Expert System for the Prediction of Metabolic Transformations* by Klopman and Sedykh (5), as well as three data driven systems: *MC4PC—An Artificial Intelligence Approach to the Discovery of Quantitative Structure Toxic Activity Relationships (QSTAR)* by Klopman et al. (10), *PASS: Prediction of*

Biological Activity for Substances by Poroikov and Filimonov (10), and *lazar: Lazy Structure–Activity Relationships for Toxicity Prediction* by Helma (12). Two further expert systems (DEREK and StAR) have been presented already in *The Use of Expert Systems for Toxicology Risk Prediction* (4).

3. CONCLUDING REMARKS

Predictive toxicology is clearly an interdisciplinary science with contributions from chemistry, biology, and medicine as well as statistics and computer science. As a deep knowledge of all of these areas is probably beyond the scope of an individual scientist, interdisciplinary collaboration is crucial for the development of a successful predictive toxicology system. This requires, however, that the involved persons have at least a basic understanding of the involved disciplines to allow a successful communication.

This book is an attempt to provide an introduction to the most important techniques and disciplines that are currently involved in predictive toxicology. Although it was written with an interdisciplinary audience in mind, some readers might have trouble understanding some chapters where the author comes from a completely different professional background. For this reason, we have added a *Glossary* and one or two references to *Introductory Literature* as additional material to each chapter. The Glossary provides a definition of the most important terms in the respective chapter[b], and the introductory literature provides suggestions for the scientific background that is needed to understand the corresponding chapter.

It is obvious that a printed book can contain only a limited amount of information and will miss the latest developments after some time. For this reason, we will provide additional material, that is related to the content of this book, on the website www.predictive-toxicology.org. You can con-

[b] Please note that the specfic meaning or usage of the terms may vary from chapter to chapter.

tact me by e-mail at helma@informatik.uni-freiburg.de if you wish to comment on the book or intend to contribute to the website.

REFERENCES

1. Guba W. Description and representation of chemicals. In: Helma C, ed. Predictive Toxicology. New York: Marcel Dekker, 2005.
2. Marchal K, De Smet F, Engelen K, De Moor B. Computational biology and toxicogenomics. In: Helma C, ed. Predictive Toxicology. New York: Marcel Dekker, 2005.
3. Cronin MTD. Toxicological information for use in predictive modeling: quality, sources, and databases. In: Helma C, ed. Predictive Toxicology. New York: Marcel Dekker, 2005.
4. Parsons S, McBurney P. The use of expert systems for toxicology risk prediction. In: Helma C, ed. Predictive Toxicology. New York: Marcel Dekker, 2005.
5. Klopman G, Sedykh A. META: An expert system for the prediction of metabolic transformations. In: Helma C, ed. Predictive Toxicology. New York: Marcel Dekker, 2005.
6. Woo Y, Lai DY. OncoLogic: a mechanism-based expert system for predicting the carcinogenic potential of chemicals. In: Helma C, ed. Predictive Toxicology. New York: Marcel Dekker, 2005.
7. Eriksson L, Johansson E, Lundstedt T. Regression- and projection-based approaches in Predictive Toxicology. In: Helma C, ed. Predictive Toxicology. New York: Marcel Dekker, 2005.
8. Kramer S, Helma C. Machine learning and data mining. In: Helma C, ed. Predictive Toxicology. New York: Marcel Dekker, 2005.
9. Frasconi P. Neural networks and kernel machines for vector and structured data. In: Helma C, ed. Predictive Toxicology. New York: Marcel Dekker, 2005.
10. Klopman G, Ivanov J, Saiakhov R, Chakravarti S. MC4PC—an artificial intelligence approach to the discovery of quantita-

tive structure toxic sactivity relationships (QSTAR). In: Helma C, ed. Predictive Toxicology. New York: Marcel Dekker, 2005.

11. Poroikov VV, Filimonov D. PASS: prediction of biological activity for substance. In: Helma C, ed. Predictive Toxicology. New York: Marcel Dekker, 2005.

12. Helma C. lazer: Lazy structure–activity relationship for toxicity prediction. In: Helma C, ed. Predictive Toxicology. New York: Marcel Dekker, 2005.

13. Rosenkranz HS, Thampathy BP. Application of substructure-based SAR in toxicology. In: Helma C, ed. Predictive Toxicology. New York: Marcel Dekker, 2005.

GLOSSARY

Algorithm: A sequence of actions that accomplishes a particular task. An algorithm operates on the *input* and generates the *output.*

Data driven modeling: A procedure that generates prediction models from experimental data.

Expert system: A prediction system that uses formalized expert knowledge.

Mechanism of toxicity: A model for a toxic effect at the cellular and molecular level.

Quantitative structure–activity relationships (QSAR): Models for the relationship between chemical structures and biological activities.

Toxicology: The science about poisonous effects of chemicals on biological systems.

Validation: The estimation of the accuracy of a prediction method.

2

Description and Representation of Chemicals

WOLFGANG GUBA

F. Hoffmann-La Roche Ltd, Pharmaceuticals Division, Basel, Switzerland

1. INTRODUCTION

Biological effects are mediated by intermolecular interactions, for instance, through the binding of a ligand to a receptor, which triggers a signaling event in a signal transduction pathway. Three-dimensional (3D) structures of receptor–ligand complexes are of great value to rationalize pharmacological or toxicological effects of small molecule ligands. However, due to experimental constraints such as purity and homogeneity of the protein, crystallizability, solubility, size of the protein–ligand complex, etc., an X-ray or NMR-based structure determination is often not feasible. In those cases empirical models have to be developed that deduce biological effects from the 2D or 3D molecular structures of small molecule ligands only, and as a result structure–activity relationships (SAR) are formulated. These SARs may either be qualitative [i.e., molecular features (substructures, functional groups) are associated with activity] or quantitative SARs QSARs (defined by correlating molecular structures with biological effects via mathematical equations). QSAR models require the translation of molecular structures into numerical scales, i.e., molecular descriptors. These descriptors are used by various linear [partial least squares (PLS) (1), etc.] or non-linear [neural networks (2)] regression algorithms to predict biological effects of molecules which have not been tested or not even synthesized.

The core of empirical model building by QSAR is the similarity principle (3), which states that similar chemical structures should have similar biological activities. The converse is not true, since similar biological activities may be displayed by chemically diverse molecules (4). Within the context of QSAR, the similarity principle implies that small changes in molecular structure cause correspondingly slight variations in biological activity, which allows the interpolation of biological activity from a calibration set to a structurally related prediction set of compounds. Thus, molecular descriptors have a pivotal role for quantifying the similarity principle and their usefulness can be ranked by the following criteria:

- relevance for the biological effect to be described
- interpretability
- speed of calculation

The biological relevance of molecular descriptors can be easily checked by the stability and predictivity of the generated mathematical QSAR models and speed is (at least for up to 1000–10,000 compounds in most cases) no longer a limiting factor. Only for virtual screening campaigns with $>10^5$ compounds does this issue need to be taken into consideration. The most critical factor is the interpretability of molecular descriptors, because a clear understanding of the correlation of molecular structures with toxicological effects is crucial for correctly associating structural features with toxic liabilities and for optimizing the biological profile of a compound.

This chapter will describe how molecules are transformed into numerical descriptors. Fragment-based and whole molecule descriptor schemes will be discussed, followed by examples for 1D, 2D, and 3D molecular descriptors. The focus will not be on reviewing algorithms for descriptor generation but rather on illustrating strategies on how to deal with homogenous and diverse sets of molecules and on outlining the scope of commonly used descriptor schemes. For more detailed information about molecular descriptors and algorithms, the reader is referred to the references and to the encyclopedic *Handbook of Molecular Descriptors* (5). The quest for a universal set of descriptors which can be generally applied to structure–activity modeling is ongoing and will probably never succeed. The choice of molecular descriptors is determined by the biological phenomenon to be analyzed and very often experience in descriptor selection is a critical success factor in QSAR model building.

2. FRAGMENT-BASED AND WHOLE MOLECULE DESCRIPTOR SCHEMES

Molecular descriptors are usually classified in terms of dimensionality of the molecular representation (1D, 2D, and 3D) from which they are derived. However, before selecting the dimensionality of the molecular descriptor scheme, the following question needs to be answered. Do the molecules in the dataset contain an invariant substructure, a common scaffold, with one or more substitution sites to which variable building blocks are attached or is there no common substructure?

A QSAR analysis correlates the variation in molecular structures with the variation in biological activities. In the case of an invariant scaffold the obvious strategy is to establish a relationship between the structural variation of the substituents (R groups), the substitution site (R_1, R_2, etc.)

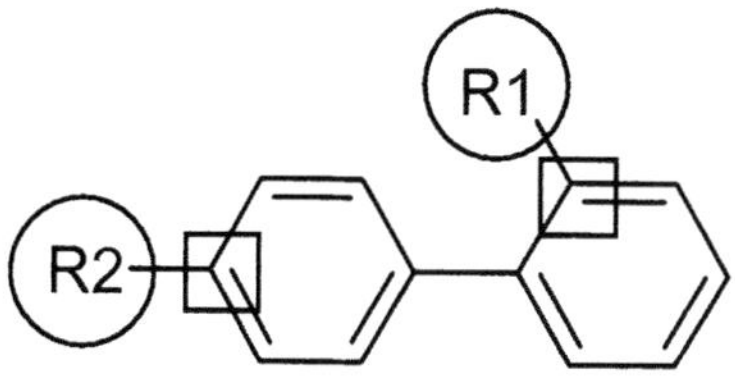

Figure 1 In datasets with a common, invariant scaffold (here, biphenyl) molecular descriptors can be generated both for the variation of substituents R (marked by circles) and for the substitution sites (marked by squares).

and the resulting biological effects. Since the R groups also influence the scaffold (e.g., via electronic effects), another approach would be to compare the effects of the substituent groups onto the common set of scaffold atoms (Fig. 1). Hybrid approaches are also feasible correlating both the variation of building blocks with respect to a substitution site and the modified properties of common scaffold atoms with biological activities.

If no common substructure can be identified, whole molecule descriptors have to be calculated. These heterogeneous datasets are more challenging than series with common scaffolds. It cannot be assumed a priori that each molecule in the dataset interacts with the biological target in the same way, and it is usually not a trivial task to identify those structural features which cause a biological effect. Later it will be illustrated how topological and 3D descriptor schemes attempt to tackle this problem.

3. FRAGMENT DESCRIPTORS

3.1. Homogeneous Dataset with a Common Scaffold

Drugs or toxic agents interact with their macromolecular targets (enzymes, receptors, ion channels, etc.) via hydrophobic, steric, polar, and electrostatic forces. The classical Hansch analysis (6) assigns physicochemical property constants to each substituent, and a correlation between the substituent constants and the biological effect is established. Commonly used fragment values are hydrophobic constants (π), molar refractivity (MR), and the electronic Hammett constant (σ) (7).

The hydrophobic constant π has been derived from the difference of octanol–water partition coefficients of a substituted molecule and the unsubstituted parent compound.

$$\pi_{\text{substituent}} = \log P_{\text{substituted compound}} - \log P_{\text{parent compound}}$$

The octanol–water partition coefficient is defined by

$$\log P = \frac{\text{Concentration of solute in octanol phase}}{\text{Concentration of solute in aqueous phase}}$$

and assumes positive values for lipophilic compounds favoring the octanol phase and negative values for polar molecules with a preference for the aqueous phase. There is a general trend between lipophilicity and toxicity of xenobiotics which is caused by an enrichment in hydrophobic body compartments (membranes, body fat, etc.) and by extensive metabolism leading to reactive species. The octanol–water partition coefficient ($\log P$) is a highly relevant measure and it can be calculated via a battery of atom- and fragment-based methods (8).

Molar refractivity is determined from the Lorentz–Lorenz equation (9) and is a function of the refractive index (n), density (d), and molecular weight (MW):

$$\text{MR} = \frac{n^2 - 1}{n^2 + 2}\frac{\text{MW}}{d}$$

Molar refractivity is related to the molar volume (MW/density), but the refractive index correction also accounts for polarizability. However, molar refractivity does not discriminate between substituents with different shapes (10); e.g., the MR values of butyl and *tert*-butyl are 19.61 and 19.62, respectively. Therefore, Verloop (11) developed the STERIMOL parameters which are based on 3D models of fragments generated with standard bond lengths and angles. Topologically based approaches to derive shape descriptors will be described below.

The Hammett constant is a measure of the electron-withdrawing or electron-donating properties of substituents and was originally derived from the effect of substituents on the ionization of benzoic acids (12,13). Hammett defined the parameter σ as follows with pK_a being the negative decadic logarithm of the ionization constant:

$$\sigma = pK_{a_{\text{benzoic acid}}} - pK_{a_{\text{meta, para-substituted benzoic acid}}}$$

Positive values of σ correspond to electron withdrawal by the substituent from the aromatic ring ($\sigma_{\text{para-nitro}} = 0.78$), whereas negative σ values indicate an electron-donating substituent ($\sigma_{\text{para-methoxy}} = -0.27$). Electronic effects can be categorized into field-inductive and resonance effects. Field-inductive effects consist of two components: 1) σ- or π-bond mediated inductive effect and 2) electrostatic field effect which is transmitted through solvent space. Resonance effects energetically stabilize a molecule by the delocalization of π electrons or by hyperconjugation (delocalization of σ electrons in a π orbital aligned with the σ bond). Swain and Lupton (14) introduced the factoring of σ values into field and resonance effects, which were more consistently redefined by Hansch and Leo (9). A compilation of electronic substituent constants has been published for 530 substituents (15). Although these tabulated electronic substituent constants are of great value, the main drawback is the limited number of available data. Therefore, quantum-chemical calculations to derive electronic whole molecule and fragment descriptors are becoming increasingly common in QSAR/QSPR modeling (16,17).

Finally, indicator variables have to be mentioned as a special case of fragment descriptors. Indicator variables show whether a particular substructure or feature is present ($I = 1$) or absent ($I = 0$) at a given substitution site on a scaffold. Typical applications of indicator variables are the description of ortho effects, cis/trans isomerism, chirality, different parent skeletons, charge state, etc. (6).

The original Hansch analysis (6) correlates the above-mentioned physicochemical descriptors of the substituents of a congeneric series with a biological activity. From a more general perspective, each substitution site (R_1, R_2, etc.) is characterized with respect to principal physicochemical properties (steric bulk, lipophilicity, hydrogen bonding, electronics), and the presence of special structural features is encoded by indicator variables. This descriptor matrix is correlated with the ligand concentration C which causes a biological effect (e.g., EC_{50}: concentration of an agonist that produces 50% of the maximal response). The regression coefficients are often determined by multiple linear regression (MLR). However, MLR assumes that the descriptors are uncorrelated, which is predominantly not the case. Therefore, PLS is recommended as the general statistical engine since it does not suffer from the drawbacks of MLR:

$$-\log C = \underbrace{a_0 + a_1\,\pi + a_2\mathrm{MR} + a_3\,\sigma}_{\mathrm{R}_1} + \cdots + \underbrace{b_0 + b_1\pi + b_2\mathrm{MR} + b_3\,\sigma}_{\mathrm{R}_2} + \cdots \underbrace{+k_1\,l_1+}_{\substack{\text{indicator}\\ \text{variables}}} \cdots$$

In addition, quadratic terms can be introduced to account for the frequently observed non-linear correlation of physicochemical properties, such as $\log P$, with bioactivity. For example, let us assume that the lipophilicity within a compound series is positively correlated with a biological response. Even if an increase in lipophilicity is accompanied by an analogous rise of activity, there will be a lipophilicity optimum beyond which activity drops. Possible causes for this deviation from a linear correlation model are a decreasing solubility in aqueous body fluids or an enrichment in biological membranes or body fat which would reduce the effective concentration of the ligand at the site of action. This can be described by the following parabolic model that defines a $\log P$ optimum beyond which activity is reduced:

$$-\log C = a_0 - a_1(\log P)^2 + a_2 \log P \cdots$$

If a stable and predictive QSAR can be derived, the analysis of the statistical model allows one to determine which substitution site or structural feature has the largest impact on activity and what physicochemical property profile is required for the optimization of activity. However, this strategy is confined to a congeneric series with a common scaffold.

3.2. Heterogeneous Dataset

In heterogeneous datasets no common molecular substructure can be defined. A QSAR model of a molecular descriptor matrix consisting of n rows (molecules) and k descriptor columns requires that the column entries of the descriptor matrix denote the same property for each molecule. For heterogeneous datasets atomic descriptors cannot be compared directly due to the different number of atoms in each molecule.

In the following two sections, van der Waals surface area descriptors and autocorrelation functions will be introduced as a means to allow for QSAR/QSPR modeling of compound sets with no common core structures.

3.2.1. van der Waals Surface Area Descriptors

The Hansch concept of correlating biological effects with principal physicochemical properties of substituents has been extended to whole molecules by Paul Labute (18). In a first step, the van der Waals surface area (VSA) is calculated for each atom of a molecule from a topological connection table with a predefined set of van der Waals radii and ideal bond lengths. Thus, V_i, the contribution of atom i to the VSA of a molecule, is a conformationally independent approximate 3D property which only requires 2D connectivity information. In a second step, steric, lipophilic, and electrostatic properties are calculated for each atom by applying the algorithms of Wildman and Crippen (19) for determining $\log P$ and molar refractivity and assigning the partial charges of Gasteiger and Marsili (20). Each of the three properties is divided into a set of predefined ranges (10 bins for $\log P$, 8 bins for MR, and 14 bins for partial charges) and for each property the atomic VSA contributions in a given property range bin are added up. Thus, the VSA descriptors correspond to a subdivision of the total molecular surface area into surface patches that are assigned to ranges of steric, lipophilic, and electrostatic properties. Summing up, each molecule is transformed into a $10+8+14=32$ dimensional vector. Linear combinations of VSA descriptors correlate well with many widely used descriptors such as connectivity indices, physicochemical, properties, atom counts, polarizability, etc. However, the interpretation of VSA-based QSAR models with respect to proposing chemical modifications for the optimization of compound properties is not straightforward.

3.2.2. Autocorrelation Transforms of 2D Descriptors

As mentioned above, the analysis of heterogeneous datasets requires the comparison of molecules with different numbers of atoms. Descriptor vectors of varying length can be transformed into directly comparable vectors of uniform length by an autocorrelation transform. Moreau and Broto (21,22) were the first to add or average atom pair properties separated by a predefined number of bonds. However, due to the summation or averaging of atom pair properties for a given distance bin, the interpretation of biological effects with respect to individual atom pair properties is no longer possible.

4. TOPOLOGICAL DESCRIPTORS

Topological descriptors are derived entirely from 2D structural formulas and, therefore, missing parameters, conformational flexibility, or molecular alignment do not have to be taken into account. The pros and cons of 2D vs. 3D descriptors will be briefly discussed in the following section. Whereas topological descriptors can be easily calculated from molecular graphs, the interpretation of topological indices with respect to molecular structures is often far from obvious. There is still a highly controversial debate about the utility of topological indices which peaked in provocative statements like "connectivity parameters are artificial parameters that are worthless in real quantitative structure–activity relationships" (23). Nevertheless, the interested reader should develop his/her own opinion and, therefore, the electrotopological state (E-state) indices developed by Kier and Hall (24) will be introduced as one of the more intuitive examples of topological indices.

The general concept of the E-state indices is to characterize each atom of a molecule in terms of its potential for electronic interactions which is influenced by the bound neighboring atoms. Kier and Hall describe the topological environment of an atom by the δ-value, which is defined as the number of adjacent atoms minus the number of bound hydrogens

$$\delta = \sigma - h$$

In other words, the δ-parameter characterizes the number of sigma electrons or bonds around each non-hydrogen atom.

In addition, the valence delta value, δ^v, is introduced as

$$\delta^v = \sigma + \pi + n - h$$

with π being the number of electrons in pi orbitals and n being the number of lone-pair electrons. Thus, the valence delta value δ^v indicates the total number of sigma, pi and lone pair electrons for each atom excluding hydrogen atoms. As an example, an sp^3 hybridized ether oxygen has a δ value of 2 (2 sigma bonds) and δ^v equals 6 (2 sigma and 4 lone pair electrons). An sp^2 hybridized carbonyl oxygen, however, has a δ value of 1 (1 sigma bond) and δ^v equals 6 (1 sigma, 1 pi and 4 lone pair electrons).
From the parameters δ and δ^{v} the term $\delta^v - \delta$ is derived:

$$\delta^v - \delta = \pi + n$$

Thus, the term $\delta^v - \delta$ is the total count of pi and lone-pair electrons for each atom in a molecule. It provides quantitative information about the potential of an atom for intermolecular interactions and, in addition, it is correlated with electronegativity (25).

The intrinsic state I combines the information about the topological environment of an atom and the availability of electrons for intermolecular interactions. This is achieved by multiplying the electronegativity-related term $\delta^v - \delta$ with the accessibility $1/\delta$ (the more sigma bonds, δ, there are around a given atom, the less accessible it is):

$$I = \frac{\delta^v - \delta}{\delta}$$

In other words, the intrinsic state I is the proportion of the total number of pi and lone-pair electrons with respect to the number of sigma bonds for a given atom. Further algebraic modifications (e.g., adding 1 to $\delta^v - \delta$ to discriminate between sp^3 carbon hydrides and scaling with the principal quantum number N to account for the diminished electronegativity of higher than second-row atoms) yield the general definition of the intrinsic state value I as:

$$I = \frac{(2/N)^2 \delta^v + 1}{\delta}$$

Finally, the influence of the molecular environment onto the I-states of each atom within a molecule is determined by summing up pairwise atomic interactions. These interactions represent the perturbation of the I-state of a given atom by the differences in I-states with all the other atoms in the same molecule. Since more distant atoms exert less perturbation than neighboring atoms, these pairwise differences in I-states are scaled by the inverse squared distances between atom pairs. Thus, the electrotopological state S_i or "E-state" of an atom i is defined as

$$S_i = I_i + \Sigma \Delta I_{ij}, \qquad \Delta_{ij} = (I_i - I_j)/r^2{}_{ij}$$

Summing up, the E-state concept quantifies steric and electronic effects. Both the topological accessibility and the electronegativity of an atom as well as the field effect of the local intramolecular environment are captured by a single parameter.

Although both I-states and E-states cannot be translated back to molecular structures directly, they can, nevertheless, be interpreted in terms of electronic and topological features. For instance, I-states for sp^3 carbon atoms decrease from primary to quaternary carbon atoms (2.000–1.250), which reflects the reduced steric accessibility. The sp^3 hybridized terminal groups $-F$, $-OH$, $-NH_2$ and $-CH_3$ have decreasing *I*-states of 8.0, 6.0, 4.0, and 2.0, respectively, which correlates with the corresponding electronegativities. In general, it is found that *E*-states >8 indicate a strong electrostatic and H-bonding effect, values between 3 and 8 represent weak H-bonding and dipolar forces, *E*-states in the range from 1 to 3 are associated with van der Waals and hydrophobic interactions and values below 1 are typical of low electronegativity and topologically buried atoms (24).

Recently, the *E*-state concept has also been extended to hydrogen atoms (26). Since hydrogen atoms always occupy a terminal position, topology has no influence on the *E*-state of a hydrogen atom. The H-bond donor strength is directly proportional to the electronegativity of the attached heavy atom and, therefore, the hydrogen *E*-state is entirely based on the Kier–Hall relative electronegativities (KHE) (25):

$$\mathrm{KHE} = \frac{\delta^v - \delta}{N^2} = \frac{\pi + n}{N^2}$$

Thus, the perturbation term for the calculation of the hydrogen *E*-state is:

$$\Delta HI_{ij} = \frac{\mathrm{KHE}_i - \mathrm{KHE}_j}{r_{ij}^2}$$

with a predefined KHE of -0.2 for hydrogen atoms. Small numerical *E*-state values for polar hydrogen atoms indicate a low electron density on the hydrogen atom and, therefore, a high polarity of the hydrogen bond.

After having discussed the definition and the interpretation of the *E*-state indices, the focus will now shift to setting up the descriptor matrices. For congeneric series with a common scaffold, individual *E*-state (atom-level) values are calculated for topologically equivalent scaffold atoms. Thus, the impact of the substituents onto the electron accessibility of the scaffold atoms is correlated with biological effects, which allows one to identify those atoms in a molecule that are most important for activity. For heterogeneous datasets, an atom-type classification scheme is applied where each atom is assigned to a set of classes which are generally defined by the element, the bond types and the number of bonded hydrogen atoms. The occurrence of each atom type and the sum of E-states for all groups of the same type are recorded.

5. 3D MOLECULAR INTERACTION FIELDS

Molecular recognition occurs in 3D space and, therefore, 3D descriptors seem to be the most natural choice for QSAR studies. However, in order to generate stable and predictive 3D-QSAR models, two hurdles need to be overcome. First, a biophore pattern needs to be identified for aligning the corresponding parts of each molecule, which is a challenging task for heterogeneous datasets. Second, for each molecule in the dataset, the biologically relevant conformer has to be selected, which is very demanding for conformationally flexible molecules with thousands or even millions of low energy conformations. Two approaches will be exemplified that are not dependent on finding a common alignment rule. Both descriptor schemes are derived from GRID (27,28), where the interaction energies between chemical probes and the target molecule(s) are calculated at each single grid point in a 3D cage. These probes represent van der Waals, H-bonding, electrostatic, and hydrophobic properties, and commonly used probes are H_2O (H-bond donor/acceptor), carbonyl oxygen (H-bond acceptor only), amide nitrogen (H-bond donor only), and the hydrophobic DRY probe. Each grid point within this cage is labelled with the interaction energy of the GRID probe with the target molecule, i.e., a molecular field is generated. These molecular interaction fields can either be visualized directly by computing contour maps or be transformed into descriptors as described below.

5.1. VolSurf

The VolSurf (28–30) procedure transforms polar and hydrophobic 3D interaction maps into a quantitative scale by calculating the volume or the surface of the interaction contours. This is illustrated in Fig. 2, where the H_2O and

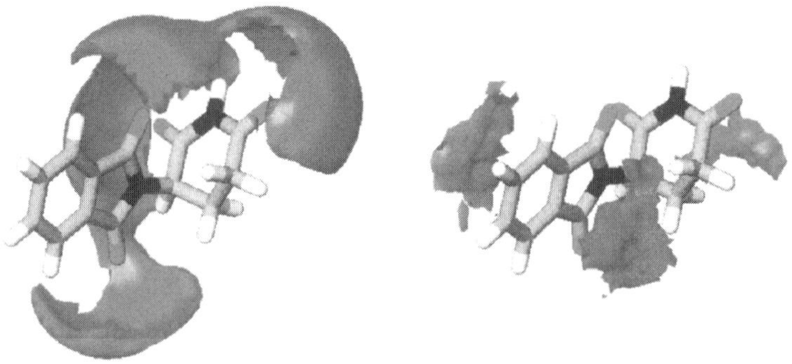

Figure 2 The GRID probes H_2O (left) and DRY (right) have been used to sample energetically favorable polar and hydrophobic interaction sites around thalidomide. These interaction contours are used by VolSurf to generate molecular surface descriptors which are relevant for intermolecular interactions.

hydrophobic DRY probes have been used to characterize polar and hydrophobic interactions sites around thalidomide. Besides molecular volume, surface, and shape, the concentration of polar interactions on the molecular surface and the size of polar and hydrophobic interaction sites are calculated at eight energy levels. The "integy" (interaction energy) moment is defined in analogy to the dipole moment and describes the distance of the center of mass to the barycenter of polar or hydrophobic interaction sites at a given energy level. If the integy moment is high, polar or hydrophobic interaction sites are not evenly distributed around the molecule, but they are concentrated in regions far away from the center of mass. If the integy moment is small, the polar or hydrophobic moieties are either close to the center of mass or they are located symmetrically on the periphery of the molecule. Further descriptors characterize the balance of hydrophilic and hydrophobic regions, membrane packing and the amphiphilic moment. Summing up, the 3D molecular structure is translated into a set of physicochemically meaningful descriptors without the need for alignment. Most of the VolSurf descriptors, with the exception of the integy moment and other descriptors not mentioned here, are only moderately dependent on conformational flexibility. Thus, size, shape, H-bonding, and hydrophobicity can be quantitatively differentiated within a series of molecules and correlated with biological effects. Several successful applications of VolSurf descriptors in the field of QSAR and QSPR have been published (31–33).

5.2. Almond

The GRID molecular interaction fields (MIFs) are also the basis for a novel, alignment-free 3D QSAR methodology called *Almond* (28,34). Whereas in standard 3D-QSAR methods like *Comparative Molecular Field Analysis* (35) (CoMFA) conformational flexibility and multiple possibilities for pharmacophore alignment have to be considered, Almond is not dependent on a molecular alignment, since it operates on pairwise pharmacophoric distances only. For this purpose, one or several low energy conformations for each molecule are characterized with the hydrophobic, amide, and carbonyl GRID probes to sample energetically favorable interaction sites representing potential hydrophobic and H-bond donor/acceptor contacts with a macromolecular target. For each probe, a representative set of interaction sites in 3D space is identified and the product of interaction energies for each pairwise combination of GRID nodes is calculated. In contrast to a conventional correlation transform only the largest product for each distance bin is recorded. This procedure is illustrated in Fig. 3, where energetically favorable interaction sites with a hydrogen bond donor have been sampled around thalidomide. Auto- and cross-correlograms are computed for single probes as well as for combination of probes and are used as descriptor profiles in QSAR studies.

In standard auto- and cross-correlation functions, all products would be retained and averaged for each distance bin, thus, individual pharmacophoric distances would be no longer present. In Almond, however, the selection of maximum products of interaction energies for each distance bin allows to identify those pharmacophoric distances which are relevant for bioactivity and to map them back onto the molecular structures. Thus, Almond is an innovative procedure for rapidly generating relevant 3D descriptors in QSAR models.

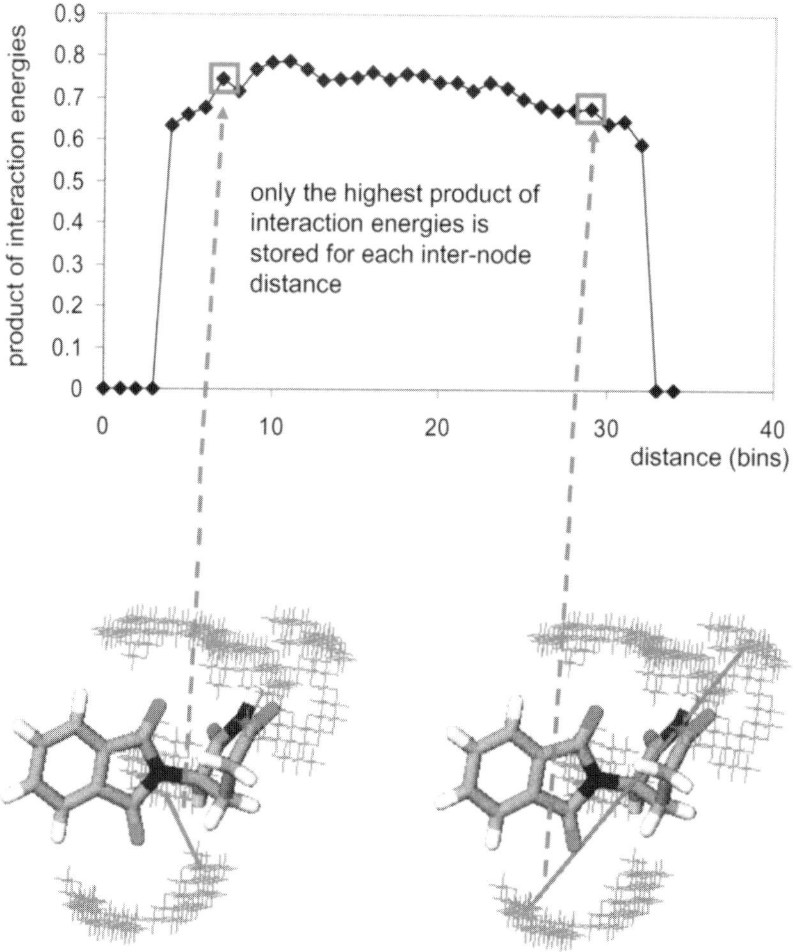

Figure 3 The alignment-independent 3D-QSAR approach *Almond* is illustrated with thalidomide as an example. First, a representative set of energetically favorable interaction sites with a hydrogen bond donor group (amide probe in GRID) is selected. Then, for each inter-node distance, the highest product of interaction energies is stored and plotted in a correlogram (top). The points in the correlogram can be traced back to the molecular structure as it is shown for a small (lower left) and a large (lower right) distance. The correlograms are computed for all compounds in the dataset with uniform length, and a descriptor matrix is generated for QSAR studies.

6. OTHER APPROACHES

The aim of this chapter was to introduce the reader to a representative set of information-rich, intuitive, and readily interpretable descriptor schemes that are applicable to homogeneous as well as heterogeneous datasets. Such a selection is, of course, subjective and, in addition, the interdisciplinary character of this book requires a limit on the presentation of the topic. This chapter will be concluded by a short summary of other important approaches, and apologies are offered for any omissions.

The reactivity of toxic agents concerning redox and both nucleophilic and electrophilic reactions is the realm of quantum-mechanical descriptors. Molecular orbital parameters (HOMO, LUMO energies) and partial charges are frequently encountered as the most relevant parameters in structure–toxicity correlations (36). Depending on the level of theory, the calculation of these parameters may be quite lengthy. However, on a semiempirical level, even large numbers of molecules can be routinely handled on present day computing platforms (17).

Structural fragments (37) or connectivity pathways (38,39) can be transformed into a fingerprint by binary encoding (1/0) the presence or absence of substructures. Alternatively a molecular hologram can be created by counting the occurrences of fragments up to a predefined length (40). A drawback is the huge number of possible fragments, i.e., descriptor columns, resulting in sparse matrices, since each molecule only populates a fraction of such a descriptor matrix. Therefore, either the fingerprints are folded back ("hashing") with each bin assigned to several fragments, or fragment descriptors are clustered into structurally similar groups.

For the characterization of 2D pharmacophoric patterns, atom pairs (41) and topological torsions (42) were introduced by scientists from Lederle. Atom pairs are defined as two atoms and the length of the shortest bond path between them. Typically all fragments with bond paths of 2–20 atoms are recorded and their occurrences are counted. Atoms may be assigned to pharmacophoric classes (H-bond donor or acceptor, hydrophobic, positively or negatively charged, other), and the distribution of pharmacophoric pairs is described with respect to pairwise distances (e.g., three instances of a donor–acceptor pair separated by five bonds). Topological torsion descriptors are the 2D analog of the torsion angle since they are computed from linear sequences of four consecutively bonded heavy atoms. The topological correlation of generalized atom types is also the conceptual basis of the program CATS (43) that generates molecular size-independent pharmacophore descriptors.

Summing up, there is an overwhelming number of molecular descriptors available and it is, of course, tempting to calculate as many descriptors as possible [e.g., with the popular Dragon package (44)] and to let a "black box" program select the most informative variables during model generation. However, it is the responsibility of the QSAR practitioner to gain a thorough understanding of the biological effect to be mathematically modelled and to choose both a relevant descriptor set and a statistical procedure that allow one to identify and to rationally modify molecular features causing toxic liabilities.

ACKNOWLEDGMENT

The author would like to thank Gabriele Cruciani (University of Perugia), Manuel Pastor (IMIM/UPF, Barcelona), Karl-Heinz Baringhaus (Aventis, Frankfurt), Hans Matter (Aventis, Frankfurt), Roy Vaz (Aventis, Bridgewater), Olivier Roche (Roche, Basel), and Harald Mauser (Roche, Basel) for stimulating scientific discussions about molecular descriptors and their applications in QSAR.

REFERENCES

1. Wold S, Sjostrom M, Eriksson L. PLS-regression: a basic tool of chemometrics. Chemometrics Intell Lab Syst 2001; 58: 109–130.

2. Zupan J, Gasteiger J. Neural Networks for Chemists. Weinheim: VCH, 1993.

3. Maggiora GM, Johnson MA, eds. Concepts and Applications of Molecular Similarity. New York: Wiley, 1990.
4. Martin YC, Kofron JL, Traphagen LM. Do structurally similar molecules have similar biological activity? J Med Chem 2002; 45:4350–4358.
5. Todeschini R, Consonni V. Handbook of Molecular Descriptors. Weinheim: Wiley-VCH, 2000.
6. Kubinyi, H. QSAR: Hansch Analysis and Related Approaches. Weinheim: VCH, 1993.
7. Hansch C, Leo A, Hoekman D. Exploring QSAR: Hydrophobic, Electronic and Steric Constants. Washington, DC: ACS, 1995.
8. Mannhold R, van de Waterbeemd H. Substructure and whole molecule approaches for calculating log *P*. J Comput Aided Mol Des 2001; 15:337–354.
9. Hansch C, Leo A. Substituent Constants for Corrrelation Analysis in Chemistry and Biology. New York: Wiley, 1979.
10. Livingstone DJ. The characterization of chemical structures using molecular properties. A survey. J Chem Inf Comput Sci 2000; 40:195–209.
11. Verloop A. The STERIMOL Approach to Drug Design. New York: Marcel Dekker, 1987.
12. Hammett LP. Reaction rates and indicator acidities. Chem Rev 1935; 17:67–79.
13. Hammett LP. The effect of structure upon the reactions of organic compounds Benzene derivatives. J Am Chem Soc 1937; 59:96–103.
14. Swain CG, Lupton EC Jr. Field and resonance components of substituent effects. J Am Chem Soc 1968; 90:4328–4337.
15. Hansch C, Leo A, Taft RW. A survey of Hammett substituent constants and resonance and field parameters. Chem Rev 1991; 91:165–195.
16. Karelson M, Lobanov VS, Katritzky AR. Quantum-chemical descriptors in QSAR/QSPR studies. Chem Rev 1996; 96:1027–1043.

17. Beck B, Horn A, Carpenter JE, Clark T. Enhanced 3D-databases: a fully electrostatic database of AM 1-optimized structures. J Chem Inf Comput Sci 1998; 38: 1214–1217.

18. Labute P. A widely applicable set of descriptors. J Mol Graphics Mod 2000; 18:464–477.

19. Wildman SA, Crippen GM. Prediction of physicochemical parameters by atomic contributions. J Chem Inf Comput Sci 1999; 39:868–873.

20. Gasteiger J, Marsili M. Iterative partial equalization of orbital electronegativity: a rapid access to atomic charges. Tetrahedron 1980; 36:3219–3222.

21. Moreau G, Broto P. The autocorrelation of a topological structure: a new molecular descriptor. Nouv J Chim 1980; 4: 359–360.

22. Moreau G, Broto P. Autocorrelation of molecular structures. Application to SAR studies. Nouv J Chim 1980; 4:757–764.

23. Kubinyi H. The physicochemical significance of topological parameters. A rebuttal. Quant Struct-Act Relat 1995; 14:149–150.

24. Kier LB, Hall LH. Molecular Structure Description. The Electrotopological State. San Diego: Academic Press, 1999.

25. Kier LB, Hall LH. Molecular Connectivity in Structure–Activity Analysis. New York: Wiley, 1986.

26. Hall LH, Vaughn TA. QSAR of phenol toxicity using electrotopological state and kappa shape indices. Med Chem Res 1997; 7:407–416.

27. Goodford PJ. Computational procedure for determining energetically favourable binding sites on biologically important macromolecules. J Med Chem 1985; 28:849–857.

28. Molecular Discovery Ltd, www.moldiscovery.com.

29. Cruciani G, Pastor M, Guba W. VolSurf: a new tool for the pharmacokinetic optimization of lead compounds. Eur J Pharm Sci 2000; 11:S29–S39.

30. Cruciani G, Crivori P, Carrupt PA, Testa B. Molecular fields in quantitative structure–permeation relationships: the VolSurf approach. THEOCHEM 2000; 503:17–30.

31. Crivori P, Cruciani G, Carrupt PA, Testa B. Predicting blood–brain barrier permeation from three-dimensional molecular structure. J Med Chem 2000; 43:2204–2216.

32. Zamora I, Oprea T, Cruciani G, Pastor M, Ungell AL. Surface descriptors for protein–ligand affinity prediction. J Med Chem 2003; 46:25–33.

33. Oprea T, Zamora I, Ungell AL. Pharmacokinetically based mapping device for chemical space navigation. J Comb Chem 2002; 4:258–266.

34. Pastor M, Cruciani G, McLay I, Pickett S, Clementi S. GRid-INdependent Descriptors (GRIND): a novel class of alignment-independent three-dimensional molecular descriptors. J Med Chem 2000; 43:3233–3243.

35. Cramer RD III, Patterson DE, Bunce JD. Comparative molecular field analysis (CoMFA). 1. Effect of shape on binding of steroids to carrier proteins. J Am Chem Soc 1988; 110: 5959–5967.

36. Trohalaki S, Pachter R. Quantum descriptors for predictive toxicology of halogenated aliphatic hydrocarbons. SAR QSAR Environ Res 2003; 14:131–143.

37. MACCS keys: www.mdli.com.

38. UNITY fingerprints: www.tripos.com.

39. Daylight fingerprints: www.daylight.com.

40. Cui S, Wang X, Liu S, Wang L. Predicting toxicity of benzene derivatives by molecular hologram derived quantitative structure–activity relationships (QSARs). SAR QSAR Environ Res 2003; 14:223–231.

41. Carhart RE, Smith DH, Venkataraghavan R. Atom pairs as molecular features in structure–activity studies: definition and applications. J Chem Inf Comput Sci 1985; 25:64–73.

42. Ramaswamy N, Bauman N, Dixon JS, Venkataraghavan R. Topological torsion: a new molecular descriptor for SAR

applications. Comparison with other descriptors. J Chem Inf Comput Sci 1987; 27:82–85.

43. Schneider G, Neidhart W, Giller T, Schmid G. "Scaffold-Hopping" by topological pharmacophore search: a contribution to virtual screening. Angew Chemie Int Ed 1999; 38: 2894–2896.

44. http://www.disat.unimib.it/chm/dragon.htm

ADDITIONAL INTRODUCTORY READING

1. Böhm H-J, Klebe G, Kubinyi H. Wirkstoffdesign. Heidelberg: Spektrum, Akademischer Verlag, 1996.

2. Leach AR, Gillet VJ. An Introduction to Cheminformatics. Dordrecht: Kluwer Academic Publishers, 2003.

GLOSSARY

The following terms have been defined by the IUPAC Medicinal Chemistry Section Committee: van de Waterbeemd H, Carter RE, Grassy G, Kubinyi H, Martin YC, Tute MS, Willett P. Chapter 37. Glossary of Terms Used in Computational Drug Design (IUPAC Recommendations 1997). Annu Rep Med Chem 1998; 33:397–409.

3D-QSAR: 3D quantitative structure–activity relationships (3D-QSAR) involve the analysis of the quantitative relationship between the biological activity of a set of compounds and their 3D properties using statistical correlation methods.

Ab initio calculations: Ab initio calculations are quantum chemical calculations using exact equations with no approximations which involve the whole electronic population of the molecule.

Chemometrics: Chemometrics is the application of statistics to the analysis of chemical data and design of chemical experiments and simulations.

CLOGP: CLOGP values are calculated 1-octanol/ water partition coefficients. Comparative Molecular Field Analysis (CoMFA): CoMFA is a 3D-QSAR method that uses statistical correlation techniques for the analysis of the quantitative relationship between the biological activity of a set of compounds with a specified alignment and their 3D electronic and steric properties.

GRID: GRID is a program which calculates interaction energies between probes and target molecules at interaction points on a 3D grid.

Hammett constant σ: The Hammett constant is an electronic substituent descriptor reflecting the electron-donating or -accepting properties of a substituent.

Hansch analysis: Hansch analysis is the investigation of the quantitative relationship between the biological activity of a series of compounds and their physicochemical substituent or global parameters representing hydrophobic, electronic, steric, and other effects using multiple regression correlation methodology.

Hansch–Fujita π constant: The Hansch–Fujita π constant describes the contribution of a substituent to the lipophilicity of a compound.

Hydrophobicity: Hydrophobicity is the association of non-polar groups or molecules in an aqueous environment which arises from the tendency of water to exclude non-polar molecules.

Indicator variable: An indicator variable is a descriptor that can assume only two values indicating the presence (=1) or absence (=0) of a given condition. It is often used to indicate the absence or presence of a substituent or substructure.

Lipophilicity: Lipophilicity represents the affinity of a molecule or a moiety for a lipophilic environment. It is commonly measured by its distribution behavior in a biphasic system, either liquid–liquid (e.g., partition coefficient in 1-octanol/water) or solid–liquid (retention on reversed-phase high-performance liquid chromatography (RP-HPLC) or thin-layer chromatography (TLC) system.

Molar refractivity (MR): The molar refractivity is the molar volume corrected by the refractive index. It represents size and polarizability of a fragment or molecule.

Molecular connectivity index: A molecular connectivity index is a numeric descriptor derived from molecular topology.

Molecular descriptors: Molecular descriptors are terms that characterize a specific aspect of a molecule.

Molecular topology: Molecular topology is the description of the way in which the atoms of a molecule are bonded together.

Multivariate statistics: Multivariate statistics is a set of statistical tools to analyze data (e.g., chemical and biological) matrices using regression and/or pattern recognition techniques.

Partial least squares (PLS): Partial least squares projection to latent structures (PLS) is a robust multivariate generalized regression method using projections to summarize multitudes of potentially collinear variables.

Pharmacophore (pharmacophoric pattern): A pharmacophore is the ensemble of steric and electronic features that is necessary for the interaction of a ligand with the target receptor and for triggering or blocking a biological response.

Pharmacophoric descriptors: Pharmacophoric descriptors are used to define a pharmacophore, including H-bonding, hydrophobic and electrostatic interaction sites, defined by atoms, ring centers, and virtual points.

Principal components analysis (PCA): Principal components analysis is a data reduction method using mathematical techniques to identify patterns in a data matrix. The main element of this approach consists of the construction of a small set of new orthogonal, i.e., non-correlated, variables derived from a linear combination of the original variables.

Quantitative structure–activity (property) relationships (QSAR/QSPR): Quantitative structure–activity or structure–property relationships (QSAR/QSPR) are mathematical relationships linking chemical structure and pharmacological activity in a quantitative manner for a series of compounds. Methods which can be used in QSAR include various regression and pattern recognition techniques. QSAR is often taken to be equivalent to chemometrics or multivariate statistical data analysis. It is sometimes used in a more limited sense as equivalent to Hansch analysis.

Regression analysis: Regression analysis is the use of statistical methods for modeling a set of dependent variables, *Y*, in terms of combinations of predictors, *X*. It includes methods such as multiple linear regression (MLR) and partial least squares (PLS).

Swain–Lupton parameters (F and R): The Swain and Lupton parameters (F and R) are electronic field and resonance descriptors derived from Hammett constants.

Verloop STERIMOL parameters: The STERIMOL parameters defined by Verloop are a set of substituent length and width parameters.

3

Computational Biology and Toxicogenomics

KATHLEEN MARCHAL, FRANK DE SMET,
KRISTOF ENGELEN, and BART DE MOOR
ESAT-SCD, K.U. Leuven,
Leuven, Belgium

1. INTRODUCTION

Unforeseen toxicity is one of the main reasons for the failure of drug candidates. A reliable screening of drug candidates on toxicological side effects in early stages of the lead component development can help in prioritizing candidates and avoiding the futile use of expensive clinical trials and animal tests. A better understanding of the underlying cause of toxicological and pharmacokinetic responses will be useful to develop such screening procedure (1).

Pioneering studies (such as Refs. 2–5) have demonstrated that observable/classical toxicological endpoints are reflected in systematic changes in expression level. The observed endpoint of a toxicological response can be expected to result from an underlying cellular adaptation at molecular biological level. Until a few years ago studying gene regulation during toxicological processes was limited to the detailed study of a small number of genes. Recently, high-throughput profiling techniques allow us to measure expression at mRNA or protein level of thousands of genes simultaneously in an organism/tissue challenged with a toxicological compound (6). Such global measurements facilitate the observation not only of the effect of a drug on intended targets (on-target), but also of side effects on untoward targets (off-target) (7). Toxicogenomics is the novel discipline that studies such large scale measurement of gene/protein expression changes that result from the exposure to xenobiotics or that are associated with the subsequent development of adverse health effects (8,9). Although toxicogenomics covers a larger field, in this chapter we will restrict ourselves to the use of DNA arrays for mechanistic and predictive toxicology (10).

1.1. Mechanistic Toxicology

The main objective of mechanistic toxicology is to obtain insight in the fundamental mechanisms of a toxicological response. In mechanistic toxicology, one tries to unravel the pathways that are triggered by a toxicity response. It is, however, important to distinguish background expression changes of genes from changes triggered by specific mechanistic or adaptive responses. Therefore, a sufficient number of repeats and a careful design of expression profiling measurements are essential. The comparison of a cell line that is challenged with a drug to a negative control (cell line treated with a nonactive analogue) allows discriminating general stress from drug specific responses (10). Because the triggered pathways can be dose- and condition-dependent, a large number of experiments in different conditions are typically needed. When an in vitro model system is used (e.g., tissue culture) to assess the influence of a drug on gene expression, it is of paramount importance that the model system accurately encapsulates the relevant biological in vivo processes.

With dynamic profiling experiments one can monitor adaptive changes in the expression level caused by administering the xenobiotic to the system under study. By sampling the dynamic system at regular time intervals, short-, mid- and long-term alterations (i.e., high and low frequency changes) in xenobiotic-induced gene expression can be measured. With static experiments, one can test the induced changes in expression in several conditions or in different genetic backgrounds (gene knock out experiments) (10).

Recent developments in analysis methods offer the possibility to derive low-level (sets of genes triggered by the toxicological response) as well as high-level information (unraveling the complete pathway) from the data. However, the feasibility of deriving high-level information depends on the quality of the data, the number of experiments, and the type of biological system studied (11). Therefore, drug triggered pathway discovery is not straightforward and, in addition, is expensive so that it cannot be applied routinely. Nevertheless, when successful, it can completely describe the effects elicited by representative members of certain classes of compounds. Well-described agents or compounds, for which both the toxicological endpoints and the molecular mechanisms resulting in them are characterized, are optimal candidates for the construction of a reference database and for subsequent predictive toxicology (see Sec. 1.2). Mechanistic insights can also help in determining the relative health risk and guide the discovery program toward safer compounds. From a statistical point of view, mechanistic toxicology does not require any prior knowledge on the molecular biological aspects of the system studied. The analysis is based on what is called unsupervised techniques. Because it is not known in advance which genes will be involved in the studied response, arrays used for mechanistic toxicology are exhaustive; they contain cDNAs representing as much coding sequences of the genome as possible. Such arrays are also referred to as diagnostic or investigative arrays (12).

1.2. Predictive Toxicology

Compounds with the same mechanism of toxicity are likely to be associated with the alteration of a similar set of elicited genes. When tissues or cell lines subjected to such compounds are tested on a DNA microarray, one typically observes characteristic expression profiles or fingerprints. Therefore, reference databases can be constructed that contain these characteristic expression profiles of reference compounds. Comparing the expression profile of a new compound with such a reference database allows for a classification of the novel compound (2,5,7,9,13,14). From the known properties of the class to which the novel substance was classified, the behavior of the novel compound (toxicological endpoint) can be predicted. The reference profiles will, however, depend to a large extent on the endpoints that were envisaged (used the cell lines, model organisms, etc.). By a careful statistical analysis (feature extraction) of the profiles in such a compendium database, markers for specific toxic endpoints can be identified. These markers consist of genes that are specifically induced by a class of compounds. They can then be used to construct dedicated arrays [toxblots (12,15), rat hepato chips (13)]. Contrary to diagnostic arrays, the number of genes on a dedicated array is limited resulting in higher throughput screening of lead targets at a lower cost (12,15). Markers can also reflect diagnostic expression changes of adverse effects. Measuring such diagnostic markers in easily accessible human tissues (blood samples) makes it possible to monitor early onset of toxicological phenomena after drug administration, for instance, during clinical trials (5). Moreover, markers (features) can be used to construct predictive models. Measuring the levels of a selected set of markers on, for instance, a dedicated array can be used to predict with the aid of a predictive model (classifier) the class of compounds to which the novel xenobiotic belongs (predictive toxicology). The impact of predictive toxicology will grow with the size of the reference databases. In this respect, the efforts made by several organizations (such as the International Life Science Institute (ILSI) http://www.ilsi.org/) to make public repositories of microarray data that are compliant with certain standards (MIAMI) are extremely useful (10,16).

1.3. Other Applications

There are plenty of other topics where the use of expression profiling can be helpful for toxicological research, including the identification of interspecies or in vitro in vivo discrepancies. Indeed, results based on the determination of dose responses and on the predicted risk of a xenobiotic for humans are often extrapolated from studies on surrogate animals. Measuring the differences in effect of administering well-studied compounds to either model animals or cultured human cells, could certainly help in the development of more systematic extrapolation methods (10).

Expression profiling can also be useful in the study of structure activity relationships (SAR). Differences in pharmacological or toxicological activity between structural related compounds might be associated with corresponding differences in expression profiles. The expression profiles can thus help distinguish active from inactive analogues in SAR (7).

Some drugs need to be metabolized for detoxification. Some drugs are only metabolized by enzymes that are encoded by a single pleiothropic gene. They involve the risk of drug accumulation to toxic concentrations in individuals carrying specific polymorphisms of that gene (17). With mechanistic toxicology, one can try to identify the crucial enzyme that is involved in the mechanism of detoxification. Subsequent genetic analysis can then lead to an a priori prediction to determine whether a xenobiotic should be avoided in populations with particular genetic susceptibilities.

2. MICROARRAYS

2.1. Technical Details

Microarray technology allows simultaneous measurement of the expression levels of thousands of genes in a single hybridization assay (7). An array consists of a reproducible pattern of different DNAs (primarily PCR products or oligonucleotides—also called probes) attached to a solid support. Each spot on an array represents a distinct coding sequence of the genome of interest. There are several microarray platforms that can be distinguished from each other in the way that the DNA is attached to the support.

Spotted arrays (18) are small glass slides on which presynthesized single stranded DNA or double-stranded DNA is spotted. These DNA fragments can differ in length depending on the platform used (cDNA microarrays vs. spotted oligoarrays). Usually the probes contain several hundred of base pairs and are derived from expressed sequence tags (ESTs) or from known coding sequences from the organism under study. Usually each spot represents one single ORF or gene. A cDNA array can contain up to 25,000 different spots.

GeneChip oligonucleotide arrays [Affymetrix, Inc., Santa Clara (19)] are high-density arrays of oligonucleotides synthesized in situ using light-directed chemistry. Each gene is represented by 15–20 different oligonucleotides (25-mers), that serve as unique sequence-specific detectors. In addition, mismatch control oligonucleotides (identical to the perfect match probes except for a single base-pair mismatch) are added. These control probes allow the estimation of cross-hybridization. An Affymetrix array represents over 40,000 genes.

Besides these customarily used platforms, other methodologies are being developed [e.g., fiber optic arrays (20)].

In every cDNA-microarray experiment, mRNA of a reference and agent-exposed sample is isolated, converted into cDNA by an RT-reaction and labeled with distinct fluorescent dyes (Cy3 and Cy5, respectively the "green" and "red" dye). Subsequently, both labeled samples are hybridized simultaneously to the array. Fluorescent signals of both channels (i.e., red and green) are measured and used for further analysis (for more extensive reviews on microarrays refer to Refs. 7,21–23. An overview of this procedure is given in Fig. 1.

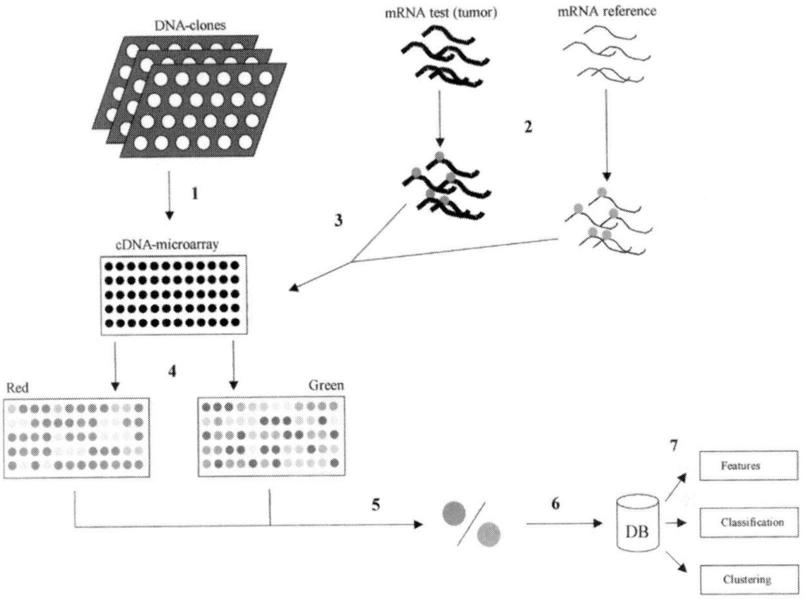

Figure 1 Schematic overview of an experiment with a cDNA microarray. 1) Spotting of the presynthesized DNA-probes (derived from the genes to be studied) on the glass slide. These probes are the purified products from PCR-amplification of the associated DNA-clones. 2) Labeling (via reverse transcriptase) of the total mRNA of the test sample (red = Cy5) and reference sample (green = Cy3). 3) Mixing of the two samples and hybridization. 4) Read-out of the red and green intensities separately (measure for the hybridization by the test and reference sample) of each probe. 5) Calculation of the relative expression levels (intensity in the red channel/intensity in the green channel). 6) Storage of results in a database. 7) Data mining.

2.2. Sources of Variation

In a microarray experiment, changes in gene expression level are being monitored. One is interested in knowing how much the expression of a particular gene is affected by the applied condition. However, besides this effect of interest, other experimental factors or sources of variation contribute to the measured change in expression level. These sources of variation prohibit direct comparison between measurements.

That is why preprocessing is needed to remove these additional sources of variation, so that for each gene, the corrected "preprocessed" value reflects the expression level caused by the condition tested (effect of interest). Consistent sources of variation in the experimental procedure can be attributed to gene, condition/dye, and array effects (24–26).

Condition and dye effects reflect differences in mRNA isolation and labeling efficiencies between samples. These effects result in a higher measured intensity for certain conditions or for either one of both channels.

When performing multiple experiments (i.e., by using more arrays), arrays are not necessarily being treated identically. Differences in hybridization efficiency result in global differences in intensities between arrays, making measurements derived from different arrays incomparable. This effect is generally called the array effect.

The gene effect explains that some genes emit a higher or lower signal than others. This can be related to differences in basal expression level, or to sequence-specific hybridization or labeling efficiencies.

A last source of variation is a combined effect, the array–gene effect. This effect is related to spot-dependent variations in the amount of cDNA present on the array. Since the observed signal intensity is not only influenced by differences in the mRNA population present in the sample, but also by the amount of spotted cDNA, direct comparison of the absolute expression levels is unreliable.

The factor of interest, which is the condition-affected change in expression of a single gene, can be considered to be a combined gene–condition (GC) effect.

2.3. Microarray Design

The choice of an appropriate design is not trivial (27–29). In Fig. 2 distinct designs are represented. The simplest microarray experiments compare expression in two distinct conditions. A test condition (e.g., cell line triggered with a lead compound) is compared to a reference condition (e.g., cell line triggered with a placebo). Usually the test is labeled with Cy5 (red dye),

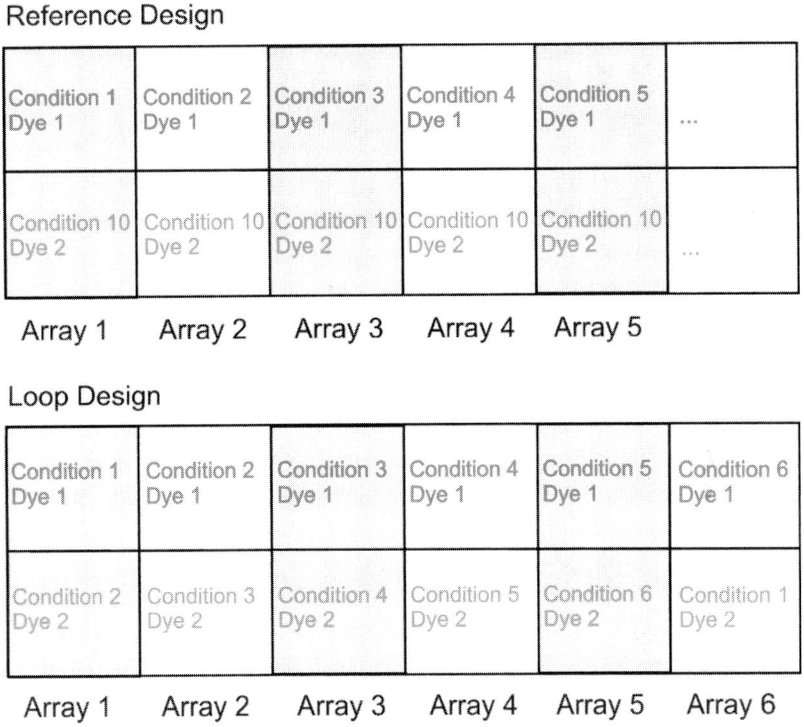

Figure 2 Overview of two commonly used microarray designs. (A) Reference design; (B) loop design. Dye 1 = Cy5; Dye 2 = Cy3; two conditions are measured on a single array.

while the reference is labeled with Cy3 (green dye). Performing replicate experiments is mandatory to infer relevant information on a statistically sound basis. However, instead of just repeating the experiments exactly in the way described above, a more reliable approach here would be to perform dye reversal experiments (dye swap). As a repeat on a second array: The same test and reference conditions are measured once more but the dyes are swapped; i.e., on this second array, the test condition is labeled with Cy3 (green dye), while the corresponding reference condition is labeled with Cy5 (red dye). This allows intrinsically compensating for dye-specific differences.

When the behavior of distinct compounds is compared or when the behavior triggered by a compound is profiled during the course of a dynamic process, more complex designs are required. Customarily used, and still preferred by molecular biologists, is the reference design: Different test conditions (e.g., distinct compounds) are compared to a similar reference condition. The reference condition can be artificial and does not need to be biologically significant. Its main purpose is to have a common baseline to facilitate mutual comparison between samples. Every reference design results in a relatively higher number of replicate measurements of the condition (reference) in which one is not primarily interested than of the condition of interest (test condition). A loop design can be considered as an extended dye reversal experiment. Each condition is measured twice, each time on a different array and labeled with a different dye (Fig. 2). For the same number of experiments, a loop design offers more balanced replicate measurements of each condition than a reference design, while the dye-specific effects can also be compensated for.

Irrespective of the design used, the expression levels of thousands of genes are monitored simultaneously. For each gene, these measurements are usually arranged into a data matrix. The rows of the matrix represent the genes while the columns are the tested conditions (toxicological compounds, timepoints). As such one obtains gene expression profiles (row vectors) and experiment profiles (column vectors) (Fig. 3).

3. ANALYSIS OF MICROARRAY EXPERIMENTS

Some of the major challenges for mechanistic and predictive toxicogenomics are in data management and analysis (5,10). A later chapter gives an overview of the state of the art methodologies for the analysis of high-throughput expression profiling experiments. The review is not comprehensive as the field of microarray analysis is rapidly evolving. Although there will be a special focus on the analysis of cDNA arrays, most of the described methodologies are generic and applicable to data derived from other high-throughput platforms.

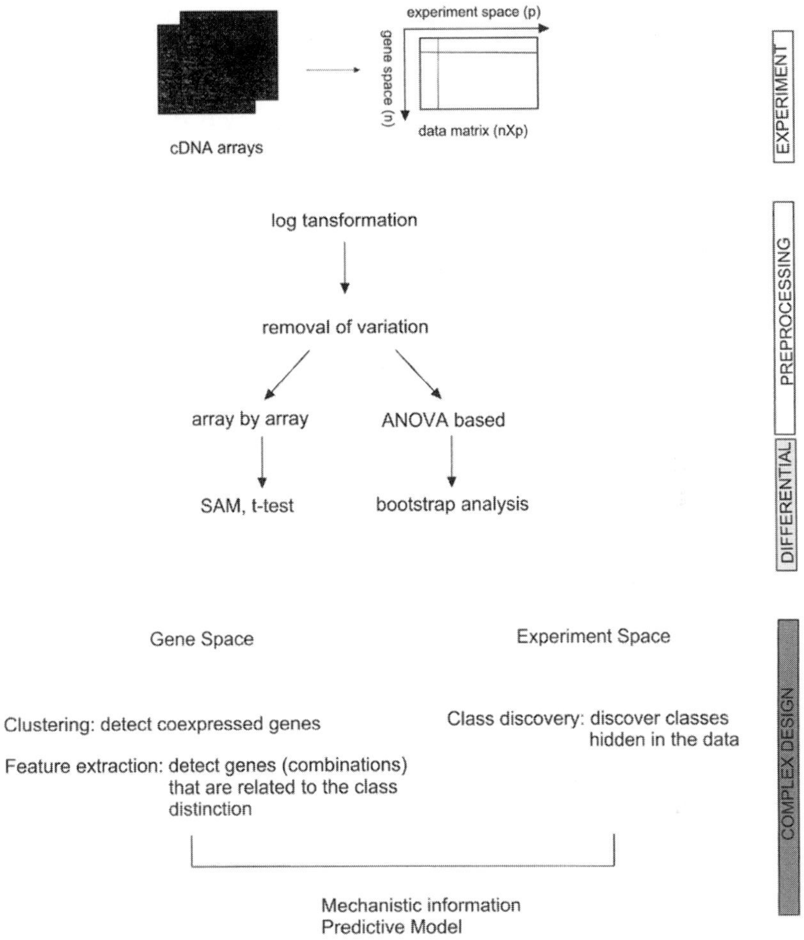

Figure 3 Schematic overview of the analysis flow of cDNA-microarray data.

3.1. Preprocessing: Removal of Consistent Sources of Variation

As mentioned before, preprocessing of the raw data is needed to remove consistent and/or the systematic sources of variation from the measured expression values. As such, the preprocessing has a large influence on the final result of the analysis. In the following, we will give an overview of the commonly used approaches for preprocessing: the array by array approach and the procedure based on analysis of variance (ANOVA) (Fig. 3). The array by array approach is a multistep procedure comprising log transformation, normalization, and identification of differentially expressed genes by using a test statistic. The ANOVA-based approach consists of a log transformation, linearization, and identification of differentially expressed genes based on bootstrap analysis.

3.1.1. Mathematical Transformation of the Raw Data: Need for a Log Transformation

The effect of the log transformation as an initial preprocessing step is illustrated in Fig. 4. In Fig. 4A, the expression levels of all genes measured in the test sample were plotted against the corresponding measurements in the reference sample. Assuming that the expression of only a restricted

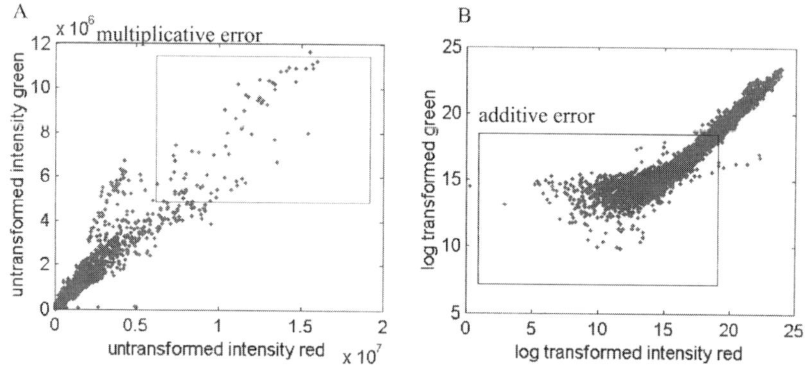

Figure 4 Illustration of the influence of log transformation on the multiplicative and additive errors. Panel A: representation of untransformed raw data. *X*-axis: intensity measured in the red channel, *Y*-axis: intensity measured in the green channel. Panel B: representation of $\log_2$ transformed raw data. *X*-axis: intensity measured in the red channel ($\log_2$ value), *Y*-axis: intensity measured in the green channel ($\log_2$ value). Assuming that only a small number of the genes will alter their expression level under the different conditions tested, the measurements of most genes in the green channel can be considered as replica's of the corresponding measurements in the red channel.

number of genes is altered (global normalization assumption, see below), measurements of the reference and the test condition can be considered to be comparable for most of the genes on the array. Therefore, the residual scattering as observed in Fig. 4A reflects the measurement error. As often observed, the error in microarray data is a superposition of a multiplicative error and an additive one. Multiplicative errors cause signal-dependent variance of residual scattering, which deteriorates the reliability of most statistical tests. Log transforming the data alleviates this multiplicative error, but usually at the expense of an increased error at low expression levels (Fig. 4B). Such an increase of the measurement error with decreasing signal intensities, as present in the log transformed data, is, however, considered to be intuitively plausible: low expression levels are generally assumed to be less reliable than high levels (24,30).

An additional advantage of log transforming the data is that differential expression levels between the two channels are represented by $\log_{(\text{test})} - \log_{(\text{reference})}$ (see Sec. 3.1.2). This brings levels of under- and overexpression to the same scale, i.e., values of underexpression are no longer bound between 0 and 1.

3.1.2. Array by Array Approach

In the array by array approach, each array is compensated separately for dye/condition and spot effects. A $\log_{(\text{test/reference})} = \log_{(\text{test})} - \log_{(\text{reference})}$ is used as an estimate of the relative expression. Using ratios (relative expression levels) instead of absolute expression levels allows compensating intrinsically for spot effects. The major drawback of the ratio approach is that when the intensity measured in one of the channels is close to 0, the ratio attains extreme values that are unstable as the slightest change in the value close to 0 has a large influence on the ratio (30,31).

Normalization methods aim at removing consistent condition and dye effects (see above). Although the use of spikes (control spots, external control) and housekeeping genes (genes not altering their expression level under the conditions tested) for normalization have been described in the literature, global normalization is commonly used (32). The global normalization principle assumes that only of a small fraction of the total number of genes on the array, the expression level is altered. It also assumes that symmetry exists in the number of genes for which the expression is increased vs. decreased. Under this assumption, the average intensity of the genes in the test condition should be equal to the average intensities of the genes in the reference condition. Therefore, for the bulk of the genes, the log-ratios should equal 0. Regardless of the procedure used, after normalization, all log-ratios will be centered around 0. Notice that the assumption of global normalization applies only to microarrays that contain a random set of genes and not to dedicated arrays.

Linear normalization assumes a linear relationship between the measurements in both conditions (test and reference). A common choice for the constant transformation factor is the mean or median of the log intensity ratios for a given gene set. As shown in Fig. 5, most often the assumption of a linear relationship between the measurements in both conditions is an oversimplification, since the relationship between dyes depends on the measured intensity. These observed nonlinearities are most pronounced at extreme intensities (either high or low). To cope with this problem, Yang et al. (32) described the use of a robust scatter plot smoother, called Lowess, that performs local linear fits. The results of this fit can be used to simultaneously linearize and normalize the data (Fig. 5).

The array by array procedure uses the global properties of all genes on the array to calculate the normalization factor. Other approaches have been described that subdivide an array into, for instance, individual print tip groups, which are normalized separately (32). Theoretically, these approaches perform better than the array by array approach in removing position-dependent "within array" variations. The drawback, however, is that the number of measurements to calculate the fit is reduced, a pitfall that can be overcome by the use of ANOVA (see Sec. 3.1.3). SNOMAD offers a free online implementation of the array by array normalization procedure (33).

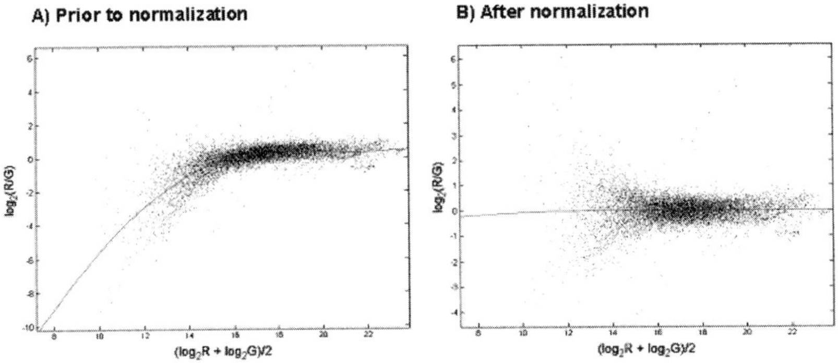

Figure 5 Illustration of the influence of an intensity-dependent normalization. Panel A: Representation of the log-ratio $M = \log_2(R/G)$ vs. the mean log intensity $A = [\log_2(R) + \log_2(G)]/2$. At low average intensities, the ratio becomes negative indicating that the green dye is consistently more intense as compared to the intensity of the red dye. This phenomena is referred to as the non-linear dye effect. Solid line represents the Lowess fit with an f value of 0.02 (R = red; G = green). Panel B: Representation of the ratio $M = \log_2(R/G)$ vs. the mean log intensity $A = [\log_2(\mathrm{R}) + \log_2(G)]/2$ after performing a normalization and linearization based on the Lowess fit. Solid line represent the new Lowess fit with an f value of 0.02 on the normalized data (R = red; G = green).

3.1.3. ANOVA-based Preprocessing

ANOVA can be used as an alternative to the array by array approach (24,27). In this case, it can be viewed as a special case of multiple linear regression, where the explanatory variables are entirely qualitative. ANOVA models the measured expression level of each gene as a linear combination of the explanatory variables that reflect, in the context of microarray analysis, the major sources of variation. Several explanatory variables representing the condition, dye and array effects (see above) and combinations of these effects are taken into account in the models (Fig. 6). One of the combined effects, the GC effect, reflects the expression of a gene solely depending on the tested condition (i.e., the condition-specific expression or the effect of interest). Of the other

$$I_{ijklmn} = \mu + B_m + D_l + A_{k(m)} + (AD)_{kl(m)} + P_{j(k(m))} + G_{i(j(o))} + E_{in(j(m))} + \varepsilon_{ijklmn}$$

Figure 6 Example of an ANOVA model. I is the measured intensity, D is the dye effect, A is the array effect, G is the gene effect, B is the batch effect (the number of separate arrays needed to cover the complete genome if the cDNAs of the genome do not fit on a single array), P is the pin effect, E is the expression effect (factor of interest). AD is the combined array–dye effect, ε is the residual error, m is the batch number, l is the dye number, j is the spot number on an array spotted by the same pin, and i is the gene number. The measured intensity is modeled as a linear combination of consistent sources of variation and the effect of interest. Note that in this model condition effect C has been replaced by the combined AD effect.

combined effects, only those having a physical meaning in the process to be modeled are retained. Reliable use of an ANOVA model requires a good insight into the experimental process. Several ANOVA models have been described for microarray preprocessing (24,34,35).

The ANOVA approach can be used if the data are adequately described by a linear ANOVA model and if the residuals are approximately normally distributed. ANOVA obviates the need for using ratios. It offers as an additional advantage that all measurements are used simultaneously for statistical inference and that the experimental error is implicitly estimated (36). Several web applications that offer an ANOVA-based preprocessing procedure have been published [e.g., MARAN (34), GeneANOVA (37)].

3.2. Microarray Analysis for Mechanistic Toxicology

The purpose of mechanistic toxicology consists of unraveling the genomic responses of organisms exposed to xenobiotics. Distinct experimental setups can deliver the required information. The most appropriate data analysis method depends both on the biological question to be answered and the experimental design. For the purpose of clarity, we make a distinction between three types of design. This subdivision is somewhat artificial and the distinction is not always clearcut. The simplest design compares two conditions to identify differentially expressed genes. (Techniques developed for this purpose are reviewed in Sec. 3.2.1.) Using more complex designs, one can try to reconstruct the regulation network that generates a certain behavior. Dynamic changes in expression can be monitored as function of time. For such a dynamic experiment, the main purpose is to find genes that behave similarly during the time course, where often an appropriate definition of similarity is one of the problems. Such coexpressed genes are identified by cluster analysis (Sec. 3.2.2). On the other hand, the expression behavior can be tested under distinct experimental conditions (e.g., the effect induced by distinct xenobiotics). One is interested not only in finding coexpressed genes, but also in knowing the experimental conditions that group together based on their experiment profiles. This means that clustering is performed both in the space of the gene variables (row vectors) and in the space of the condition variables (column vectors). Although such designs can also be useful for mechanistic toxicology, they are usually performed in the context of class discovery and predictive toxicology and will be further elaborated in Sec. 3.3. The objective of clustering is to detect low-level information. We describe this information as low-level because the correlations in expression patterns between genes are identified, but all causal relationships (i.e., the high-level information) remains undiscovered. Genetic network inference (Sec. 3.2.3), on the other hand, tries to infer this high-level information from the data.

3.2.1. Identification of Differentially Expressed Genes

When preprocessed properly, consistent sources of variation have been removed and the replicate estimates of the (differential) expression of a particular gene can be combined. To search for differentially expressed genes, statistical methods are used that test whether two variables are significantly different. The exact identity of these variables depends on the question to be answered. When expression in the test condition is compared to expression in the reference condition, it is generally assumed that for most of the genes no differential expression occurs (global normalization assumption). Thus, the zero hypothesis implies that expression of both test and reference sample is equal (or that the log of the relative expression equals 0). Because in a cDNA experiment the measurement of the expression of the test condition and reference condition is paired (measurement of both expression levels on a single spot), the paired variant of the statistical test is used.

When using a reference design, one is not interested in knowing whether the expression of a gene in the test condition is significantly different from its expression in the reference condition since the reference condition is artificial. Rather, one wants to know the relative differences between the two compounds tested on different arrays using a single reference. Assuming that the ratio is used to estimate the relative expression between each condition and a common reference, the zero hypothesis now will be equality of the average ratio in both conditions tested. In this case, the data are no longer paired. This application is related to feature extraction and will be further elaborated in Sec. 3.3.1.

A major emphasis will be on the description of selection procedures to identify genes that are differentially expressed in the test vs. reference condition.

The fold test is a nonstatistical selection procedure that makes use of an arbitrary chosen threshold. For each gene, an average ratio is calculated based on the different ratio estimates of the replicate experiments (log-ratio $= \log_{(\text{test})} - \log_{(\text{reference})}$). Average ratios of which the expression ratio exceeds a threshold (usually twofold) are retained. The fold test is based on the assumption that a larger observed fold change can be more confidently interpreted as a stronger response to the environmental signal than smaller observed changes. A fold test, however, discards all information obtained from replicates (30). Indeed, when either one of the measured channels obtains a value close to 0, the log-ratio estimate usually obtains a high but inconsistent value (large variance on the variables). Therefore, more sophisticated variants of the fold test have been developed. These methods simultaneously construct an error model of the raw measurements that incorporates multiplicative and additive variations (38–40).

A plethora of novel methods to calculate a test statistic and the corresponding significance level have recently been proposed, provided replicates are available. Each of these methods first calculates a test statistic and subsequently determines the significance of the observed test statistic. Distinct *t*-test like methods are available that differ from each other in the formula that describes the test statistic and in the assumptions regarding the distribution of the null hypothesis. *t*-Test methods are used for detecting significant changes between repeated measurements of a variable in two groups. In the standard *t*-test, it is assumed that data are sampled from a normal distribution with equal variances (zero hypothesis). For microarray data, the number of repeats is too low to assess the validity of this assumption of normality. To overcome this problem, methods have been developed that estimate the distribution of the zero hypothesis from the data itself by permutation or bootstrap analysis (36,41). Some methods avoid the necessity of estimating a distribution of the zero hypothesis by using order statistics (41). For an exhaustive comparison between the individual performances of each of these methods, we refer to Marchal et al. (31) and for the technical details, we refer to the individual references and Pan (2002) (42).

When ANOVA is used to preprocess the data, significantly expressed genes are often identified by bootstrap analysis (Gaussian statistics are often inappropriate, since normality assumptions are rarely satisfied). Indeed, fitting the ANOVA model to the data allows the estimation of the residual error which can be considered as an estimate of the experimental error. By adding noise (randomly sampled from the residual error distribution) to the estimated intensities, thousands of novel bootstrapped datasets, mimicking wet lab experiments, can be generated. In each of the novel datasets, the difference in GC effect between two conditions is calculated as a measure for the differential expression. Based on these thousands of estimates of the difference in GC effect, a bootstrap confidence interval is calculated (36).

An extensive comparison of these methods showed that a *t*-test is more reliable than a simple fold test. However, the *t*-test suffers from a low power due the restricted number of replicate measurements available. The method of Long et al. (43) tries to cope with this drawback by estimating the population variance as a posterior variance that consists of a contribution of the measured variance and a prior variance. Because they assume that the variance is intensity-dependent, this prior variance is estimated based on the measurements of other genes with similar expression levels as the gene of interest. ANOVA-based methods assume a constant error variance for the entire range of intensity measurements (homoscedasticity). Because the calculated confidence intervals are based on a linear model and microarray data suffer from nonlinear intensity-dependent effects and large additive effects at low expression levels (Sec. 3.1.1), the estimated confidence intervals are usually too restrictive for elevated expression levels and too small for measurements in the low intensity range. In our experience, methods that did not make an explicit assumption on the distribution of the zero hypotheses, such as Statistical Analysis of Microarrays (SAM) (41), clearly outperformed the other methods for large datasets.

Another important issue in selecting significantly differentially expressed genes is correction for multiple testing. Multiple testing is crucial since hypotheses are calculated for thousands of genes simultaneously. Standard Bonferroni correction seems overrestrictive (30,44). Therefore, other corrections for multiple testing have been proposed (45). Very promising for microarray analysis seems the application of the False Discovery Rate (FDR) (46). A permutation-based implementation of this method can be found in the SAM software (41).

3.2.2. Identification of Coexpressed Genes

3.2.2.1. Clustering of the Genes

As mentioned previously, normalized microarray data are collected in a data matrix. For each gene, the (row) vector leads to what is generally called an expression profile. These expression profiles or vectors can be regarded as (data) points in a high-dimensional space. Genes involved in a similar biological pathway or with a related function often exhibit a similar expression behavior over the coordinates of the expression profile/vector. Such similar expression behavior is reflected by a similar expression profile. Genes with similar expression profiles are called coexpressed. The objective of cluster analysis of gene expression profiles is to identify subgroups (= clusters) of such coexpressed genes (47,48). Clustering algorithms group together genes for which the expression vectors are "close" to each other in the high-dimensional space based on some distance measure. A first generation of algorithms originated in research domains other than biology (such as the areas of "pattern recognition" and "machine learning"). They have been applied successfully to microarray data. However, confronted with the typical characteristics of biological data, recently a novel generation of algorithms has emerged. Each of these algorithms can be used with one or more distance metrics (Fig. 7). Prior to clustering, microarray data usually are filtered, missing values are replaced, and the remaining values are rescaled.

3.2.2.2. Data Transformation Prior to Clustering

The "Euclidean distance" is frequently used to measure the similarity between two expression profiles. However, genes showing the same relative behavior but with diverging absolute behavior (e.g., gene expression profiles with a different baseline and/or a different amplitude but going up and down at the same time) will have a relatively high Euclidean distance. Because the purpose is to group expression profiles that have the same relative behavior, i.e., genes that are up- and downregulated together, cluster algorithms based on the Euclidean distance will therefore erroneously assign

Minkowski distance

$$d(x,y) = \sqrt[r]{\sum_{i=1}^{p} |x_i - y_i|^r}$$

r = 1: Manhattan distance
r = 2: Euclidean distance

Pearson correlation distance

$$s(x,y) = \frac{\sum_{i=1}^{p}(x_i - \bar{x})(y_i - \bar{y})}{\sqrt{\sum_{i=1}^{p}(x_i - \bar{x})^2 \times \sum_{i=1}^{p}(y_i - \bar{y})^2}}$$

$$\bar{x} = \frac{1}{p}\sum_{i=1}^{p} x_i$$

$$\bar{y} = \frac{1}{p}\sum_{i=1}^{p} y_i$$

Figure 7 Overview of commonly used distance measures in cluster analysis. x and y are points or vectors in the p-dimensional space. x_i and y_i ($i = 1, \ldots, p$) are the coordinates of x and y. p is the number of experiments.

the genes with different absolute baselines to different clusters. To overcome this problem, expression profiles are standardized or rescaled prior to clustering. Consider a gene expression profile $g(g_1, g_2, \ldots, g_p)$ of dimension p (i.e., p time points or conditions) with average expression level μ and standard deviation σ. Microarray data are commonly rescaled by replacing every expression level g_i by

$$\frac{g_i - \mu}{\sigma}$$

This operation results in a collection of expression profiles all being 0 mean and with standard deviation 1 (i.e., the absolute differences in expression behavior have largely been removed). The Pearson correlation coefficient, a second customarily used distance measure, inherently performs this rescaling as it is basically equal to the cosine of the angle between two gene expression profile vectors.

As previously mentioned, a set of microarray experiments in which gene expression profiles have been generated frequently contains a considerable number of genes that do not contribute to the biological process that is being studied. The expression values of these profiles often show little variation over the different experiments (they are called constitutive with respect to the biological process studied). By applying the rescaling procedure, these profiles will be inflated and will contribute to the noise of the dataset. Most existing clustering algorithms attempt to assign each gene expression profile, even the ones of poor quality to at least one cluster. When also noisy and/or random profiles are assigned to certain clusters, they will corrupt these clusters and hence the average profile of the clusters. Therefore, filtering prior to the clustering is advisable. Filtering involves removing gene expression profiles from the dataset that do not satisfy one or possibly more very simple criteria (49). Commonly used criteria include a minimum threshold for the standard deviation of the expression values in a profile (removal of constitutive genes). Microarray datasets regularly contain a considerable number of missing values. Profiles containing too many missing values have to be omitted (filtering step). Sporadic missing values can be replaced by using specialized procedures (50,51).

3.2.2.3. Cluster Algorithms

The first generation of cluster algorithms includes standard techniques such as *K*-means (52), self-organizing maps (53,54), and hierarchical clustering (49). Although biologically meaningful results can be obtained with these algorithms, they often lack the fine-tuning that is necessary for biological problems. The family of hierarchical clustering algorithms was and is probably still the method preferred by biologists (49) (Fig. 8). According to a certain measure, the distance between every couple of clusters is calculated (this is called the pairwise distance matrix). Iteratively, the two closest clusters are merged giving rise to a tree structure, where the height of the branches is proportional to the pairwise distance between the clusters. Merging stops if only one cluster

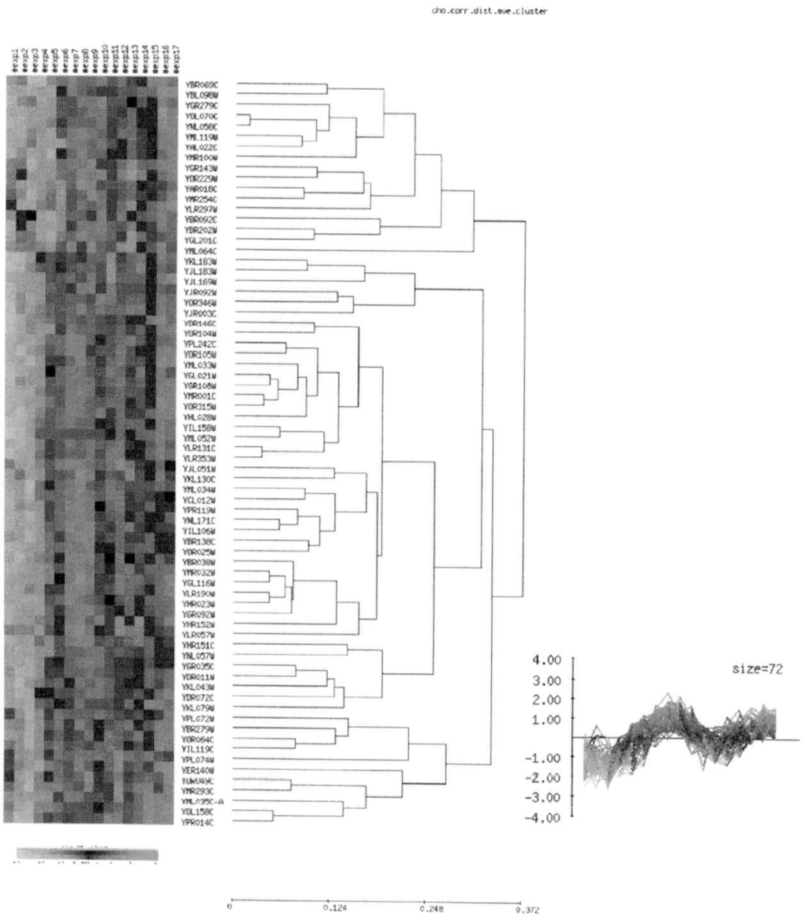

Figure 8 Hierarchical clustering. Hierarchical clustering of the dataset of Cho et al. (119) representing the mitotic yeast cell cycle. A selection of 3000 genes was made as described in Ref. 51. Hierarchical clustering was performed using the Pearson correlation coefficient and an average linkage distance (UPGMA) as implemented in EPCLUST (65). Only a subsection of the total tree is shown containing 72 genes. The columns represent the experiments, the rows the gene names. A green color indicates downregulation, while a red color represents upregulation, as compared to the reference condition. In the complete experimental setup, a single reference condition was used (reference design).

is left. However, the final number of clusters has to be determined by cutting the tree at a certain level or height. Often it is not straightforward to decide where to cut the tree as it is typically rather difficult to predict which level will give the most valid biological results. Secondly, the computational complexity of hierarchical clustering is quadratic in the number of gene expression profiles, which can sometimes be limiting considering the current (and future) size of the datasets.

Centroid methods form another attractive class of algorithms. The K-means algorithm for instance starts by assigning at random all the gene expression profiles to one of the N clusters (where N is the user-defined number of clusters). Iteratively, the center (which is nothing more than the average expression vector) of each cluster is calculated, followed by a reassignment of the gene expression vectors to the cluster with the closest cluster center. Convergence is reached when the cluster centers remain stationary. Self-organizing maps can be considered as a variation on centroid methods that also allow samples to influence the location of neighboring clusters. These centroid algorithms suffer from similar drawbacks as hierarchical clustering: The number of clusters is a user-defined parameter with a large influence on the outcome of the algorithm. For a biological problem, it is hard to estimate in advance how many clusters can be expected. Both algorithms assign each gene of the dataset to a cluster. This is from a biological point of view counterintuitive, since only a restricted number of genes are expected to be involved in the process studied. The outcome of these algorithms appears to be very sensitive to the chosen parameter settings [number of clusters for K-means (Fig. 9)], the distance measure that is used and the metrics to determine the distance between clusters (average vs. complete linkage for hierarchical clustering). Finding the biological most relevant solution usually requires extensive parameter fine-tuning and is based on arbitrary criteria (e.g., clusters look more coherent) (55).

Besides the development of procedures that help to estimate some of the parameters needed for the first generation of algorithms [e.g., like the number of clusters present in the data (56–58)], a panoply of novel algorithms have been

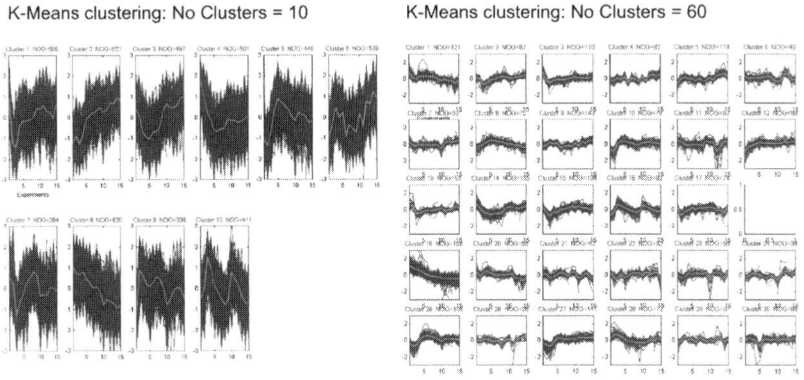

Figure 9 Illustration of the effect of using different parameter settings on the end result of a *K*-means clustering of microarray data. Data were derived from Ref. 119 and represent the dynamic profile of the cell cycle. The cluster number is the variable parameter of the *K*-means clustering. By underestimating the number of clusters, genes within a cluster will have a very heterogeneous profile. Since *K*-means assigns all genes to a cluster (no inherent quality criterion is imposed), genes with a noisy profile disturb the average profile of the clusters. When increasing the number of clusters, the profiles of genes that belong to the same cluster become more coherent and the influence of noisy genes is less exacerbating. However, when too high the cluster number, genes belonging biologically to the same cluster might be assigned to separate clusters with very similar average profiles.

designed that cope with the problems mentioned above in different ways: Self-organizing tree algorithm or SOTA (59) combines self-organizing maps and divisive hierarchical clustering; quality-based clustering (60) only assigns genes to a cluster that meet a certain quality criterion; adaptive quality-based clustering (51) is based on a principle similar to quality-based clustering, but offers a strict statistical meaning to the quality criterion; gene shaving (61) is based on Principal Component Analysis (PCA). Other examples include model-based clustering (56,58), clustering based on simulated annealing (57) and CAST (62). For a more extensive overview of these algorithms, refer to Moreau et al. (47).

Some of these algorithms determine the number of clusters based on the inherent data properties (51,58–60,63). Quality criteria have been developed to minimize the number of false positives. Only those genes are retained, in the clusters, that satisfy a quality criterion. This results in clusters that contain genes with tightly coherent profiles (51,60). Fuzzy clustering algorithms allow a gene to belong to more than one cluster (61). Distinct publicly available implementations of these novel algorithms are freely available for academic users (INCLUSive (64), EPCLUST (65), AMADA (66), Cluster (49), etc.).

3.2.2.4. Cluster Validation

Depending on the algorithms and the distance measures used, clustering will give different results. Therefore validation, either statistically or biologically, of the cluster results is essential. Several methods have been developed to assess the statistical relevance of a cluster. Intuitively, a cluster can be considered reliable if the within cluster distance is small (i.e., all genes retained are tightly coexpressed) and the cluster has an average profile well delineated from the remainder of the dataset (maximal intercluster distance). This criterion is formalized by Dunn's validity index (67). Another desirable property is cluster stability: gene expression levels can be considered as a superposition of real biological signals and small experimental errors. If true biological signals are more pronounced than the experimental variation, repeating the experiments should not interfer with the identification of the biological true clusters. Following this reasoning, cluster stability is assessed by creating new in silico replicas (i.e., simulated replicas) of the dataset of interest by adding a small amount of artificial noise to the original data. The noise can be estimated from a reasonable noise model (68,69) or by sampling the noise distribution directly from the data (36). These newly generated datasets are preprocessed and clustered in the same way as the original dataset. If the biological signal is more pronounced than the noise signal in the measurements of one particular gene, adding small artificial variations (in the range of the experimental noise present in the dataset) to the expression profile of such gene will not influence its overall profile and cluster membership. The result (cluster membership) of that particular gene is robust towards what is called a sensitivity analysis and a reliable confidence can be assigned to the cluster result of that gene.

An alternative approach of validating clusters is by assessing the biological relevance of the cluster result. Genes exhibiting a similar behavior might belong to the same biological process. This is reflected by enrichment of functional categories within a cluster (51,55). Also, for some clusters, the observed coordinate behavior of the gene expression profiles might be caused by transcriptional coregulation. In such case, detection of regulatory motifs is useful as a biological validation of cluster results (55,70–72).

3.2.3. Genetic Network Inference

The final goal of mechanistic toxicology is the reconstruction of the regulatory networks that underlie the observed cell responses. A complete regulatory network consists of proteins interacting with each other, with DNA or with metabolites to constitute a complete signaling pathway (73). The action of regulatory networks determines how well cells can react or adapt to novel conditions. From this perspective, a cellular reaction against a xenobiotic compound can be considered as a stress response that triggers a number of specialized regulation pathways and induces the essential survival machinery. A regulatory network viewed at the level of transcriptional regulation is called a genetic network. This genetic network can be monitored by microarray experiments. In contrast to clustering that searches for correlation in the data, genetic network inference goes one step beyond and tries to reconstruct the causal relationships between the genes. Although methods for genetic network inference are being developed, the sizes of the currently available experimental datasets do not yet meet the extensive data requirements of most of these algorithms. In general, the number of experimental data is still much smaller than the number of parameters that is to be estimated (i.e., the problem is underdetermined). The low signal to noise level of microarray data and the inherent stochasticity of biological systems (74,75) aggravates the problem of underdetermination. Combining expression data with additional sources of information (prior information) can possibly offer a solution (76–79). Most of the current inference algorithms already make use of general knowledge on the characteristics of biological networks, such as the presence of hierarchical network structures (77,80), a powerlaw distribution of the number of connections (81), sparsness of a network (82,83), and a maximal indegree (maximal number of incoming and outgoing edges).

In order to unravel pathways, both dynamic and static experiments can be informative. However, most of the developed algorithms can only handle static data. Dynamic data can always be converted to static data by treating the transition from a previous time point to a consecutive time point as a single condition. However, this is at the expense of losing the specific information that can be derived from the dynamical characteristics of the data. Treating this biological time signals as responses of a dynamical system is one of the big challenges of the near future.

Networks are either represented graphically or by a matrix representation. In a matrix representation, each column and row represent a gene and the matrix elements represent causal relationships. In a graph, the nodes represent the genes and the edges between the nodes reflect the interactions between the genes. To each edge corresponds an interaction table (matrix representation) that expresses the type and strength of the interaction between the nodes it connects.

A first group of inference methods explicitly uses the graphical network representation. As such algorithms based on Boolean models have been proposed (84,85). Interactions are modeled by Boolean rules and expression levels are described by two discrete values. Although such discrete representations require relatively few data, the discretization leads to a considerable loss of information that was present in the original expression data. Most Boolean models cannot cope with the noise of the experimental data or with the stochasticity of the biological system although certain attempts have been made (86).

Bayesian networks (or belief networks) are from that perspective more appropriate (87). Because of their probabilistic nature, they cope with stochasticity automatically. Also, in this probabilistic framework, additional sources of information can easily be taken into account (76). With a few exceptions that can handle continuous data (88,89), most of the inference implementations based on Bayesian networks require data discretization. Bayesian networks can also cope with hidden variables (90). Hidden variables represent essential network components for which no changes in expression can be observed, either because of measurements error (then called missing variables), or because of biological reasons, e.g., the compound acts at posttranslational level. Inference algorithms based on Bayesian networks have been developed both for static data (76,88,89,91,92) and dynamic data (87,93,94).

The probabilistic nature of Bayesian networks certainly offers an advantage over the deterministic characteristics of Boolean networks. The downside, however, is the extensive data requirement that is much less explicit in the simpler Boolean models than in Bayesian networks. To combine the best of both methods, a hybrid model based on the use of Bayesian Boolean networks has been proposed. This method combines the rule-based reasoning of the Boolean models with probabilistic characteristics of Bayesian networks (95). A second group of methods uses the matrix, representation of a network. These methods are based on linear or nonlinear models. In linear models, each gene transcription level depends linearly on the expression level of its parents, for instance represented by linear differential equations (96,97). Nonlinear models make use of black box representations such as neural networks (98), nonlinear differential equations (99), or nonlinear differential equations based on empirical rate laws of enzyme kinetics (100). Nonlinear optimization methods are used to fit the model equations to the data and to estimate the model parameters. Estimating all of the parameters requires an unrealistic large amount of data. The matrix method of singular value decomposition (SVD) has been proposed to solve linear models more efficiently and to generate a family of possible candidate networks for the undetermined problem (101–104).

To this day, genetic network inference is, given the relatively small number of available experiments, an undetermined problem. The solution of any algorithm will therefore pinpoint a number of possible solutions, i.e., networks that are equally consistent with the data. To further reduce the number of possible networks, design methods have been developed (105). These methods predict, based on a first series of experiments, the consecutive set of experiments that will be most informative. Close collaboration between data-analysts and molecular biologists using experiment design procedures and consecutive series of experiments will be indispensable for biological relevant inference. Practical examples where genetic network inference has resulted in the reconstruction of at least part of a network are rare. Most of the successful studies use heuristic methods that are based on biological intuition and that combine expression data with additional prior knowledge (e.g., 77,106).

3.3. Microarray Analysis for Predictive Toxicology

Every toxicological compound affects the expression of genes in a specific way. Every gene represented on the array, therefore, has a characteristic expression level triggered by the compound. All these characteristic gene expression levels contribute to a profile that is specifically associated with a certain compound (typical fingerprint or reference profile or experiment profile). Each reference profile thus consists of a vector with thousands of components (one component for each probe present on the array) and corresponds to a certain column of the expression matrix (see Sec. 2.3). Assuming that compounds with a similar mechanism of toxicity are associated with the alteration of a similar set of genes, they should exhibit similar reference profiles. In our setup, a class or a group of compounds corresponds to the set of compounds that have a similar characteristic profile.

Based on this reasoning, reference databases are constructed. For each class of compounds, representatives, for which the toxicological response is well-characterized mechanistically are selected. For these representatives, reference profiles are assessed. The main goal of predictive toxicology is to determine the class to which a novel compound belongs by comparing its experiment profile to the reference profiles present in the database. However, due to its huge dimension (thousands of components), it is impossible to use the complete experiment profile at once in predictive toxicology. Prediction is based on a selected number of features (genes or combination of genes) that are most correlated with the class differences between the compounds (that are most discriminative). Identification of such features relies on feature extraction methods (Sec. 3.3.1). Sometimes the number of classes and the exact identity of classes present in the data are not known, i.e., it is not known in advance which of the tested compounds belong to the same class of compounds. Class discovery (or clustering of experiments) is an unsupervised technique that tries to detect these hidden classes and the features associated with them (Sec. 3.3.2). Eventually, once the classes and related features have been identified in the reference database, classifiers can be constructed that predict the class to which a novel compound belongs (class prediction or classification Sec. 3.3.3).

3.3.1. Feature Selection

Due to its high dimensionality, using the complete experiment profile to predict the class membership of a novel compound is infeasible. Dimensions need to be reduced, e.g., the profile consisting of the expression levels of 10,000 genes will be reduced to a profile that only consists of a restricted number of most discriminative features (such as 100). The problem of dimensionality reduction thus relates to the identification of the genes for which the expression profile is most correlated with the distinction between the different classes of compounds. Several approaches for feature selection exist, some of which will be elaborated below.

3.3.1.1. Selection of Individual Genes

The aim is to identify single genes the expression of which is correlated with the class distinction one is interested in. Features then correspond to these individual genes (i.e., single gene features). Because not all genes have an expression that contains information about a certain class distinction, some genes can be omitted when studying these classes. Contrary to class discovery, feature extraction as described here requires that the class distinction is known in advance (i.e., it is a supervised method). For this simple method of feature selection, standard statistical tests to identify two variables that are significantly different from each other are applicable (*t*-test, Wilcoxon rank-sum test, etc; see Sec. 3.2.1). Other specialized methods have been developed such as the nonparameter rank based methods of Park et al. (107) or the measure of correlation described by Golub et al. (108).

Methods for multiple testing are also required here (see Sec. 3.2.1). Indeed, a statistical test has to be calculated for every single gene in the dataset (several thousands!). As a consequence, several genes will be selected coincidentally (they will have a high score or low *p*-value without having any true correlation with the class distinction, i.e., they are false positives).

Although frequently applied in predictive applications (109,110), using single gene features might not result in the best predictive performance. Indeed, in general, a class distinction is not determined by the activity of a single gene, but rather by the interaction of several genes. Therefore, using a combination of genes as a single feature, is a more realistic approach (see Sec. 3.3.1.2).

3.3.1.2. Selection of a Combination of Genes

In this section, methods for dimensionality reduction are described that are based on the selection of different combinations (linear or nonlinear) of gene expression levels as features.

Principal Component Analysis (PCA) is one of the methods that can be used in this context (111). PCA finds linear combinations of the gene expression levels of a microarray experiment in such a way that these linear combinations have maximal spread (or standard deviation) for a certain collection of microarray experiments. In fact, PCA searches for the combinations of gene expression levels that are most informative. These (linear) combinations are called the principal components for a particular collection of experiments and they can be found by calculating the eigenvectors of Σ (covariance matrix of A—note that in this formula A has to be centralized, i.e., the mean column vector of A has to lie in the origin):

$$\Sigma = \frac{1}{p-1} A \cdot A'$$

where A is the expression matrix ($n \times p$ matrix—collection of p microarray experiments where n gene expression levels were measured). The eigenvectors or principal components with the largest eigenvalues also correspond to the linear combinations with the largest spread for the collection of microarray experiments represented by A. For a certain experiment, the linear combinations (or features) themselves can be calculated by projecting the expression vector (for that experiment) onto the principal components. In general, only the principal components with the largest eigenvalues will be used. So when 1) E ($n \times 1$) is the expression vector for a certain microarray experiment (where also n gene expression levels were measured), 2) the columns of P ($n \times m$ matrix) contain the m principal components corresponding to the m largest eigenvalues of A, and 3) F ($m \times 1$) is given by

$$F = P' \cdot E$$

then the m components of F contain the m features or linear combinations for the microarray experiment with expression vector E according to the first m principal components of the collection of microarray experiments represented by A.

As an unsupervised method, PCA can also be used in combination with, for example, class discovery or clustering. Also nonlinear versions of PCA (that use nonlinear combinations—kernel PCA (112) and PCA-similar methods such as PLS (partial least squares) (113)–are available.

3.3.1.3. Feature Selection by Clustering Gene Expression Profiles

As discussed in Sec. 3.2.2, genes can be subdivided into groups (clusters) based on the similarity in their gene expression profile. These clusters might contain genes that contribute similarly to the distinction between the different classes of compounds. If the latter is the case, genes within a cluster of gene expression profiles can be considered as one single feature (mathematically represented by the mean expression in this cluster).

3.3.2. Class Discovery

Compounds or drugs can, according to their effects in living organisms, be subdivided in different classes. These effects are reflected in the characteristic expression profiles of cells exposed to a certain compound (fingerprints, reference profile). The knowledge of these different classes enables classification of new substances. However, the current knowledge of these different classes might still be imperfect. The current taxonomy may contain classes that include substances with a high variability in expression profile. Also current class borders might be suboptimal. All this suggests that a refinement of the classification system and a rearrangement of the classes might improve predicting the behavior of new compounds.

Unsupervised methods such as clustering allow automatically finding the different classes/clusters in a group of microarray experiments, without knowing the properties of these classes in advance (i.e., the classification system of the compounds to which the cells were exposed to is unknown). A cluster, in general, will group microarray experiments (or the associated xenobiotics) with a certain degree of similarity in their experiment expression profile or fingerprint. The distinct clusters identified by the clustering procedure will—at least partially—match with the existing classification used for grouping compounds. However, it is not excluded that novel yet unknown entities or classes might originate from these analyses.

Several methods [e.g., hierarchical clustering (114), *K*-means clustering (115), self-organizing maps (108)] discussed in Sec. 3.2.2.3 can also be used in this context (i.e., clustering of the experiment expression profiles or columns of the expression matrix instead of clustering the gene expression profiles or rows of the expression matrix). For some methods (e.g., *K*-means that are not able to cluster limited sets of high-dimensional data points), clustering of the experiment profiles must be preceded by unsupervised feature extraction or dimensionality reduction (Sec. 3.3.1) (Fig. 10).

When clustering gene expression profiles is performed concurrently with or in preparation of the cluster analysis of the experiment profiles, this is called biclustering. For

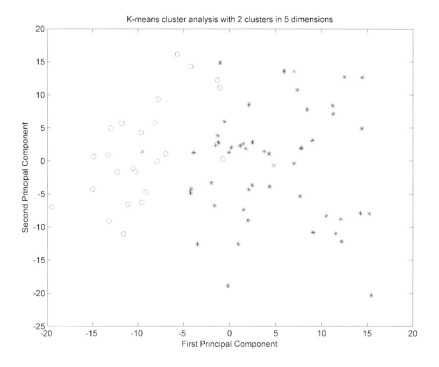

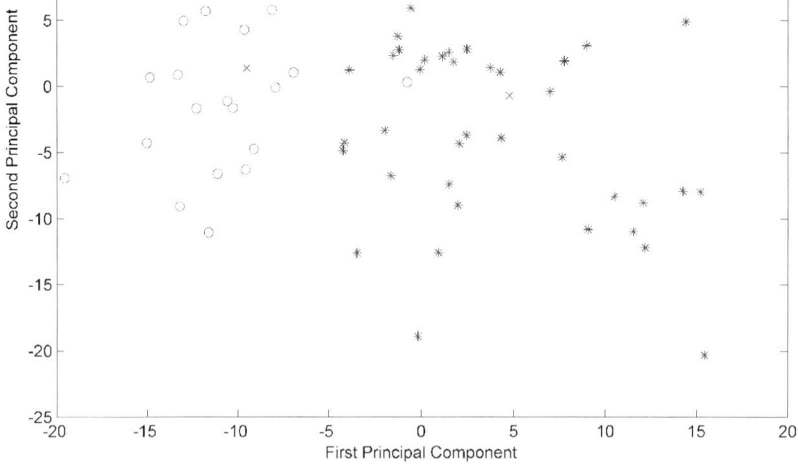

instance, hierarchical clustering simultaneously calculates a tree structure for both columns (experiments) and rows (genes) of the data matrix. One can also start with the cluster analysis of the gene expression profiles. Subsequently, one or a subset of these clusters (that seem biologically relevant) is selected. Cluster analysis of the experiments is based on this selection (114). Another technique is to find what is called "a bicluster" (106). A bicluster is defined as a subset of genes that shows a consistent expression profile over a subset of microarray experiments (and vice versa); i.e., one looks for a homogeneous submatrix of the expression matrix (116).

Figure 10 Illustration of class discovery by cluster analysis. The use of microarrays in toxicological gene expression is taking a lead from the work that has been carried out in the field of cancer research. From this field also the following example was taken because of its illustrative value. The dataset derived is from the study of Golub et al. (108) and describes a comparison between mRNA profiles of blood or bone marrow cells extracted from 72 patients suffering from two distinct types of acute leukemia (ALL or AML). Class labels (ALL or AML) were known in advance. In this example, it was demonstrated that the predefined classes could be rediscovered based on unsupervised learning techniques. Patients were clustered based on their experiment profiles (column vectors). Since each experiment profile consisted of the expression levels of thousands of genes (it represents a point in the n-dimensional space), its dimensionality was too high to use K-means clustering without prior dimensionality reduction. Dimensionality was reduced by PCA. The five principal components with the largest eigenvalues were retained and K-means clustering (two clusters) was performed in the five-dimensional space. Patients assigned to the first cluster are represented by circles, patients belonging to the second cluster by stars. Patients with ALL are in blue, and patients with AML are in red. Cluster averages are indicated by black crosses. For the ease of visualization, the experiments (patients) are plotted on the first two principal components. Note that all patients of the first cluster have AML and that almost all patients (with one exception) of the second cluster have ALL.

3.3.3. Class Prediction

Predictive toxicogenomics tries to predict the toxicological endpoints of compounds, with unknown properties or side-effects, by using high-throughput measurements, such as microarrays. This implicates that first the class membership of the novel compound needs to be predicted. Subsequently, the properties of the unknown compound will be derived through extrapolation of the characteristics of the reference members of the class of compounds to which the unknown compound was predicted to belong.

To be able to predict the class membership of novel compounds, a classifier has to be built. Based on a set of features and a training set (reference database), a classifier model [like neural networks (111), support vector machines (112), linear discriminant analysis (111), Bayesian networks (117,118), etc.] will be trained. This means that the parameters of the model will be determined using the data in the training set (Fig. 11). This classifier is subsequently used to predict the class membership of a novel compound.

4. CONCLUSIONS AND PERSPECTIVES

Conclusively, the use high-throughput molecular biological data have much to offer the mechanistic and predictive toxicologist. The impact of these data on toxicological research will grow with the size of public datasets and reference databases. The combination and interpretation of all the data generated will be a major computational challenge for the future that can only be tackled by an integrated effort of both experts in toxicology and data analysis.

ACKNOWLEDGMENTS

Bart De Moor is a full professor at the K.U. Leuven. Kathleen Marchal is a postdoctoral researcher of the Belgian Fund for Scientific Research (FWO-VLaanderen); Frank De Smet is a postdoctoral research assistant of the K.U. Leuven. Kristof

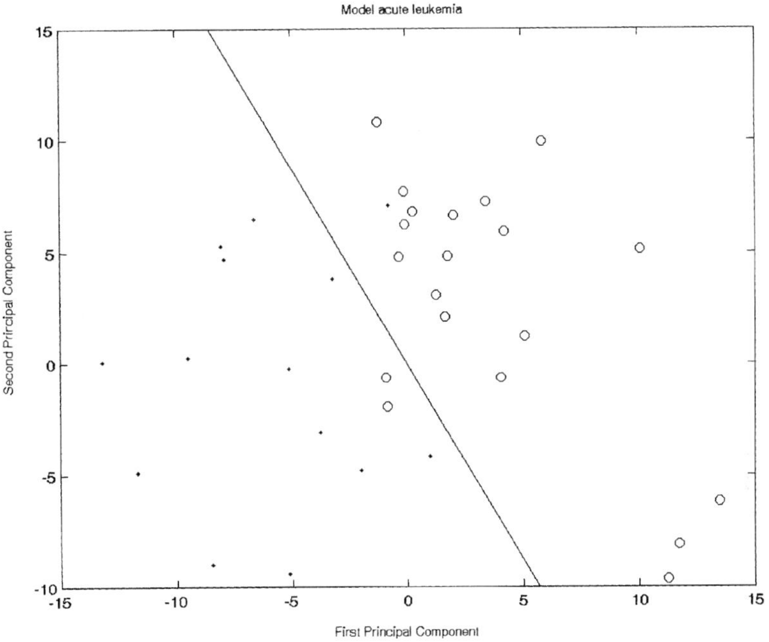

Figure 11 Example of a predictive method. This example resumes the example of Fig. 10 and illustrates the application of a classification model to predict the class membership of patients with acute leukemia based on their experiment profile. A linear classification model was built using Linear Discriminant Analysis based on the first two ($m=2$) principal components of the patients of a training set containing 38 patients. The line in this figure represents the linear classifier for which the parameters were derived using the patients of the training set. Only the patients of the test set (remaining 34 patients) are shown (after projection onto the principal components of the training set). The patients above the line are classified as ALL and below as AML. Note that this resulted in three misclassifications. Test set: $\circ$ = ALL, $\cdot$ = AML.

Engelen is research assistant of the IWT. B. This work is partially supported by: (1) IWT projects: STWW-00162, GBOU-SQUAD-20160; (2) Research Council KULeuven: GOA AMBIORICS, IDO genetic networks; (3) FWO projects: G.0115.01, G.0413.03, and G.0388.03; (4) IUAP V-22 (2002–2006); (5) FP6-NCE Biopattern.

REFERENCES

1. Ulrich R, Friend SH. Toxicogenomics and drug discovery: will new technologies help us produce better drugs? Nat Rev Drug Discov 2002; 1:84–88.
2. Gerhold D, Lu M, Xu J, Austin C, Caskey CT, Rushmore T. Monitoring expression of genes involved in drug metabolism and toxicology using DNA microarrays. Physiol Genomics 2001; 5:161–170.
3. Waring JF, Ciurlionis R, Jolly RA, Heindel M, Ulrich RG. Microarray analysis of hepatotoxins in vitro reveals a correlation between gene expression profiles and mechanisms of toxicity. Toxicol Lett 2001; 120:359–368.
4. Waring JF, Gum R, Morfitt D, Jolly RA, Ciurlionis R, Heindel M, Gallenberg L, Buratto B, Ulrich RG. Identifying toxic mechanisms using DNA microarrays: evidence that an experimental inhibitor of cell adhesion molecule expression signals through the aryl hydrocarbon nuclear receptor. Toxicology 2002; 181:537–550.
5. Amin RP, Hamadeh HK, Bushel PR, Bennett L, Afshari CA, Paules RS. Genomic interrogation of mechanism(s) underlying cellular responses to toxicants. Toxicology 2002; 181:555–563.
6. Greenbaum D, Luscombe NM, Jansen R, Qian J, Gerstein M. Interrelating different types of genomic data, from proteome to secretome: 'oming in on function. Genome Res 2001; 11:1463–1468.
7. Clarke PA, te Poele R, Wooster R, Workman P. Gene expression microarray analysis in cancer biology, pharmacology, and drug development: progress and potential. Biochem Pharmacol 2001; 62:1311–1336.
8. Nuwaysir EF, Bittner M, Trent J, Barrett JC, Afshari CA. Microarrays and toxicology: the advent of toxicogenomics. Mol Carcinog 1999; 24:153–159.
9. Hamadeh HK, Amin RP, Paules RS, Afshari CA. An overview of toxicogenomics. Curr Issues Mol Biol 2002; 4:45–56.

10. Pennie WD, Kimber I. Toxicogenomics; transcript profiling and potential application to chemical allergy. Toxicol In Vitro 2002; 16:319–326.

11. Naudts B, Marchal K, De Moor B, Verschoren A. Is it realistic to infer a gene network from a small set of microarray experiments. Internal Report ESAT/SCD K.U.Leuven. http://www. esat.kuleuven. ac.be/~sistawww/cgi-bin/pub.pl.

12. Pennie WD. Use of cDNA microarrays to probe and understand the toxicological consequences of altered gene expression. Toxicol Lett 2000; 112:473–477.

13. de Longueville F, Surry D, Meneses-Lorente G, Bertholet V, Talbot V, Evrard S, Chandelier N, Pike A, Worboys P, Rasson JP, Le Bourdelles B, Remacle J. Gene expression profiling of drug metabolism and toxicology markers using a low-density DNA microarray. Biochem Pharmacol 2002; 64:137–149.

14. Gant TW. Classifying toxicity and pathology by gene-expression profile—taking a lead from studies in neoplasia. Trends Pharmacol Sci 2002; 23:388–393.

15. Pennie WD. Custom cDNA microarrays; technologies and applications. Toxicology 2002; 181–182:551–554.

16. Brazma A, Hingamp P, Quackenbush J, Sherlock G, Spellman P, Stoeckert C, Aach J, Ansorge W, Ball CA, Causton HC, Gaasterland T, Glenisson P, Holstege FC, Kim IF, Markowitz V, Matese JC, Parkinson H, Robinson A, Sarkans U, Schulze-Kremer S, Stewart J, Taylor R, Vilo J, Vingron M. Minimum information about a microarray experiment (MIAME)—toward standards for microarray data. Nat Genet 2001; 29:365–371.

17. Gerhold DL, Jensen RV, Gullans SR. Better therapeutics through microarrays. Nat Genet 2002; 32(suppl):547–551.

18. Duggan DJ, Bittner M, Chen Y, Meltzer P, Trent JM. Expression profiling using cDNA microarrays. Nat Genet 1999; 21:10–14.

19. Lipshutz RJ, Fodor SP, Gingeras TR, Lockhart DJ. High density synthetic oligonucleotide arrays. Nat Genet 1999; 21:20–24.

20. Epstein JR, Leung AP, Lee KH, Walt DR. High-density, microsphere-based fiber optic DNA microarrays. Biosens Bioelectron 2003; 18:541–546.

21. Southern EM. DNA microarrays. History and overview. Methods Mol Biol 2001; 170:1–15.

22. Blohm DH, Guiseppi-Elie A. New developments in microarray technology. Curr Opin Biotechnol 2001; 12:41–47.

23. Brown PO, Botstein D. Exploring the new world of the genome with DNA microarrays. Nat Genet 1999; 21:33–37.

24. Kerr MK, Martin M, Churchill GA. Analysis of variance for gene expression microarray data. J Comput Biol 2000; 7:819–837.

25. Schuchhardt J, Beule D, Malik A, Wolski E, Eickhoff H, Lehrach H, Herzel H. Normalization strategies for cDNA microarrays. Nucleic Acids Res 2000; 28:E47.

26. Yue H, Eastman PS, Wang BB, Minor J, Doctolero MH, Nuttall RL, Stack R, Becker JW, Montgomery JR, Vainer M, Johnston R. An evaluation of the performance of cDNA microarrays for detecting changes in global mRNA expression. Nucleic Acids Res 2001; 29:E41–E41.

27. Kerr MK, Churchill GA. Experimental design for gene expression microarrays. Biostatistics 2001; 2:183–201.

28. Yang YH, Speed T. Design issues for cDNA microarray experiments. Nat Rev Genet 2002; 3:579–588.

29. Churchill GA. Fundamentals of experimental design for cDNA microarrays. Nat Genet 2002; 32(suppl):490–495.

30. Baldi P, Long AD. A Bayesian framework for the analysis of microarray expression data: regularized *t*-test and statistical inferences of gene changes. Bioinformatics 2001; 17:509–519.

31. Marchal K, Engelen K, De Brabanter J, Aerts S, De Moor B, Ayoubi T, Van Hummelen P. Comparison of different methodologies to identify differentially expressed genes in two-sample cDNA microarrays. J Biol Syst 2002; 10:409–430.

32. Yang YH, Dudoit S, Luu P, Lin DM, Peng V, Ngai J, Speed TP. Normalization for cDNA microarray data: a robust

composite method addressing single and multiple slide systematic variation. Nucleic Acids Res 2002; 30:el5.

33. Colantuoni C, Henry G, Zeger S, Pevsner J. SNOMAD (Standardization and NOrmalization of MicroArray Data): web-accessible gene expression data analysis. Bioinformatics 2002; 18:1540–1541.

34. Engelen K, Coessens B, Marchal K, De Moor B. MARAN: a web-based application for normalizing micro-array data. Bioinformatics 2003; 19:893–894.

35. Wolfinger RD, Gibson G, Wolfinger ED, Bennett L, Hamadeh H, Bushel P, Afshari C, Paules RS. Assessing gene significance from cDNA microarray expression data via mixed models. J Comput Biol 2001; 8:625–637.

36. Kerr MK, Churchill GA. Bootstrapping cluster analysis: assessing the reliability of conclusions from microarray experiments. Proc Natl Acad Sci USA 2001; 98:8961–8965.

37. Didier G, Brezellec P, Remy E, Henaut A. GeneANOVA—gene expression analysis of variance. Bioinformatics 2002; 18:490–491.

38. Newton MA, Kendziorski CM, Richmond CS, Blattner FR, Tsui KW. On differential variability of expression ratios: improving statistical inference about gene expression changes from microarray data. J Comput Biol 2001; 8:37–52.

39. Ideker T, Thorsson V, Siegel AF, Hood LE. Testing for differentially-expressed genes by maximum-likelihood analysis of microarray data. J Comput Biol 2000; 7:805–817.

40. Rocke DM, Durbin B. A model for measurement error for gene expression arrays. J Comput Biol 2001; 8:557–569.

41. Tusher VG, Tibshirani R, Chu G. Significance analysis of microarrays applied to the ionizing radiation response. Proc Natl Acad Sci USA 2001; 98:5116–5121.

42. Pan W. A comparative review of statistical methods for discovering differentially expressed genes in replicated microarray experiments. Bioinformatics 2002; 18:546–554.

43. Long AD, Mangalam HJ, Chan BY, Tolleri L, Hatfield GW, Baldi P. Improved statistical inference from DNA microarray

data using analysis of variance and a Bayesian statistical framework. Analysis of global gene expression in *Escherichia coli* K12. J Biol Chem 2001; 276:19937–19944.

44. Troyanskaya OG, Garber ME, Brown PO, Botstein D, Altman RB. Nonparametric methods for identifying differentially expressed genes in microarray data. Bioinformatics 2002; 18:1454–1461.
45. Dudoit S, Yang YH, Callow MJ, Speed TP. Statistical methods for identifying differentially expressed genes in replicated cDNA microarray experiments. Technical Report #578, Stanford University, 2000:1–38.
46. Storey JD, Tibshirani R. Statistical significance for genome-wide studies. Proc Natl Acad Sci USA 2003; 100:9440–9445.
47. Moreau Y, De Smet F, Thijs G, Marchal K, De Moor B. Functional bioinformatics of microarray data: from expression to regulation. Proceedings of the IEEE 2002; 30:1722–1743.
48. De Moor B, Marchal K, Mathys J, Moreau Y. Bioinformatics: organisms from Venus, technology from Jupiter, algorithms from Mars. Eur J Control 2003; 9:237–278.
49. Eisen MB, Spellman PT, Brown PO, Botstein D. Cluster analysis and display of genome-wide expression patterns. Proc Natl Acad Sci USA 1998; 95:14863–14868.
50. Troyanskaya O, Cantor M, Sherlock G, Brown P, Hastie T, Tibshirani R, Botstein D, Altaian RB. Missing value estimation methods for DNA microarrays. Bioinformatics 2001; 17:520–525.
51. De Smet F, Mathys J, Marchal K, Thijs G, De Moor B, Moreau Y. Adaptive quality-based clustering of gene expression profiles. Bioinformatics 2002; 18:735–746.
52. Tou JT, Gonzalez RC. Pattern classification by distance functions. Pattern Recognition Principles. Adison-Wesley, 1979:75–109.
53. Kohonen T. Self-Organizing Maps. Berlin, Germany: Springer-Verlag, 1997.
54. Tamayo P, Slonim D, Mesirov J, Zhu Q, Kitareewan S, Dmitrovsky E, Lander ES, Golub TR. Interpreting patterns of

gene expression with self-organizing maps: methods and application to hematopoietic differentiation. Proc Natl Acad Sci USA 1999; 96:2907–2912.

55. Tavazoie S, Hughes JD, Campbell MJ, Cho PJ, Church GM. Systematic determination of genetic network architecture. Nat Genet 1999; 22:281–285.

56. Ghosh D, Chinnaiyan AM. Mixture modelling of gene expression data from microarray experiments. Bioinformatics 2002; 18:275–286.

57. Lukashin AV, Fuchs R. Analysis of temporal gene expression profiles: clustering by simulated annealing and determining the optimal number of clusters. Bioinformatics 2001; 17:405–414.

58. Yeung KY, Fraley C, Murua A, Raftery AE, Ruzzo WL. Model-based clustering and data transformations for gene expression data. Bioinformatics 2001; 17:977–987.

59. Herrero J, Valencia A, Dopazo J. A hierarchical unsupervised growing neural network for clustering gene expression patterns. Bioinformatics 2001; 17:126–136.

60. Heyer LJ, Kruglyak S, Yooseph S. Exploring expression data: identification and analysis of coexpressed genes. Genome Res 1999; 9:1106–1115.

61. Hastie T, Tibshirani R, Eisen MB, Alizadeh A, Levy R, Staudt L, Chan WC, Botstein D, Brown P. 'Gene shaving' as a method for identifying distinct sets of genes with similar expression patterns. Genome Biol 2000; 1:RESEARCH0003.

62. Ben Dor A, Shamir R, Yakhini Z. Clustering gene expression patterns. J Comput Biol 1999; 6:281–297.

63. Sharan R, Shamir R. CLICK: a clustering algorithm with applications to gene expression analysis. Proc Int Conf Intell Syst Mol Biol 2000; 8:307–316.

64. Thijs G, Moreau Y, De Smet F, Mathys J, Lescot M, Rombauts S, Rouze P, De Moor B, Marchal K. INCLUSive: INtegrated Clustering, Upstream sequence retrieval and motif Sampling. Bioinformatics 2002; 18:331–332. http://www.esat.kuleuven.ac.be/~dna/BioI/Software.html.

65. Brazma A, Parkinson H, Sarkans U, Shojatalab M, Vilo J, Abeygunawardena N, Holloway E, Kapushesky M, Kemmeren P, Lara GG, Oezcimen A, Rocca-Serra P, Sansone SA. ArrayExpress—a public repository for microarray gene expression data at the EBI. Nucleic Acids Res 2003; 31:68–71.

66. Xia X, Xie Z. AMADA: analysis of microarray data. Bioinformatics 2001; 17:569–570.

67. Azuaje F. A cluster validity framework for genome expression data. Bioinformatics 2002; 18:319–320.

68. Bittner M, Meltzer P, Chen Y, Jiang Y, Seftor E, Hendrix M, Radmacher M, Simon R, Yakhini Z, Ben Dor A, Sampas N, Dougherty E, Wang E, Marincola F, Gooden C, Lueders J, Glatfelter A, Pollock P, Carpten J, Gillanders E, Leja D, Dietrich K, Beaudry C, Berens M, Alberts D, Sondak V. Molecular classification of cutaneous malignant melanoma by gene expression profiling. Nature 2000; 406:536–540.

69. McShane LM, Radmacher MD, Freidlin B, Yu R, Li MC, Simon R. Methods for assessing reproducibility of clustering patterns observed in analyses of microarray data. Bioinformatics 2002; 18:1462–1469.

70. Marchal K, Thijs G, De Keersmaecker S, Monsieurs P, De Moor B, Vanderleyden J. Genome-specific higher-order background models to improve motif detection. Trends Microbiol 2002; 11:61–66.

71. Thijs G, Marchal K, Lescot M, Rombauts S, De Moor B, Rouzé P, Moreau Y. A Gibbs sampling method to detect overrepresented motifs in the upstream regions of coexpressed genes. J Comput Biol 2002; 9:447–464.

72. Thijs G, Lescot M, Marchal K, Rombauts S, De Moor B, Rouzé P, Moreau Y. A higher-order background model improves the detection of promoter regulatory elements by Gibbs sampling. Bioinformatics 2001; 17:1113–1122.

73. Brazhnik P, de la Fuente A, Mendes P. Gene networks: how to put the function in genomics. Trends Biotechnol 2002; 20:467–472.

74. Elowitz MB, Levine AJ, Siggia ED, Swain PS. Stochastic gene expression in a single cell. Science 2002; 297:1183–1186.

75. Rao CV, Wolf DM, Arkin AP. Control, exploitation and tolerance of intracellular noise. Nature 2002; 420:231–237.

76. Hartemink AJ, Gifford DK, Jaakkola TS, Young RA. Combining location and expression data for principled discovery of genetic regulatory network models. Pac Symp Biocomput 2002; 437–449.

77. Lee TI, Rinaldi NJ, Robert F, Odom DT, Bar-Joseph Z, Gerber GK, Hannett NM, Harbison CT, Thompson CM, Simon I, Zeitlinger J, Jennings EG, Murray HL, Gordon DB, Ren B, Wyrick JJ, Tagne JB, Volkert TL, Fraenkel E, Gifford DK, Young RA. Transcriptional regulatory networks in *Saccharomyces cerevisiae*. Science 2002; 298:799–804.

78. Banerjee N, Zhang MQ. Functional genomics as applied to mapping transcription regulatory networks. Curr Opin Microbiol 2002; 5:313–317.

79. Zhang Z, Gerstein M. Reconstructing genetic networks in yeast. Nat Biotechnol 2003; 21:1295–1297.

80. Laub MT, McAdams HH, Feldblyum T, Fraser CM, Shapiro L. Global analysis of the genetic network controlling a bacterial cell cycle. Science 2000; 290:2144–2148.

81. Guelzim N, Bottani S, Bourgine P, Kepes F. Topological and causal structure of the yeast transcriptional regulatory network. Nat Genet 2002; 31:60–63.

82. Rung J, Schlitt T, Brazma A, Freivalds K, Vilo J. Building and analysing genome-wide gene disruption networks. Bioinformatics 2002; 18(suppl 2):S202–S210.

83. Thieffry D, Salgado H, Huerta AM, Collado-Vides J. Prediction of transcriptional regulatory sites in the complete genome sequence of *Escherichia coli* K-12. Bioinformatics 1998; 14:391–400.

84. Liang S, Fuhrman S, Somogyi R. Reveal, a general reverse engineering algorithm for inference of genetic network architectures. Pac Symp Biocomput 1998; 18–29.

85. Akutsu T, Miyano S, Kuhara S. Algorithms for identifying Boolean networks and related biological networks based on matrix multiplication and fingerprint function. J Comput Biol 2000; 7:331–343.

86. Akutsu T, Miyano S, Kuhara S. Inferring qualitative relations in genetic networks and metabolic pathways. Bioinformatics 2000; 16:727–734.

87. Murphy K, Mian I. Modelling gene expression data using dynamic Bayesian networks. Technical Report 1999, Computer Science Division, University of California, Berkeley, CA. http://www.cs .berkeley.edu/~murphyk/publ.html.

88. Yoo C, Thorsson V, Cooper GF. Discovery of causal relationships in a gene-regulation pathway from a mixture of experimental and observational DNA microarray data. Pac Symp Biocomput 2002; 498–509.

89. Imoto S, Goto T, Miyano S. Estimation of genetic networks and functional structures between genes by using Bayesian networks and nonparametric regression. Pac Symp Biocomput 2002; 175–186.

90. Friedman N. Learning belief networks in the presence of missing values or hidden variables. Proceedings of the 14th International Conference on Machine Learning (ICML) 1997.

91. Pe'er D, Regev A, Elidan G, Friedman N. Inferring subnetworks from perturbed expression profiles. Bioinformatics 2001; 17(suppl 1):S215–S224.

92. Friedman N, Nachman I, Linial M, Pe'er D. Using Bayesian networks to analyze expression data. J Comput Biol 2000; 7:601–620.

93. Ong IM, Glasner JD, Page D. Modelling regulatory pathways in *E. coli* from time series expression profiles. Bioinformatics 2002; 18(suppl 1):S241–S248.

94. Smith VA, Jarvis ED, Hartemink AJ. Evaluating functional network inference using simulations of complex biological systems. Bioinformatics 2002; 18(suppl 1):S216–S224.

95. Shmulevich I, Dougherty ER, Kim S, Zhang W. Probabilistic Boolean networks: a rule-based uncertainty model for gene regulatory networks. Bioinformatics 2002; 18:261–274.

96. D'Haeseleer P, Liang S, Somogyi R. Genetic network inference: from co-expression clustering to reverse engineering. Bioinformatics 2000; 16:707–726.

97. Chen T, He HL, Church GM. Modeling gene expression with differential equations. Pac Symp Biocomput 1999; 29–40.

98. Wahde M, Hertz J. Coarse-grained reverse engineering of genetic regulatory networks. Biosystems 2000; 55:129–136.

99. Akutsu T, Miyano S, Kuhara S. Algorithms for inferring qualitative models of biological networks. Pac Symp Biocomput 2000; 293–304.

100. Kato M, Tsunoda T, Takagi T. Merring genetic networks from DNA microarray data by multiple regression analysis. Genome Inform Ser Workshop Genome Inform 2000; 11:118–128.

101. Alter O, Brown PO, Botstein D. Singular value decomposition for genome-wide expression data processing and modeling. Proc Natl Acad Sci USA 2000; 97:10101–10106.

102. Yeung MK, Tegner J, Collins JJ. Reverse engineering gene networks using singular value decomposition and robust regression. Proc Natl Acad Sci USA 2002; 99:6163–6168.

103. Holter NS, Mitra M, Maritan A, Cieplak M, Banavar JR, Fedoroff NV. Fundamental patterns underlying gene expression profiles: simplicity from complexity. Proc Natl Acad Sci USA 2000; 97:8409–8414.

104. Raychaudhuri S, Stuart JM, Altman RB. Principal components analysis to summarize microarray experiments: application to sporulation time series. Pac Symp Biocomput 2000; 455–466.

105. Ideker TE, Thorsson V, Karp RM. Discovery of regulatory interactions through perturbation: inference and experimental design. Pac Symp Biocomput 2000; 305–316.

106. Tanay A, Sharan R, Shamir R. Discovering statistically significant biclusters in gene expression data. Bioinformatics 2002; 18(suppl 1):S136–S144.

107. Park PJ, Pagano M, Bonetti M. A nonparametric scoring algorithm for identifying informative genes from microarray data. Pac Symp Biocomput 2001; 52–63.

108. Golub TR, Slonim DK, Tamayo P, Huard C, Gaasenbeek M, Mesirov JP, Coller H, Loh ML, Downing JR, Caligiuri MA, Bloomfield CD, Lander ES. Molecular classification of cancer:

class discovery and class prediction by gene expression monitoring. Science 1999; 286:531–537.

109. Xu J, Stolk JA, Zhang X, Silva SJ, Houghton RL, Matsumura M, Vedvick TS, Leslie KB, Badaro R, Reed SG. Identification of differentially expressed genes in human prostate cancer using subtraction and microarray. Cancer Res 2000; 60:1677–1682.

110. Wang K, Gan L, Jeffery E, Gayle M, Gown AM, Skelly M, Nelson PS, Ng WV, Schummer M, Hood L, Mulligan J. Monitoring gene expression profile changes in ovarian carcinomas using cDNA microarray. Gene 1999; 229:101–108.

111. Bishop CM. Neural Networks for Pattern Recognition. New York: Oxford University Press, 1995.

112. Suykens J, Van Gestel T, De Brabanter J, De Moor B, Vandewalle J. Least Squares Support Vector Machines. Singapore: World Scientific, 2002.

113. Johansson D, Lindgren P, Berglund A. A multivariate approach applied to microarray data for identification of genes with cell cycle-coupled transcription. Bioinformatics 2003; 19:467–473.

114. Alizadeh AA, Eisen MB, Davis RE, Ma C, Lossos IS, Rosenwald A, Boldrick JC, Sabet H, Tran T, Yu X, Powell JI, Yang L, Marti GE, Moore T, Hudson J Jr, Lu L, Lewis DB, Tibshirani R, Sherlock G, Chan WC, Greiner TC, Weisenburger DD, Armitage JO, Warnke R, Standt LM. Distinct types of diffuse large B-cell lymphoma identified by gene expression profiling. Nature 2000; 403:503–511.

115. Quackenbush J. Computational analysis of microarray data. Nat Rev Genet 2001; 2:418–427.

116. Sheng Q, Moreau Y, De Moor B. Biclustering microarray data by Gibbs sampling. Bioinformatics 2003; 19(suppl 2):II196–II205.

117. Moreau Y, Antal P, Fannes G, De Moor B. Probabilistic graphical models for computational biomedicine. Methods Inf Med 2003; 42:161–168.

118. Jordan M. Learning in Graphical Models. Cambridge, MA, London: MIT Press, 1999.

119. Cho RJ, Campbell MJ, Winzeler EA, Steinmetz L, Conway A, Wodicka L, Wolfsberg TG, Gabrielian AE, Landsman D, Lockhart DJ, Davis RW. A genome-wide transcriptional analysis of the mitotic cell cycle. Mol Cell 1998; 2:65–73.

REFERENCES AS INTRODUCTORY TO TOXICOGENOMICS

1. Clarke PA, te Poele R, Wooster R, Workman P. Gene expression microarray analysis in cancer biology, pharmacology, and drug development: progress and potential. Biochem Pharmacol 2001; 62:1311–1336.
2. Amin RP, Hamadeh HK, Bushel PR, Bennett L, Afshari CA, Paules RS. Genomic interrogation of mechanism(s) underlying cellular responses to toxicants. Toxicology 2002; 181–182:555–563.
3. Ulrich R, Friend SH. Toxicogenomics and drug discovery: will new technologies help us produce better drugs? Nat Rev Drug Discov 2002; 1:84–88.
4. Vrana KE, Freeman WM, Aschner M. Use of microarray technologies in toxicology research. Neurotoxicology 2003; 24:321–332.

REFERENCES AS INTRODUCTORY TO METHODOLOGICAL REVIEWS

1. Quackenbush J. Computational analysis of microarray data. Nat Rev Genet 2001; 2:418–427.

GLOSSARY

Additive error: This represents the absolute error on a measurement that is independent of the measured expression level. Consequently, the relative error is inversely proportional to the measured intensity and is high for measurements with low magnitude.

Bayesian network: This represents a mathematical model that allows both a compact representation of the joint probability distribution over a large number of variables, and an efficient way of using this representation for statistical inference. It consists of a directed acyclic graph that models the interdependencies between the variables, and a conditional probability distribution for each node with incoming edges.

Class discovery: This represents the automatic identification of the hidden classes in a dataset without a priori knowledge on the class distinction. The data reduction or grouping is derived solely from the data. This can be obtained by using unsupervised learning techniques, such as, clustering.

Classification/prediction: This represents determination for a certain experiment (microarray experiment of a certain compound) of its class membership based on a classifier or predictive model: objects are classified into known groups. Classification is based on supervised learning techniques.

Clustering: This represents unsupervised learning technique that organizes multivariate data into groups with roughly similar patterns, i.e., clustering algorithms group together genes (experiments) with a similar expression profile. Similarity is defined by the use of a specific distance measure.

Coexpressed genes: These are genes with a similar expression profile. Genes of which the behavior of the expression is similar in different conditions or at different time-points.

Data matrix: This is a mathematical representation of a complex microarray experiment. Each row represents the expression vector of a particular gene. Each column of the matrix represents an experimental condition. Each entry in the matrix represents the expression level of a gene in a certain condition.

Dedicated microarrays: These contain only a restricted number of genes, usually marker genes or genes characteristic for a certain toxicological endpoint. Using dedicated arrays offers the advantage of higher throughput screening of lead targets at a lower cost.

Diagnostic or investigative microarrays: These contain probes representing as much coding sequences of a genome as possible.

DNA Microarray: This is a high-throughput technology that enables the measurement of mRNA transcript levels at a genomic scale. DNA microarrays are produced by high density depositing thousands of individual spots (called probes) of synthetic, unique oligonucleotides or cDNA gene sequences to a solid substrate such as a glass microscope slide or a membrane.

Dye reversal experiment: This is a specific type of experimental design used for cDNA arrays. On the first array the test condition is labeled with Cy5 (red dye), while the reference is labeled with Cy3 (green dye). On the second array, the dyes are swapped, i.e., reference condition is labeled with Cy5 (red dye), while the test is labeled with Cy3 (green dye).

Dynamic experiment: This is a complex microarray experiment that monitors adaptive changes in the expression level elicited by administering the xenobiotic to the system under study. By sampling the system at regular time intervals during the time course of the adaptation, short-, mid-, and long-term alterations in xenobiotic-induced gene expression are measured.

Expression profile of a gene: This is a vector that contains the expression levels of a certain gene measured in the different experimental conditions tested; corresponds to the row in the data matrix.

Expression profile of an experiment/compound (also "fingerprint" or "reference pattern"): This is a vector that contains the expression levels of all genes measured in the specific experimental condition represented by the column; corresponds to the column in the data matrix.

FDR: The FDR (false discovery rate) is considered as a sensible measure of balance between the number of false positives and true positives. The FDR is the rate that the features called significant are truly null or the number of false positives among the features called significant.

Feature: This represents a gene (single feature) or combination of genes (complex feature) of which the expression levels are associated with a class distinction of interest (e.g., of which expression is switched on in one class and switched off in the other class).

Feature extraction: This represents a mathematical or statistical methodology that identifies the features that are most correlated with a specific class distinction.

Filtering: This represents removal of genes from the dataset of which the expression does not change over the tested conditions, i.e., genes that are not involved in the process studied.

Global normalization assumption: This is a general assumption stating that, from one biological condition to the next, only of a small fraction of the total number of genes shows an altered expression level and that symmetry exists in the number of genes for which the expression is upregulated vs. downregulated.

Mechanistic toxicogenomics: This involves the use of high-throughput technologies to gain insight into the molecular biological mechanism of a toxicological response.

Missing values: These are gene expression values that could not be accurately measured and that were omitted form the data matrix.

Multiple testing: When considering a family of tests, the level of significance and power are not the same as those for an individual test. For instance, a significance of $\alpha = 0.01$ indicates a probability of 1% of falsely rejecting the null hypothesis (e.g., assuming differential expression while there is none). This means that for a family of 1000 tests, say every 1000 genes tested, 10 would be expected to pass the test although not being differentially expressed. To limit this number of false positives in a multiple test, a correction is needed (e.g., Bonferroni correction).

Multiplicative error: This means that the absolute error on the measurement increases with the measurement magnitude. The relative error is constant, but the variance between replicate measurements increases with the mean expression value. Multiplicative errors cause signal-dependent variance of the residuals.

Network inference: This represents reconstruction of the molecular biological structure of regulatory networks from high-throughput measurements, i.e., deriving the causality relationships between genetic entities (proteins, genes) from the data.

PCA: Principal Component Analysis.

Predictive model or classifier: This represents a mathematical model (neural network, Bayesian model, etc.) of which the parameters are estimated by the use of a trainingsset (i.e., the reference database). The predictive model is subsequently used to predict the class membership of a novel compound, i.e., to assign a novel compound to a predefined class of compounds based on its expression profile.

Predictive toxicogenomics: This involves the prediction of the toxicological endpoints of compounds, with yet unknown properties or side-effects by the aid of high-throughput profiling experiments such as microarrays. A reference database of expression fingerprints of known compounds and a predictive model or classifier trained on this reference database are needed.

Preprocessing: This is a pretreatment process, that removes consistent and/or systematic sources of variation from the raw data.

Power: This represents the discriminant power of a statistical test (computed as $1-\beta$) and the probability of rejecting the null hypothesis when the alternative hypothesis is true. It can be interpreted as the probability of correctly rejecting a false null hypothesis. Power is a very descriptive and concise measure of the sensitivity of a statistical test, i.e., the ability of the test to detect differences.

Probes: These are the spots/oligonucleotides on the microarray that represent the different genes of the genome.

Reference databases: This is a compendium of characteristic expression profiles or fingerprints of well-described agents or compounds, for which both the toxicological endpoints and the molecular mechanisms resulting in them are characterized.

Rescaling microarray data: This represents transformation of the gene expression profiles by subtracting the mean expression level and by dividing by the standard deviation of the profile.

Significance: This represents the significance level of a statistical test, referred to as α, and the maximum *probability* of accidentally rejecting a true *null hypothesis* (a decision known as a *Type I error*).

Static experiments: This is a complex microarray experiment that tests the induced changes in expression under several conditions or in different genetic backgrounds (gene knock out experiments). Samples are taken when the steady state expression levels are reached.

SVD: Singular value decomposition.

Target: These are the labeled transcripts, present in the mRNA sample that is hybridized to the array.

Test statistic: This value is calculated from the data points (e.g., a mean) and used to evaluate a null hypothesis against an alternative hypothesis. In the framework of testing for differentially expressed genes, the null hypothesis states that the genes are not differentially expressed.

Toxicogenomics: This is a subdiscipline of toxicology that combines large scale gene/protein expression measurements and the expanding knowledge of genomics to identify and evaluate genome-wide effects of xenobiotics.

Underdetermination: The number of parameters to be estimated exceeds the number of experimental data points. The mathematical problem has no single solution.

4

Toxicological Information for Use in Predictive Modeling: Quality, Sources, and Databases

MARK T. D. CRONIN
School of Pharmacy and Chemistry,
John Moores University, Liverpool, U.K.

1. INTRODUCTION

All modeling processes require data. For predictive toxicology, the starting place for all models is the collation of toxicological data. This chapter will describe where toxicological data may be obtained and how they should be approached. The reader is also referred to a number of other papers where the reader may obtain further information (1,2) and the excellent collections of papers published in *Toxicology*, Volume 157, Issues 1–2 (3–14) and Volume 173, Issues 1–2 (15–28). It is hoped that this chapter will provide a source of information for further investigation. It is well beyond the remit of one chapter, and probably the knowledge of one scientist, to describe all sources of information. However, it is hoped that this chapter will provide a starting place to show how data may be obtained for predictive toxicology.

1.1. Fundamentals

For the development of models, data on a single chemical are normally required. It is generally not possible to create models and make predictions on mixtures of chemicals, simply because the ethos behind the modeling process is the association of an activity with the structure and/or properties of single chemicals. There are some exceptions to this rule; quantitative structure–activity relationship (QSAR) analyses of surfactants, for example, have been developed for mixtures by assuming that surfactants may be represented as single structures. Thus, for a mixture of surfactants with varying chain lengths, an average of the chain lengths may be taken (provided all components are assumed to be at approximately the same concentration). For properties and activities where chemical structure may be approximated in this manner, this approach is relatively successful. However, all discussion in this chapter will assume that single chemicals are being modeled. A further point to note is that many compounds are toxic as a result of their metabolism, and only the parent compound, and not the metabolite, is usually considered in a QSAR.

It should also be noted that nearly all predictive toxicology relates to organic chemicals. The development of quantitative models for metal ions and organo-metallic complexes is a separate issue and seldom considered. Generally speaking, the activities and properties of metal ions cannot be modelled in conjunction with purely organic molecules. This is because most commonly used descriptors (e.g., log P, molecular weight, etc.) cannot be applied to metal ions. Instead QSARs may be developed for metal ions in isolation (29). Despite this, the information contained in toxicity data for metal ions may be captured in the form of rules for knowledge-based expert systems. In this case, rules for metals may exist in the same database as those for organic fragments. In the same manner, it is difficult to parameterize the properties of organo-metallic complexes, thus it is difficult to develop QSARs for them (unless properties are measured). The difficulty of the development of QSARs for metal ions becomes a mute point, however, when the relatively small number of possible ions, in conjunction with the wealth of existing data, is considered.

It should not need stating that data for the development of models should be accurate, reliable, and consistent (see discussion of data quality below). The model developer, as well as the model user, should appreciate the impact of the robustness, or otherwise, of the data on which the model is developed. Further both parties should be aware of the meaning of the endpoint, and its mechanistic relevance. Gone are the days when models should be accepted from workers that have trawled the literature or toxicological databases, merely to collate data for model development without recourse to the quality of the data. Neither should models be accepted where overly multivariate approaches have been used to force relationships between descriptors and activity. It is also important that modelers should note that there will be different degrees of difficulty in modeling different toxicity endpoints. Endpoints with complex mechanisms (e.g., carcinogenicity) and endpoints that cover multiple observations (e.g., developmental toxicity which combine a variety of different "sub-endpoints" such as growth retardation, structural and functional anomalies, etc.) should be handled with caution. It is recommended that consultation with an expert in these areas is made before modeling is attempted.

All biological data for use in QSAR must be presented properly. For quantitative endpoints that involve a concentration provoking a specific biological response (e.g., LC_{50}, EC_{50}, etc.), concentrations must be reported and used in molar units. This is a fundamental assumption as the relative effect of one molecule as compared to another is being modeled (i.e., a free energy relationship). The use in predictive toxicology of units of concentration by weight or ppm, etc. is not acceptable. It is also acceptable to use the relative biological response of the application of an equimolar concentration to a system, although this latter approach may be insensitive and is not favored. For categoric data (i.e., presence or absence of carcinogenicity, etc.), the classification must be consistent for the complete data set. Historically, a number of classification criteria have been available [e.g., from the European Union, United States of America, Japan, Organization of Economic Co-operation and Development, (OECD) etc.]. The modeler must ensure if categoric data are collated from a number of sources, that the classifications are all based on a single and consistent set of criteria.

The depiction of ionized structures in QSAR modeling causes much confusion. Many substances will ionize at the varying physiological pHs within a mammal for instance. Ionization may dramatically alter the properties of a compound such as its aqueous solubility and hydrophobicity. For most modeling approaches, the unionized structure is assumed and utilized. While this simplifies greatly the modeling procedure, the modeller should always be aware of the drawbacks of such assumptions. In the same manner, most substances formulated as salts are also treated as the neutral unionized molecule.

A final, and fundamental, point to make regarding data in predictive toxicology studies is that they should be reported in all publications, i.e., all data used in the study should be made available. Ideally a publication will have a straightforward table that includes the unambiguous chemical names and, if possible, Chemical Abstracts Service (CAS) number as well as the toxicity data used in the model. If it is not possible to include all data (e.g., if the table is too large for publication), then data should be made available online and/or on request from the authors. Editors and referees in peer-reviewed journals must get into the habit of rejecting papers that do not provide the primary information source. There are many reasons for the data to be published. These include the fact that it would allow the user of the model to assess the chemical domain, and other modelers to develop models. For too long commercial modelers have been able to publish reports on predictive models without quoting their data sources, in an attempt to maintain the commercial advantage, and then use the publication (which sometimes contains neither data nor model) to boost the commercial credibility of the software.

1.2. Data, Information, and Knowledge (And Do Not Forget Wisdom and Understanding)

"Knowledge management" is a science in itself and separate from predictive toxicology. It is, however, useful to apply some of the terms coined in "knowledge management" to predictive toxicology. More specifically, we may define the relationship between knowledge, information, and data.

- *Data* are a collection of facts, preferably stored in a usable format.
- *Information* is the data in addition to an understanding of its meaning.
- *Knowledge* is the information in addition to an understanding of its meaning.
- *Wisdom* is knowledge in addition to experience (of a user).

Knowledge management theory asserts that each level has more "value" associated with it than the levels above it. In the context of toxicology, this author interprets the above definitions as follows:

- *Data* are the raw facts regarding the toxicological assessment of a chemical. These may be the dose-response concentrations and effects for a lethality endpoint, etc. These are normally stored in an (electronic) form suitable for analysis and manipulation, e.g., Probit analysis.
- *Information* is the interpretation of the data. Thus, a concentration causing a 50% effect (LC_{50}, EC_{50}, etc.) or a categoric description of activity (toxic or non-toxic).
- Given information for a series of chemicals, *knowledge* can be obtained by assessing the relationship between the biological effects and physico-chemical and/or structural features. This may result in a method to predict toxic events.

- *Wisdom* allows a user to apply a predictive technique (knowledge) successfully. This may include when a prediction is likely to be accurate and, from experience, what factors will influence the prediction.

It should be obvious from the above definitions and their applications in predictive toxicology that data, information, and knowledge can be obtained from machines. Wisdom is more indefinable, and normally requires subtle human judgment, on the basis of expertise.

2. REQUIREMENTS FOR TOXICOLOGICAL DATA FOR PREDICTIVE TOXICITY

For predictive toxicology, all data must be in a usable format. An unambiguous endpoint, or description of activity, must be clearly associated with a chemical structure.

2.1. Where to Find Data

As will become obvious from this chapter, there are many different sources of toxicity data. At the outset the investigator must be clear what is being searching for. There are clear distinctions that must be drawn, and each type of data source has advantages for different uses. From the opinion of the author (a modeler), the following uses of data may be envisaged:

- Data for development (and validation) of general models, i.e., for a heterogeneous group of compounds, that could be used to predict the toxicity of a very broad range of toxicants. Compilations of data [e.g., fathead minnow database, carcinogenic potency database—(see Sec. 3)] are suitable sources of information.
- Data for development (and validation) of models for specific groups of compounds, such as a chemical class or congeneric series. Small groups of data are required.
- Data for the toxicological assessment of particular compounds or groups of compounds (i.e., not modeling per se). The online databases (e.g., TOXNET) provide good starting points for such exercises.

The sources of toxicological information should come as no surprise to the majority of scientific researchers, e.g., journal articles, books, product information sheets, databases, etc. Sources and typical examples are given in Table 1 .

2.2. Definition of Quality of Data

2.2.1. Toxicological Data

Data for modeling purposes must be of a high quality. The assessment of the quality of any item is normally a subjective decision, often biased by one's own experiences and preferences. It is certainly no easy task to describe the quality of toxicological information. Some, general, attempts have been made, especially with regard to regulatory information. Klimisch et al. (30), for example, evaluated the quality of data for use in hazard and risk assessment. Reliability was differentiated into four categories as described in Table 2. Many of the issues raised by Klimisch et al. (30) provide a useful starting point for the assessment of data quality in predictive toxicology.

With regard to predictive toxicology, the quality of individual data must be assessed (i.e., the toxicity value for an individual chemical) as well as the quality of the data set itself (i.e., the reliability of the toxicity values for a selection or group of chemicals). Intrinsically, this makes the definition of quality more complex than that described by Klimisch et al. (30). For predictive methods, several criteria with which to define quality have been defined for the biological data sets, and it should also be remembered that the assessment of quality should also be applied to the physico-chemical descriptors as noted below (31,32). Suggested criteria by which to assess biological data sets are summarized below:

Reliability: Data must be reliable, in other words they must accurately represent the toxic endpoint being assessed.

Consistent: Biological data must be consistent, i.e., the test result should be repeatable any number of times with low error.

Reproducible: The biological endpoint must be reproducible; in other words if the test or assay is repeated in a number of laboratories, then the result should be the same.

Table 1 General Sources of Toxicological Information and Typical Examples

Source of information	Example	Comment
Single (or small number of tests)—typically in a journal article	Reports of test results published in journals. These may be found, for instance, by a literature search	Searching primary literature is a time-consuming process with no guarantee of finding a significant number of data
Large numbers of data in a journal/book article. These normally may be from a single laboratory for a single endpoint	Fathead minnow database (37)	Generally high quality data. However there are very few such databases in existence!
Compilations of toxicity data. Normally for the same endpoint, but values from many different laboratories	*Vibrio fischeri* (formerly *Photobacterium phosphoreum*) toxicity compilation (65)	Data quality is variable, and often little assessed. Some compounds may have multiple entries, meaning a decision is required as to which to use
Publicly available databases	TOXNET	Large amounts of data, most of which may not be suitable for predictive modeling. Often databases may be searched only one compound at a time

Closed (corporate) databases	Most large pharmaceutical and agrochemical companies willhave their own databases	These may be an excellent source of toxicological information, but will be available only within the company
Commercial databases	MDL Toxicity Database	Commercial databases may be expensive for modeling and contain data of variable (and unchecked) quality
Internet (search engines)	Searching the internet may provide information	Time-consuming with no guarantee of finding data. Data retrieved will be of a highly variable quality
Handbooks, Pharmacoepias, etc.	Pesticide Manual, British Pharmacopoaeia	Large amounts of data, however, the data may be of variable quality

Table 2 Categories of Reliability that May Be Associated with Toxicological Information for Risk and Hazard Assessment Purposes

Code	Category	Brief definition
1	Reliable without restriction	Performed according to an accepted international test guideline (and GLP)
2	Reliable with restriction	Well-documented studies, not necessarily to an accepted guideline or GLP
3	Not reliable	Use of non-acceptable methods, or exposure by a non-relevant route
4	Not assignable	Insufficient experimental details provided

From Ref. 30.

Meaningful and relevant to desired endpoint: The design of the test must provide data and information that is appropriate for the use of the information. Thus, for risk assessment of chemicals in the environmental, a no-observable effect concentration (NOEC) may be more relevant than an LC_{50} (however, it should be noted that an NOEC is less likely to be a high quality datum as it is often obtained by relatively fewer experimental observations than an LC_{50} which requires a more stringent dose–response relationship). Also a pertinent species must be used, e.g., data for a marine fish are of less use to address risk assessment in freshwaters.

Standardized, validated, and statistically relevant assay: High quality data will be associated with well defined tests. For instance, those performed to OECD test guidelines and Good Laboratory Practice (GLP) should be considered to be high quality data.

Low biological error: A high quality test should produce a result with low variability and thus only a small degree of biological error. It should be remembered in modeling that we require all data (biological and physicochemical) to have low error. The modeling process should describe the relationship between the biological and chemical information and go no further. Over-fitting of models often means that error is being modeled.

2.2.2. Physico-Chemical and Structural Data

Similar principles to the assessment of the biological activity may be applied to physicochemical and/or structural property data used in developing predictive models (31,32). All chemical measurements (e.g., octanol–water partitioning) will have an experimental error associated with them. Calculations of properties (e.g., logarithm of octanol–water partition coefficient, log K_{ow}) will also have calculation errors associated with them. Conformationally dependent properties used in predictive modeling have also been shown to be susceptible to error. Most molecules at body temperature are extremely flexible structures and will have many energetically possible conformations. Thus, properties such as molecular dimensions which are dependent on conformation will therefore be affected. Calculated molecular orbital properties are also highly dependent of conformation (33). It is the view of this author that some properties from molecular orbital calculations (dipole moment in particular) vary so greatly depending on conformation, that their practical use in predictive toxicology is questionable.

Despite the problems associated with physicochemical properties, some types of structural properties and indices have no error or uncertainty associated with them. Simple empirical descriptors of molecules, such as indicator variables for specific functional groups, are implicitly unambiguous. Topological indices derived from graph theory (such as molecular connectivities) also fall into this category.

2.3. Impact of Poor Quality Data

Poor quality toxicological or physicochemical and structural data will introduce error and uncertainty into any model derived from them. This, is turn, means that models developed on poor quality data must be expected to have a lower statistical fit and be less predictive than models developed on higher quality data for the same endpoint. Thus, to develop an accurate, predictive and meaningful model, high quality data are a prerequisite.

3. HIGH QUALITY DATA SOURCES FOR PREDICTIVE MODELING

All modelers desire high quality data with which to develop their algorithms. While high quality data may be available in isolation throughout the scientific literature, the reality for predictive toxicology is that there are very few sources of large data sets of high quality toxicity values (as defined in the previous section). In the area of human health effects, high quality data include those from the Carcinogenic Potency DataBase (i.e., a database which includes those compiled by the United States National Toxicology Program— see Sec. 5.1 for more details); as well as the specific endpoints such as for skin sensitization (34) and eye and skin irritation (35,36). In the environmental area, the key data set that has underpinned much research in this area is the 96 hr fathead minnow LC_{50} database (37). Another source of high quality data sets is the 40 hr inhibition of growth of *Tetrahymena pyrifonnis* (38–41). Also worthy of mention in this regard are the data available for in vitro endocrine disruption (typically estrogenicity) (42–45). Despite the fact that some data sets meet the criteria for high quality, care must still be taken in their use. In the comparison of two carcinogenicity data sets that are normally to be considered to be of high quality, Gottmann et al. (46) estimated that for chemicals that were common to both data sets, there was only a 57% concordance (i.e., little more than chance) between the data. In addition, Helma et al. (47) pointed out a number of errors, inconsistencies, and ambiguities in these databases.

4. DATABASES PROVIDING GENERAL SOURCES OF TOXICOLOGICAL INFORMATION

There are a number of useful databases a modeler may wish to search to obtain information. The excellent articles by Wright (10) and Wukovitz (12) provide very useful summaries of resources and an indication of how to search them to obtain the most information for minimum outlay.

4.1. MDL Toxicity Database (Formerly Registry of Toxic Effects of Chemical Substances)

The Registry of Toxic Effects of Chemical Substances (RTECS) is one of the more well known and established toxicological data repositories. The RTECS constitutes a compendium of data extracted from the open scientific literature. The data are arranged in alphabetical order by prime chemical name. Six types of toxicity data are included in the database: primary irritation; mutagenic effects; reproductive effects; tumourigenic effects; acute toxicity; and other multiple dose toxicity. Specific numeric toxicity values such as LD_{50}, LC_{50}, TDL_0, and TCL_{50} are noted as well as the species studied and route of administration used. The bibliographic source is listed for each citation. It must be stressed that no attempt has been made to evaluate the studies cited in RTECS and the assessment of the quality (or otherwise) of the data is at the discretion of the user.

Until December 2001 RTECS was compiled, maintained, and updated by the United States National Institute of Occupational Safety and Health (NIOSH). Mandated through United States legislation, the first list, known as the "Toxic Substances List," was published on 28 June, 1971. It included toxicological data for approximately 5000 chemicals. Since that time, the list has continuously grown and been updated, and its name changed to RTECS. As of January, 2001, the last update of the database by NIOSH, RTECS contained information on 152,970 chemicals. Since the last update of RTECS by NIOSH in January 2001 a "PHS Trademark Licensing Agreement" has been put in place for RTECS. This non-exclusive licensing agreement provides for the transfer and continued development of the "RTECS Database and its Trademark" to MDL Information Systems, Inc. (www.mdli.com), a wholly owned subsidiary of Elsevier Science, Inc. Under this agreement, MDL will be responsible for updating, licensing, and marketing and distributing RTECS. The database is now available as the MDL Toxicity Database.

4.2. TOXNET

TOXNET is the United States National Library of Medicine's (NLM) reference sources for toxicology and environmental health (sis.nlm.nih.gov/Tox/ToxMain.html). It is an open and freely available web-based facility to search for toxicological information and is well introduced by Wexler (15). It provides an extremely powerful tool to obtain toxicological data, particularly on discreet chemicals. This facility is actually a cluster of databases on toxicology, hazardous chemicals, and related areas. The databases within TOXNET may be searched individually, or all simultaneously. When searching all databases simultaneously, the number of records retrieved for each of the TOXNET databases is displayed and the individual results may be viewed subsequently. The databases in TOXNET may be searched using any combination of words, chemical names, and numbers, including CAS numbers.

The TOXNET cluster of databases includes the following specific reference sources. (For more details follow the links from the TOXNET web-page (toxnet.nlm.nih.gov).)

4.2.1. Hazardous Substances Data Bank

The Hazardous Substances Data Bank (HSDB) is a wide ranging database containing over 4500 records including human and animal toxicity data, safety and handling, environmental fate, and other information. The data bank focuses on the toxicology of potentially hazardous chemicals. Specifically, it collates information relating to human exposure, industrial hygiene, emergency handling procedures, environmental fate, regulatory requirements, and related areas. All data are referenced and derived from a core set of books, government documents, technical reports, and selected primary journal literature. The HSDB is peer-reviewed by a committee of experts in the major subject areas within the scope of the data bank.

4.2.2. Integrated Risk Information System

Integrated Risk Information System (IRIS) has been compiled by the U.S. Environmental Protection Agency (EPA) and contains over 500 chemical records. These data support human health risk assessment and focus particularly on hazard identification and dose–response assessment. Further, the toxicity data are reviewed by work groups of EPA scientists and represents EPA consensus. Among the key data provided in IRIS are EPA carcinogen classifications, unit risks, slope factors, oral reference doses, and inhalation reference concentrations.

4.2.3. GENE-TOX

The GENE-TOX program was established to select assay systems for evaluation, review data in the scientific literature, and recommend proper testing protocols and evaluation procedures for these systems. The resulting GENE-TOX database was created by the U.S. EPA and contains mutagenicity test data on over 3000 chemicals. It was established following an expert peer review of the open scientific literature.

4.2.4. Chemical Carcinogenesis Research Information System

Chemical Carcinogenesis Research Information System (CCRIS) is a scientifically evaluated and fully referenced data bank, developed and maintained by the U.S. National Cancer Institute (NCI). It contains over 8000 chemical records with carcinogenicity, mutagenicity, tumor promotion, and tumor inhibition test results. The test results recorded in CCRIS have been reviewed by experts in carcinogenesis and mutagenesis. Data are derived from studies cited in primary journals, current awareness tools, NCI reports, and other special sources.

4.2.5. TOXLINE

TOXLINE is the data base of U.S. National Library of Medicine (NLM) providing information on the biochemical, pharmacological, physiological, and toxicological effects of drugs and other chemicals. It contains more than 3 million bibliographic citations, almost all with abstracts and/or indexing terms and CAS numbers. TOXLINE references are drawn from various sources grouped into two major parts: (i) TOXLINE Core and (ii) TOXLINE Special. Both parts of TOXLINE allow versatile searching and have a number of capabilities to display information.

4.2.6. Developmental and Reproductive Toxicology and Environmental Teratology Information Center

Developmental and Reproductive Toxicology and Environmental Teratology Information Center (DART/ETIC) is a database that includes current and older literature on developmental and reproductive toxicology. More specifically, it covers teratology and other aspects of developmental and reproductive toxicology. DART/ETIC is funded by the U.S. EPA, the National Institute of Environmental Health Sciences, the National Center for Toxicological Research of the Food and Drug Administration, and the NLM. It contains over 100,000 references to literature published since 1965.

4.2.7. U.S. EPA's Toxic Chemical Release Inventory

Toxic Chemical Release Inventory (TRI) is an annually compiled series of databases that constitute the toxic releases files. It contains information on the annual estimated releases of toxic chemicals to the environment and is based upon data collected by the U.S. EPA. The data within TRI cover air, water, land, and underground injection releases, as well as transfers to waste sites, and waste treatment methods and efficiency, as reported by industrial facilities around the United States.

4.2.8. ChemIDplus

ChemIDplus is a free, web-based search system that provides access to structure and nomenclature authority files used for the identification of chemical substances cited in NLM databases. The database contains over 367,000 chemical records, of which over 142,000 include chemical structures, and is searchable by name, synonym, CAS number, molecular formula, classification code, locator code, and structure. ChemIDplus also provides structure searching and direct links to many biomedical resources at NLM and on the internet for chemicals of interest.

4.3. European Union International Uniform Chemical Information Database

Partly in response to regulatory requirements within the European Union, an electronic data reporting and management tool has been developed by the European Commission. This tool, the International Uniform ChemicaL Information Database (IUCLID) is very well described by Heidorn et al. (48,49). The IUCLID provides a platform to support all three steps of the risk assessment process, i.e., data collection, priority-setting, and risk assessment. IUCLID is distributed by the European Chemicals Bureau (European Commission Joint Research Centre, Ispra, Italy)—more (up to date) details on this database, how to obtain it and pricing may be obtained from ecb.ei.jrc.it/IUCLID/. In addition a comprehensive IUCLID Guidance Document has been prepared by the European Chemical Industry Council (Brussels) and is freely downloadable from www.cefic.be/activities/hse/mgt/hpv/Iuclid/iuclid.htm. It must be noted that while IUCLID is a useful regulatory tool for assessing chemicals its use as a source of chemical information or to obtain toxicity data for predictive toxicology is limited.

4.4. Commercial Sources of Toxicity Data

There are a number of subscription (fee-paying) sources of toxicity data which may be useful to be the predictive toxicologist. It is not possible to list all of these in detail, but the following may provide useful places to start. The C-QSAR package developed by Prof. Corwin Hansch is available from the Biobyte Corporation (www.biobyte.com). The C-QSAR package contains two main databases of QSAR equations relating biological and physicochemical activities to structural parameters (these models can be applied to chemicals from the same chemical class). The "BIO" database currently contains over 5600 equations for biological endpoints, and the "PHYS" database over 7500 equations for physicochemical properties. While the "BIO" database contains information on all biological activities, there is a significant proportion of toxicological data that may be retrieved. A further advantage is that modeling and interpretation have already been performed on the data. More details of the databases and their philosophy are available from Hansch and Gao (50) and Hansch et al. (51).

Another company specializing in toxicological databases with respect to predictive toxicology is TerraBase Inc. (see www.terrabase-inc.com for more details). TerraBase Inc. markets a number of databases including TerraTox Explorer, which contains physicochemical properties and acute toxicity endpoints on more than 15,000 substances in almost 100 species of aquatic and terrestrial organisms. The TerraTox–Pesticides database has information on more than 1500 pesticides, pesticide metabolites, and degradation products for almost 100 species of aquatic and terrestrial organisms.

5. DATABASES PROVIDING SOURCES OF TOXICOLOGICAL INFORMATION FOR SPECIFIC ENDPOINTS

As well as the databases providing information on a range of toxic effects, there are a number of databases for individual toxic effects that may be of use to the predictive toxicologist. A number of such databases are described below.

5.1. Carcinogenicity and Mutagenicity

Sources of data and information for carcinogenicity and mutagenicity are well reviewed by Junghans et al. (17), Richard and Williams (2), and Young (23); this section merely summarizes the information provided in those papers. Richard and Williams (2) separate sources of carcinogenicity and mutagenicity data broadly into those that are publicly available and those commercially available. This section of this chapter cannot go into any of the detail and insight provided by Richard and Williams (2), to which the reader is referred for a more complete review in this area. This chapter summarizes the main publicly available databases, with an emphasis on those applicable to predictive toxicology.

A number of sources of mutagenicity and carcinogenicity data are available through TOXNET (see Sec. 4.2). In particular, the GENE-TOX and CCRIS databases provide a large amount of information. A further source of information is the Carcinogenic Potency Database (CPDB), which is available online at potency.berkeley.edu/cpdb.html. This has been a well utilized source of data for predictive toxicology. It summarizes the results of chronic, long-term animal cancer tests. The CPDB provides a single, standardized, and easily accessible (i.e., online) database. Both qualitative and quantitative information on positive and negative experiments are reported, including all bioassays from the National Cancer Institute/National Toxicology Program (NCI/NTP) and experimental results from the general literature that meet a set of inclusion criteria. Analyses of 5152 experiments on 1298 chemicals are presented. Full information regarding the test is reported including the species, strain, and sex of test animal; features of experimental protocol such as route of administration, duration of dosing, dose level(s) in milligrams or kilograms body weight/day, and duration of experiment; histopathology and tumor incidence; carcinogenic potency (i.e., TD_{50} value); and literature citation. Issues with regard to data quality in carcinogenic databases are dealt with by Gottmann et al. (46).

5.2. Teratogenicity

A large number of resources are available to obtain information regarding the developmental toxicity (teratogenicity) of chemicals. Many of these resources are reviewed by Polifka and Faustman (18). Of these the Teratogen Information

System (TERIS) (http://depts.washin.gton.edu/druginfo/Vaccine/TERIS.html) is a computerized database designed to assist physicians or other health care professionals in assessing the risks of possible teratogenic exposures in pregnant women. TERIS includes data on teratogenicity, transplacental carcinogenesis, embryonic or fetal death, and fetal and perinatal pharmacologic effects of drugs and selected environmental agents. The database consists of a series of agent summaries, each of which are based on a thorough review of published clinical and experimental literature. In TERIS, analysis of each agent's teratogenicity has been made on the basis of the reproducibility, consistency, and biological plausibility of available clinical, epidemiological, and experimental data. Reproducibility is judged by whether similar findings have been obtained in independent studies. Concordance is considered to be particularly important if the studies are of different design and if the types of anomalies observed in various studies are consistent. Effects seen in animal investigations are weighed more heavily if the exposure is similar in dosage and route to that encountered clinically and if the species tested are closely related to humans phylogenetically.

A further database REPROTOX (http://reprotox.org/) provides on-line information regarding the reproductive effects of prescription, over-the-counter, and recreational drugs as well as industrial and environmental chemicals. The summaries, inclusive of human, animal, and in vitro data, cover the available information on every aspect of human reproduction including fertility, male exposure, and lactation. Information derived from medical and scientific sources is documented in selected references.

5.3. Other (Human Health) Endpoints

Useful databases of information (of varying quality) are available for a range of human health endpoints. As with all endpoints covered in this chapter, this is not an exhaustive collation but biased towards significant data collections and those of use to the predictive toxicologist. An excellent source of human hazard information is "*Sax's Dangerous Properties of Industrial Materials*" (52). Now in its 10th edition, it provides data on toxicological, fire, reactivity, explosive potential, and regulatory properties of chemicals. Sax's contains over 23,500 entries with information on names (with 108,000 domestic and international synonyms); physical properties, including solubility and flammability data; toxicity data such as those for skin and eye irritation, as well as mutagenic, teratogenic, reproductive, carcinogenic, human, and acute lethal effects and occupational classifications. Sax's has remained for many years the starting place for toxicologists to obtain fundamental information regarding commercial chemicals. The data it contains are, however, less used in predictive toxicology, probably due to the difficulty in retrieving the data (Sax's is in hardcopy only and all three volumes comprise about 4000 pages), and the issues with varying data quality and variations in test methods.

While there must be a large number of acute and chronic toxicity data in existence, very few databases are openly available for modeling purposes. Of those available, for the reasons mentioned above, few, if any, could be considered as "high quality" compilations. The largest compilation of, for instance, rat toxicity data are those in the MDL Toxicity (formerly RTECS) database. There have been several attempts to develop models from these data (53,54). It must be stressed, however, that only low quality models must be expected from data derived from the former RTECS database due to the lack of consistency of the data. Other published compilations of mammalian toxicity data are available including data on over 2900 no-observed effect levels for more than 600 chemical substances (55).

The European Centre for Ecotoxicology and Toxicology of Chemicals (ECETOC) (www.ecetoc.org) has produced a number of reports that bring together significant collections of data. Generally these reports have been the result of industry-based Task Forces. For instance Bagley et al. (35) provided a data bank of 149 in vivo rabbit eye irritation data for 132 chemicals. Care was taken to select only high quality data, and the chemicals tested were known to be available at high and consistent purity and were expected to be stable in storage. All in vivo data compiled had been generated since 1981 in studies carried out according to OECD Test Guideline 405 and following the principles of Good Laboratory Practice. The data were obtained from tests normally using at least three animals evaluated at the same time, involving instillation of 0.1 mL (or equivalent weight) into the conjunctival sac, and in which observations were made at least 1, 2, and 3 days after instillation. The chemicals are ranked for eye irritation potential on the basis of a "modified maximum average score."

On a similar basis to the eye irritation data, Bagley et al. (36) listed in vivo rabbit skin irritation data for 176 chemicals. All chemicals were known to be of high or consistent purity and stable on storage. The chemicals were tested undiluted in in vivo studies, apart from those chemicals where high concentrations could be expected to cause severe effects. In vivo data were generated in studies carried out since 1981 according to OECD Test Guideline 404 and following the principles of Good Laboratory Practice. The data were obtained from tests normally using at least three rabbits evaluated at the same time, involving application of 0.5 g or 0.5 mL to the flank under semiocclusive patches for 4 hr, and in which observations were made at least 24, 48, and 72 hr after removal of the patch. The chemicals represented a wide range of chemical classes acids, acrylates/methacrylates, alcohols, aldehydes, alkalis, amines, brominated derivatives, chlorinated solvents, esters, ethers, fatty acids and mixtures, fragrance oils, halogenated aromatics, hydrocarbons (unsaturated), inorganics, ketones, nitriles, phenolic derivatives, S-containing compounds, soaps/surfactants, and triglycerides and different degrees of irritancy. They are ranked for skin irritation potential on the basis of a "primary irritation index."

With regard to skin sensitization, Cronin and Basketter (34) published the results of over 270 in vivo skin sensitization tests (mainly from the guinea pig maximization test). All data were obtained in the same laboratory and represent one of the few occasions when large amounts of information from corporate databases have been released into the open literature. A larger database of animal and human studies for

1034 chemicals is described by Graham et al. (56). While the data are not reported in the manuscript, it is noted in the methods that they may be obtained on application to the authors.

Also related to skin toxicity, published skin permeability data for over 90 chemicals have been compiled and are available in databases compiled by workers such as Flynn (57) and Wilschut et al. (58). These data are mainly in vitro assessments of percutaneous absorption using excised human skin. The quality and accuracy of the data in these compilations must be treated with extreme caution, however, as they are collations of historic data from the literature (31,59).

Many other data exist regarding the effects of drugs on animals and humans. One of the most convenient sources of pharmacokinetic data is the compilation from Hardman et al. (60). The subject matter relating to predictive ADME is outside of the remit and scope of this chapter. Good reference sources for further information are Duffy (61), Testa et al. (62), and van de Waterbeemd et al. (63). Also relating to drugs, web-resources for drug toxicity are described by Wolfgang and Johnson (19) and for veterinary toxicologists by Poppenga and Spoo (28).

5.4. Environmental Endpoints

5.4.1. United States EPA ECOTOX Database

The U.S. EPA allows free on-line public access to the ECOTOX database. This provides ecotoxicological information regarding single chemical substances for aquatic and terrestrial life forms. The primary source of information is the peer-reviewed literature. It also includes independently compiled data files provided by various U.S. and international governmental agencies. At the time of writing, the most recent update of the ECOTOX database was 20 December, 2002. The database is available at http://www.epa.gov/ecotox. It contains ecotoxicological information on several thousand chemical substances and may be searched by name, CAS number, observed effect group, and publication year. Advanced searches allow the user to focus on more specific criteria such as study site type (e.g., laboratory, field), exposure media (e.g., freshwater, soil), route of chemical exposure (e.g., oral, diet), and statistically derived endpoints (e.g., LD_{50}, NOEL). Search results can be downloaded as an ASCII delimited file format.

5.4.2. Miscellaneous Other Sources of Environmental Data

There is a considerable quantity of data available for environmental endpoints, e.g., toxicity and fate, that is well reviewed by Russom (20). Probably the largest source of environmental information is the ECOTOX database (described above). Within that database are a number of collections of high quality data, the most notable is the fathead minnow (*Pimephales promelas*) 96 hr LC_{50} for over 600 chemicals. A broad range of chemicals have been tested; Russom et al. (37) have classified them into groupings of narcotics (three distinct groups), oxidative phosphorylation uncouplers, respiratory inhibitors, electrophiles and proelectropbiles, acetylcholinesterase inhibitors, and central nervous system seizure agents. All tests were performed at the Center for Lake Superior Environmental Studies, University of Wisconsin, Superior, WI, U.S.A. using the same protocol and mostly by the same workers. As such they meet the criteria for classification as high quality data.

ECETOC also established a task force to assemble a high-quality database on aquatic toxicity from published papers. The database described by Solbe et al. (64) and available from ECETOC, contains over 2200 records of effects for about 360 substances on 121 aquatic species. While there are a considerable quantity of data available within the ECETOC aquatic toxicity database, consideration must, of course, be given to the likely variable quality of these data with regard to inter-laboratory variation.

With regard to in vitro aquatic test systems, particular attention should be paid to the work of Prof. T.W. Schultz at the College of Veterinary Medicine, The University of Tennessee, Knoxville TN. Over the past two decades, Prof. Schultz has measured toxicity data for well in excess of 2000 single chemicals. The toxicity data are the inhibition of growth over 40 hr of the ciliated protozoan *T. pyriformis*. The data compiled by Prof. Schultz represent one of, if not the, largest data collections for a single test, performed under the same laboratory conditions, and under the direction of a single worker. Used correctly, it is, without doubt, a immensely important data collection for predictive toxicologists. Full methods are provided in Ref. 40; data for aromatic compounds are available in Refs. 38,41; and for aliphatic compounds in Ref. 39.

A further large compilation of in vitro toxicity data was published by Kaiser and Palabrica (65), which comprise the 5,15, and 30 min inhibition of bioluminescence from the marine bacterium *Vibrio fischeri* (formerly *Photobacterium phosphoreum*), the so-called Microtox assay. Data are available for over 1300 individual chemicals. Examination of the compilation reveals multiple entries for a number of compounds. As an example of data quality, it is obvious that there is considerable variability in the data (several orders of magnitude for some compounds). The reason for this is likely to be that, while the Microtox test is highly standardized, these data have been retrieved from at least 30 literature sources, with no regard for quality assurance. Cronin and Schultz (66) analyzed some of the data through the use of QSARs and demonstrated inconsistencies in some data values. The compilation from Kaiser and Palabrica (65) forms the basis of the commercially available TerraTox–*Vibrio fischeri* database from TerraBase Inc. (www.terrabase-inc.com).

In the area of endocrine disruption, a number of significant data that meet the criteria for high quality are available. It must be remembered that endocrine disruption is a collection of physiological effects that may be brought about by disruption of the endocrine (hormonal) system. Most comprehensive data sets are available for the action of a particular hormone, or some receptor-binding event, most commonly relating to the estrogen receptor. With an endpoint such as estrogenicity, there are two possibilities for predictive modeling. The first is whether or not a compound may bind to the estrogen receptor (and so may cause a response) and the second is the relative potency of the response. Of the more significant data sets, Fang et al. (42) reported results from a validated estrogen receptor competitive binding assay for 230 chemicals. The chemicals tested included both natural and xenoestrogens and the binding affinity covers a 10^6-fold range. Another high quality data set is for chemicals evaluated in the Glaxo–Wellcome *Saccharomyces cerevisiae*-based Lac-Z reporter. In this assay, relative gene activation is compared to 17β-estradiol. Notable amongst these data sets are the data provided by Schultz et al. (43) for 120 aromatic compounds, and 37 substituted anilines (44). The estrogenic activity of more than 500 chemicals including natural substances, medicines, pesticides, and industrial chemicals was reported by Nishihara et al. (45). These authors developed a simple and rapid screening method using the yeast two-hybrid system based on the ligand-dependent interaction of nuclear hormone receptors with coactivators. A large compilation of data is also available commercially from TerraBase Inc. (www.terrabase-inc.com). The TerraTox–Steroids–quantitative receptor binding assay (RBA) database contains information on over 3000 individual chemicals. These data are normalized to 17β-estradiol, progesterone, testosterone, mibolerone, androgen, and others.

5.5. Pesticides

Web resources for pesticide toxicity are well reviewed by Felsot (26). The classic source of information regarding pesticides still, however, remains *The Pesticide Manual* (67). *The Pesticide Manual* contains 812 detailed main entries as well as abbreviated details covering 598 superseded products. Entries cover herbicides, fungicides, insecticides, acaricides, nematicides, plant growth regulators, herbicide safeners, repellents, synergists, pheromones, beneficial microbial and invertebrate agents, rodenticides, and animal ectoparasiticides. Each entry includes information on structure, name, physicochemical properties, commercial history (e.g., patent, manufacturer, etc.), mode of action, mammalian toxicology profile, ecotoxicity data, and information regarding environmental fate. The *e-Pesticide Manual* is a fully searchable CD containing all the information from within the printed *Pesticide Manual* and includes access to more than 6000 trade names.

6. SOURCES OF CHEMICAL STRUCTURES

For researchers developing predictive toxicology models from literature data, there is often a great need to convert the chemical names into structures. When this author first started searching for structures, it was often with little more than a chemical catalogue, a good textbook on chemical nomenclature and the considerable assistance of knowledgeable colleagues. This process has become a lot easier in the past decade with increased availability of computational, and on-line, data searching techniques. The sources of chemical structure represent places where the author currently visits, and recommends students to visit, to obtain structural information.

6.1. Chemfinder

There are a number of online services available at www.chemfinder.com (or chemfinder.cambridgesoft.com/). Chemfinder itself is a free service, although there are only a limited number of searches available per day. ChemlNDEX is a commercial fee-charging service, that allows unlimited searches of the same database, plus additional services such as the ability to export data. Chemfinder is searchable by chemical name (or partial names using "*" as a wildcard), CAS number, as well as molecular formula or weight. The free database includes chemical structures (pictorial as well as 2D and 3D representations), physical properties (e.g., boiling and melting points, etc.) as well as hyperlinks to other information. The fee-paying database includes access to National Cancer Institute data.

6.2. The Merck Index

The Merck Index (68) is one of the more famous (and traditional) sources of chemical, structural, and biological information. It represents an internationally recognized, one-volume encyclopedia of industrial chemicals, drugs, and biological agents. Each monograph in the encyclopedia (each record in the database) discusses a single chemical entity or a small group of very closely related compounds. Updates contain material not yet available in print. Information available in the Merck Index includes structure, names, physical and toxicity data, therapeutic and commercial uses, caution and hazard information, as well as chemical, biomedical, and patent literature references. The Merck Index is also available online; information about this subscription service is obtainable at www.merck.cora/pubs/index/online.html.

6.3. Pharmacopoeias

Pharmacopoeias contain much information regarding drug and veterinary substances. Those commonly use or include the United States, British, and European Pharmacopoeias. The U.S. Pharmacopoeia (www.usp.org) has more than 4000 monographs featuring the standards for prescription and non-prescription drug ingredients and dosage forms, dietary supplements, medical devices, and other healthcare products. The monographs include descriptions, requirements, tests, analytical procedures, and acceptance criteria. Information on the British Pharmacopoaeia can be obtained from www.pharmacopoeia.org.uk and on the European Pharmacopoaeia from www.pheur.org.

6.4. Chemical Catalogues

There are many chemical catalogues available, usually free of charge for hard copy from chemical companies. These normally allow for fast searching providing a full name is known. A good example is Sigma-Aldrich Ltd (www.sigmaaldrich.com) This company supplies over 85,000 biochemicals and organic chemicals. The hard copy of the Aldrich catalogue has been invaluable to this author, in addition the catalogue can be searched online (including searches for partial names) at www.sigmaaldrich.com. Chemical catalogues are useful places to relate name to structure and often provide other physicochemical information (e.g., melting and boiling points) and hazard data.

6.5. SMILECAS Database

If the CAS number of a compound is known, this can very easily be converted into a Simplified Molecular Line Entry System (SMILES) string using the Syracuse Research Corporation SMILECAS database. This is free to download as part of the EPISUITE package from www.epa.gov/opptintr/exposure/docs/episuitedl.htm. The latest version (EPISUITE ver 3.11) contains well in excess of 100,000 SMILES notations. Thus, the users of the programs may enter a CAS number and obtain the SMILES notation. This function will also work in batch mode in other algorithms in EPISUITE.

6.6. Open NCI Database

A large amount of information and the ability to search for, and build, specific molecules is provided by the so-called "Erlangen/Bethesda/Frederick Data and Online Services" at cactus.nci.nih.gov. This is a collaboration between researchers at the Computer Chemistry Centre at the University of Erlangen Nuremberg, Germany, and members of the Laboratory of Medicinal Chemistry, NCI, and NEE. The intention of these services is to provide to the public structures, data, tools, programs and other useful information. No access restrictions are placed on their use. All data that are provided, as far as they originated at the NCI, are by definition public domain. A description of the services is provided at the web-site (cactus.nci.nih.gov) and in (69,70).

7. SOURCES OF FURTHER TOXICITY DATA

The requirement for databases of high quality toxicological information for predictive toxicology and associated modeling has long been recognized. However, as noted above, there are few credible sources of such data. Despite the lack of publicly available data, many high quality data are being held by companies and corporations. There have been many calls for these data to be made available to modellers (71). The reality is, however, that very few toxicity data are likely to be released from industry. Only legislative pressure may change this situation, although there will always be a compromise with the proprietary nature of such commercial data. Some efforts to compile data are noted below, although, at least in the shortterm, it is difficult to consider that they will be able to collate any more than the data described in this chapter.

7.1. International Life Sciences Institute—Health and Environmental Sciences Institute (ILSI-HESI)—Lhasa Structure–Activity Relationship Database Project

The ILSI-HESI (Washington, D.C., U.S.A.) and Lhasa Ltd. (University of Leeds, England) have established an international collaboration to develop a structure–activity relationship database [see www.ilsi.org/file/SARFactSheetWEB.pdf and (72) for more details]. The aim of this database is to collect toxicity test results, including toxicological data, as well as physicochemical and molecular structural information for predictive toxicology. It is anticipated that part of the database will be madc publicly available online, providing companies with a channel to publish toxicological data. A further part of the database will only be available to sponsors. Database management tools will be supplied with the database so that companies can use the software to store proprietary data. As yet this project is only at the data collection stage, and few details regarding the project are published.

7.2. DSSTox

A further database project is being supported and maintained at the U.S. EPA. The Distributed Structure-Searchable Toxicity (DSSTox) Database Network will operate through a common file format for publicly available toxicity data. The database will include the molecular structures of all tested chemicals. Much effort is being made to improve the structure-linked access to publicly available sources of toxicity information, outlining current web-based resources as well as two new database initiatives for standardizing and consolidating public chemical toxicity information. Regulators and researchers will be able to place toxicity data from many public sources into a single file, and access and explore these data using chemical structure. The standardized DSSTox data files, to be made available on the Web, should assist scientists in improving models for predicting the potential toxicity of chemicals from molecular structure (73).

7.3. Use of High Production Volume Chemicals Data

There are considerable quantities of toxicity, fate, physicochemical and other data available for the so-called high production volume (HPV) chemicals. In the United States, HPV chemicals are those with production or importation volumes in excess of 1 million pounds (in weight), and there are over 3000 such substances. The U.S. EPA has requested information on these chemicals. Generally speaking, the information available on these chemicals will be of a high quality and produced from tests performed to recognized standards. While many of these data are confidential, Walker et al. (74) recommend their use for the development of QSARs to predict properties of non-HPV chemicals.

8. CONCLUSIONS

Toxicology data are required for the modeling of toxicological events. To obtain high quality models of toxicological events, data set of high quality and comparable data are required. Criteria for assessing the quality of toxicological information and data sets for predictive toxicology are outlined in this chapter. The sad reality is that, while many data are potentially available for modeling, very few data sets can be considered to be of high quality. Modelers are encouraged to ascertain the quality of their data before modeling to ensure they comprehend its limitations. Also, users of models must understand issues relating to quality of data.

REFERENCES

1. Kaiser KLE. Toxicity data sources. In: Cronin MTD, Livingstone DJ, eds. Predicting Chemical Toxicity and Fate. Boca Raton FL, U.S.A.: CRC Press, 2004; 17–29.
2. Richard AM, Williams CR. Public sources of mutagenicity and carcinogenicity data: use in structure–activity relationship models. In: Benigni R, ed. Quantitative Structure–Activity Relationship (QSAR) Models of Mutagens and Carcinogens. Boca Raton, FL, USA: CRC Press, 2003:145–173.
3. Wexler P. Introduction to special issue on digital information and tools. Toxicology 2001; 157:1–2.
4. Wexler P. TOXNET: an evolving web resource for toxicology and environmental health information. Toxicology 2001; 157:3–10.
5. Poore LM, King G, Stefanik K. Toxicology information resources at the Environmental Protection Agency. Toxicology 2001; 157:11–23.
6. Brinkhuis RP. Toxicology information from US government agencies. Toxicology 2001; 157:25–49.
7. Stoss FW. Subnational sources of toxicology information: state, territorial, tribal, county, municipal, and community resources online. Toxicology 2001; 157:51–65.
8. Kehrer JP, Mirsalis J. Professional toxicology societies: web based resources. Toxicology 2001; 157:67–76.
9. Montague P, Pellerano MB. Toxicology and environmental digital resources from and for citizen groups. Toxicology 2001; 157:77–88.
10. Wright LL. Searching fee and non-fee toxicology information resources: an overview of selected databases. Toxicology 2001; 157:89–110.
11. Keita-Ouane F, Durkee L, Clevenstine E, Ruse M, Csizer Z, Kearns P, Halpaap A. The IOMC organisations: a source of chemical safety information. Toxicology 2001; 157:111–119.

12. Wukovitz LD. Using internet search engines and library catalogs to locate toxicology information. Toxicology 2001; 157:121–139.
13. Sharpe JF, Eaton DL, Marcus CB. Digital toxicology education tools: education, training, case studies, and tutorials. Toxicology 2001; 157:141–152.
14. South JC. Online resources for news about toxicology and other environmental topics. Toxicology 2001; 157:153–164.
15. Wexler P. Introduction to special issue (Part II) on digital information and tools. Toxicology 2002; 173:1.
16. Hakkinen PJ, Green DK. Alternatives to animal testing: information resources via the internet and world wide web. Toxicology 2002; 173:3–11.
17. Junghans TB, Sevin IF, Ionin B, Seifried H. Cancer information resources: digital and online sources. Toxicology 2002; 173:13–34.
18. Polifka JE, Faustman EM. Developmental toxicity: web resources for evaluating risk in humans. Toxicology 2002; 173:35–65.
19. Wolfgang GHI, Johnson DE. Web resources for drug toxicity. Toxicology 2002; 173:67–74.
20. Russom CL. Mining environmental toxicology information: web resources. Toxicology 2002; 173:75–88.
21. Winter CK. Electronic information resources for food toxicology. Toxicology 2002; 173:89–96.
22. Goldberger BA, Polettini A. Forensic toxicology: web resources. Toxicology 2002; 173:97–102.
23. Young RR. Genetic toxicology: web resources. Toxicology 2002; 173:103–121.
24. Patterson J, Hakkinen PJ, Wullenweber AE. Human health risk assessment: selected internet and world wide web resources. Toxicology 2002; 173:123–143.
25. Greenberg GN. Internet resources for occupational and environmental health professionals. Toxicology 2002; 173:145–152.

26. Felsot AS. WEB resources for pesticide toxicology, environmental chemistry, and policy: a utilitarian perspective. Toxicology 2002; 173:153–166.

27. Busby B. Radiation information and resources on-line. Toxicology 2002; 173:167–178.

28. Poppenga RH, Spoo W. Internet resources for veterinary toxicologists. Toxicology 2002; 173:179–189.

29. Walker JD, Enache M, Bearden JC. QSARs for predicting toxicity of metal ions. Environ Toxicol Chem 2003; 22:1916–1935.

30. Klimisch H-J, Andreae M, Tillmann U. A systematic approach for evaluating the quality of experimental toxicological and ecotoxicological data. Reg Toxicol Pharmacol 1997; 25:1–5.

31. Cronin MTD, Schultz TW. Pitfalls in QSAR. J Mol Struct (Theochem) 2003; 622:39–51.

32. Schultz TW, Cronin MTD. Essential and desirable characteristics of ecotoxicity quantitative structure–activity relationships. Environ Toxicol Chem 2003; 22:599–607.

33. Seward JR, Cronin MTD, Schultz TW. The effect of precision of molecular orbital descriptors on toxicity modeling of selected pyridines. SAR QSAR Environ Res 2002; 13:325–340.

34. Cronin MTD, Basketter DA. A multivariate QSAR analysis of a skin sensitization database. SAR QSAR Environ Res 1994; 2:159–179.

35. Bagley DM, Gardner JR, Holland G, Lewis RW, Vrijhof H, Walker AP. Eye irritation: updated reference chemicals data bank. Toxicol in Vitro 1999; 13:505–510.

36. Bagley DM, Gardner JR, Holland G, Lewis RW, Regnier JF, Stringer DA, Walker AP. Skin irritation: reference chemicals data bank. Toxicol in Vitro 1996; 10:1–6.

37. Russom CL, Bradbury SP, Broderius SJ, Hammermeister DE, Drummond RA. Predicting modes of toxic action from chemical structure: acute toxicity in the fathead minnow) *Pimephales promelas*). Environ Toxicol Chem 1997; 16:948–967.

38. Cronin MTD, Aptula AO, Duffy JC, Netzeva TL Rowe PH, Valkova IV, Schultz TW. Comparative assessment of methods

to develop QSARs for the prediction of the toxicity of phenols to *Tetrahymena pyriformis*. Chemosphere 2002; 49:1201–1221.

39. Schultz TW, Cronin MTD, Netzeva TL, Aptula AO. Structure–toxicity relationships for aliphatic chemicals evaluated with *Tetrahymena pyriformis*. Chem Res Toxicol 2002; 15: 1602–1609.

40. Schultz TW. TETRATOX: *Tetrahymena pyriformis* population growth impairment endpoint—a surrogate for fish lethality. Toxicol Meth 1997; 7:289–309.

41. Schultz TW, Netzeva TI. An exercise in external validation: the benzene response-surface model for *Tetrahymena* toxicity. In: Cronin MTD, Livingstone DJ, eds. Predicting Chemical Toxicity and Fate. Boca Raton, FL, U.S.A.: CRC Press, 2004; 265–284.

42. Fang H, Tong WD, Shi LM, Blair R, Perkins R, Branham W, Hass BS, Xie Q, Dial SL, Moland CL, Sheehan DM. Structure–activity relationships for a large diverse set of natural, synthetic, and environmental estrogens. Chem Res Toxicol 2001; 14:280–294.

43. Schultz TW, Sinks GD, Cronin MTD. Structure–activity relationships for gene activation estrogenicity: evaluation of a diverse set of aromatic chemicals. Environ Toxicol 2002; 17:14–23.

44. Hamblen EL, Cronin MTD, Schultz TW. Estrogenicity and acute toxicity of selected anilines using a recombinant yeast assay. Chemosphere 2003; 52:1173–1181.

45. Nishihara T, Nishikawa J, Kanayama T, Dakeyama F, Saito K, Imagawa M, Takatori S, Kitagawa Y, Hori S, Utsumi H. Estrogenic activities of 517 chemicals by yeast two-hybrid assay. J Health Sci 2000; 46:282–298.

46. Gottmann E, Kramer S, Pfahringer B, Helma, C. Data quality in predictive toxicology: reproducibility of rodent carcinogenicity experiments. Environ Health Persp 2001; 109:509–514.

47. Helma C, Kramer S, Pfahringer B, Gottmann E. Data quality in predictive toxicology: Identification of chemical structures and calculation of chemical properties. Environ Health Persp 2000; 108:1029–1033.

48. Heidorn CJA, Hansen BG, Nørager O. IUCLID: a database on chemical substances information as a tool for the EU Risk Assessment Program. J Chem Inf Comput Sci 1996; 36: 949–954.

49. Heidorn CJA, Rasmussen K, Hansen BG, Nørager O, Allanou R, Setnaeve R. Scheer S, Kappes D, Bernasconi R. IUCLID: an information management tool for existing chemicals and biocides. J Chem Int Comput Sci 2003; 43:779–786.

50. Hansch C, Gao H. Comparative QSAR: radical reactions of benzene derivatives in chemistry and biology. Chem Rev 1997; 97:2995–3059.

51. Hansch C, Gao H, Hoekman D. A generalized approach to comparative QSAR. In: Devillers J, ed. Comparative QSAR. Washington, DC: Taylor and Francis, 1998:285–368.

52. Lewis RJ Sr. Sax's Dangerous Properties of Industrial Materials. Vol. 3. 10th ed. Chichester, England: John Wiley, 2000.

53. Wang GL, Bai NB. Structure–activity relationships for rat and mouse LD50 of miscellaneous alcohols. Chemosphere 1998; 36:1475–1483.

54. Wang J, Lai L, Tang Y. Data mining of toxic chemicals: structure patterns and QSAR. J Mol Model 1999; 5:252–262.

55. Munro IC, Ford RA, Kennepohl E, Sprenger JG. Correlation of structural class with no-observed-effect levels: a proposal for establishing a threshold of concern. Fd Chem Toxicol 1996; 34:829–867.

56. Graham C, Gealy R, Macina OT, Karol MH, Rosenkranz HS. QSAR for allergic contact dermatitis. Quant Struct–Act Relat 1996; 15:224–229.

57. Flynn GL. Physicochemical determinants of skin absorption. In: Gerrity TR, Henry CJ, eds. Principles of Route-to-Route Extrapolation for Risk Assessment. New York: Elsevier, 1990: 93–127.

58. Wilschut A, Tenberge WF, Robinson PJ, Mckone TE. Estimating skin permeation—the validation of 5 mathematical skin permeation models. Chemosphere 1995; 30:1275–1296.

59. Moss GP, Dearden JC, Patel H, Cronin MTD. Quantitative structure–permeability relationships (QSPRs) for percutaneous absorption. Toxicol in Vitro 2002; 16:299–317.

60. Hardman JG, Goodman AG, Limbird LE eds., Goodman and Gilman's The Pharmacological Basis of Therapeutics. 10th ed. New York:McGraw-Hill, 2001.

61. Duffy JC. Prediction of pharmacokinetic parameters in drug design and toxicology. In: Cronin MTD, Livingstone DJ. eds. Predicting Chemical Toxicity and Fate. London, Boca Raton, FL, U.S.A.: CRC Press, 2004; 229–261.

62. Testa B, van de Waterbeemd H, Folkers G, Guy R, eds. Pharmacokinetic Optimisation in Drug Research. Basel: Wiley-VCH 2001.

63. van de Waterbeemd H, Lennernas H, Artursson P. Drug Bioavailability. Estimation of Solubility, Permeability, Absorption and Bioavailability. Basel: Wiley-VCH, 2003.

64. Solbe J, Mark U, Buyle B, Guhl W, Hutchinson T, Kloepper-Sams P, Lange R, Munk R, Scholz N, Bontinck W, Niessen H. Analysis of the ECETOC aquatic toxicity (EAT) database. 1. General introduction. Chemosphere 1998; 36:99–113.

65. Kaiser KLE, Palabrica VS. *Photobacterium phosphoreum* toxicity data index. Water poll Res J Can 1991; 26:361–431.

66. Cronin MTD, Schultz TW. Validation of *Vibrio fisheri* acute toxicity data: mechanism of action-based QSARs for non-polar narcotics and polar narcotic phenols. Sci Tot Environ 1997; 204:75–88.

67. Tomlin C, ed. The Pesticide Manual. 12th ed. Bracknell, England: British Crop Protection Council Publications, 2000.

68. O'Neil MJ, Smith A, Heckelman PE, Obenchain JR, eds. The Merck Index: an Encyclopedia of Chemicals, Drugs, and Biologicals. 13th ed. Whitehouse Station, NJ: Merck and Co Inc., 2001.

69. Voigt JH, Bienfait B, Wang S, Nicklaus MC. Comparison of the NCI open database with seven large chemical structural databases. J Chem Inf Comput Sci 2001; 41:702–712.

70. Ihlenfeldt W-D, Voigt JH, Bienfait B, Oellien F, Nicklaus MC. Enhanced CACTVS browser of the open NCI database. J Chem Inf Comput Sci 2002; 42:46–57.

71. Cronin MTD, Dearden JC. QSAR in toxicology. 1. Prediction of aquatic toxicity. Quant Struct-Act Relat 1995; 14:1–7.

72. Combes RD, Rodford RA. The use of expert systems for toxicity prediction—illustrated with reference to the DEREK program. In: Cronin MTD, Livingstone DJ. eds. Predicting Chemical Toxicity and Fate. Boca Raton FL, U.S.A.: CRC Press, 2004; 193–204.

73. Richard AM, Williams CR, Cariello NF. Improving structure-linked access to publicly available chemical toxicity information. Curr Opin Drug Disc Dev 2002; 5:136–143.

74. Walker JD, Dimitrov S, Mekenyan O. Using HPV chemical data to develop QSARs for non-HPV chemicals: opportunities to promote more efficient use of chemical testing resources. QSAR Comb Sci 2003; 22:386–395.

INTRODUCTORY LITERATURE

1. Cronin MTD, Livingstone DJ, eds. Predicting Chemical Toxicity and Fate. Boca Raton FL, U.S.A.: CRC Press, 2004.

2. Walker JD, Jaworska J, Comber MHI, Schultz TW, Dearden JC. Guidelines for developing and using quantitative structure–activity relationships. Environ Toxicol Chem 2003; 22: 1653–1665.

GLOSSARY

Chemical catalogues: Freely available chemical catalogues from chemical companies provide an excellent source of information regarding chemical substances. They may be available as hardcopy or electronically on the internet. Chemical catalogues are useful places to relate name to structure and often provide other physicochemical information (e.g., melting and boiling points) and hazard data.

Consistency of toxicity data: Toxicity data for the computational modeling of toxicity should be consistent, i.e., the test result should be repeatable any number of times with low error.

Data: Data are raw facts. With regard to toxicity data for the computational modeling of toxicity, these may be the dose–response concentrations and effects for a lethality endpoint, etc.

European Union international Uniform Chemical Information Database (IUCLID): IUCLID is an electronic data reporting and management tool that has been developed by the European Commission. It provides a platform to support all three steps of the risk assessment process, i.e., data collection, priority-setting, and risk assessment and is distributed by the European Chemicals Bureau (European Commission Joint Research Centre, Ispra, Italy).

Information: Information is the interpretation of data. With regard to toxicological information for the computational modeling of toxicity, this could be a concentration causing a 50% effect (LC_{50}, EC_{50} etc.) or a categoric description of activity (toxic or non-toxic).

Knowledge: Given information for a series of chemicals, knowledge can be obtained by assessing the relationship between the biological effects and physico-chemical and/or structural features.

Merck Index: The Merck Index is a good source of chemical, structural, and biological information. It is a one-volume encyclopedia of industrial chemicals, drugs, and biological agents. Information available in the Merck Index includes structure, names, physical and toxicity data, therapeutic and commercial uses, caution and hazard information as well as chemical, biomedical, and patent literature references.

Pharmacopoeias: Pharmacopoeias (including those from the United States, Britain, and the European Union) contain much information regarding drug and veterinary substances. The monographs within in pharmacopoeia include descriptions, requirements, tests, analytical procedures, and acceptance criteria for pharmaceutical products.

Quality of a toxicity data set: A toxicity data set may be assessed for quality in terms of its reliability;

consistency; reproducibility; relevance to the desired endpoint; whether it comes from a standardized and validated assay; and its error. It is recommended that data sets for the computational modeling of toxicity should be as of high a quality as possible.

Quality of toxicity data: An individual toxicity datum may be assessed for quality in terms of its accuracy and precision.

Relevance of toxicity data: The design of the toxicity test must provide data and information that is appropriate for the use of the information.

Reliability of toxicity data: Toxicity data for the computational modeling of toxicity should be reliable, in other words they must represent the toxic endpoint being assessed accurately.

Reproducibility of toxicity data: Toxicity data for the computational modeling of toxicity should be reproducible, in other words if the test or assay is repeated in a number of laboratories then the result should be the same.

SMILECAS database: This is a database (available from the Syracuse Research Corporation EPISUITE software) that contains in excess of 100,000 Simplified Molecular Line Entry System (SMILES) strings. SMILES strings can be obtained directly from the Chemical Abstract Service (CAS) number.

Standardized, validated, and statistically relevant toxicity test: High quality toxicity data will be associated with well-defined tests. Highly standardized and validated toxicity tests include those performed to Organisation for Economic Cooperation and Development (OECD) test Guidelines and according to Good Laboratory Practice (GLP).

Toxicity data set: A collection of toxicity data that may be used for the computational modeling of toxicity. It is usually, though not always, for closely related chemicals and ideally should be from the same test protocol.

Toxicity database (or databank): This represents a collection of toxicity data, not normally brought together for the purposes of modeling, but which may be mined to develop models. Often highly chemically heterogeneous and varied in nature, and may contain millions of entries.

Toxicity datum: This is normally a numerical value (although can be quantitative or qualitative) describing the relative toxic effect of a chemical substance. This must be determined experimentally and forms the basis of computational modeling of toxicity.

TOXLINE: TOXLINE is the United States National Library of Medicine's (NLM) database providing information on the biochemical, pharmacological, physiological, and toxicological effects of drugs and other chemicals. It contains more than 3 million bibliographic citations, almost all with abstracts and/or indexing terms and CAS numbers.

TOXNET: TOXNET is the United States National Library of Medicine's reference sources for toxicology and environmental health. It is an open and freely available web-based facility to search for toxicological information and as such provides a powerful tool to obtain toxicological data, particularly on discreet chemicals. This facility is actually a cluster of databases on toxicology, hazardous chemicals, and related areas. The databases within TOXNET may be searched individually, or all simultaneously.

United States Environmental Protection Agency ECOTOX database: The ECOTOX database is freely available and provides ecotoxicological information regarding single chemical substances for aquatic and terrestrial life forms. Ecotoxicological information is available for several thousand chemical substances and may be searched by name, CAS number, observed effect group, and publication year. The primary source of information is the peer-reviewed literature.

Wisdom: Wisdom allows a user to apply a predictive technique (knowledge) successfully. This may include when a prediction is likely to be accurate and, from experience, what factors will influence the prediction.

5

The Use of Expert Systems for Toxicology Risk Prediction

SIMON PARSONS

Department of Computer and
Information Science, Brooklyn College,
City University of New York,
Brooklyn, New York, U.S.A.

PETER McBURNEY

Department of Computer Science,
University of Liverpool,
Liverpool, U.K.

OVERVIEW

One approach to predicting the toxicology of novel compounds is to apply expert knowledge. The field of artificial intelligence has identified a number of ways of doing this, and some of these approaches are briefly described in this chapter. We also examine two expert systems—DEREK, which predicts a variety of types of toxicology, and StAR, which predicts carcinogenicity—in some detail. StAR reasons about carcinogenicity using a system of argumentation. We believe that argumentation systems have great potential in this area, and so discuss them at length.

1. INTRODUCTION

One way to build a computer system to solve a problem is to replicate the way that a human would deal with the problem. For some tasks, this is not a good solution. We would not write a program to do arithmetic by the same symbolic manipulation that humans carry out—it just would not be efficient[a]—and the same is true of any problem for which there are clear algorithmic solutions. However, for some problems, the best we can do is to try to replicate the way that humans solve the problems. Problems, such as diagnosing an illness, identifying chemical compounds from the output of a mass spectrometer, and deciding how to configure a computer, are all tasks where copying what humans do seems to be the best we can do, and the same may be true of predicting the toxicology of novel compounds.

Now, all the tasks mentioned above have at least one thing in common. These are all tasks that humans can only complete once they have completed a significant amount of training and have gained a good deal of experience. They are tasks that can only be completed by human *experts*. When we build systems that capture the aspects of the problem-solving ability of these experts, we call them *expert systems*, and it is such systems that are the subject of this chapter.[b]

The study of expert systems is a subfield of artificial intelligence, and rose to great prominence in the late 1970s. This was when the pioneers of the expert system field were building and trialing the earliest expert systems—MYCIN, which carried out medical diagnosis (2,3); DENDRAL, which identified chemical compounds from mass spectrometer readings (4,5); and R1/XCON which configured VAX computers (6).

[a] Indeed, Reverse Polish Notation was invented precisely to improve the efficiency of computer manipulation of arithmetic symbols (1).

[b] It should be stressed that the important feature of expert systems is this capturing of the ability of human experts. The aim of building an expert system is not to *mimic* a human expert, but to isolate an expert's problem-solving ability with the aim of using this ability to improve upon the problem-solving performance of that expert. Some expert systems do indeed manage to outperform the expert whose ability they capture, others are not so successful.

Expert systems seemed to offer great advantages over conventional software systems and other artificial intelligence techniques. They could, it seemed, be used to replicate, and possibly replace, expensive human experts, and bring scarce expertise into the domain of mass production. This led to a huge growth in interest in the field, both academically and commercially, and the field seemed to have a bright future. However, by the late 1980s, it had become clear that expert systems were not as widely applicable as some had claimed, and most of the interest in the area subsided.

This subsidence, in our view, was only to be expected. Expert systems were oversold, and it is only natural that this would become apparent in due course. However, underneath all the hype, the basic idea behind expert systems remains sound. For some problems, they provide a very good solution and in that kind of role they are flourishing (as we will see, flourishing remarkably widely) and will continue to.

Our aim in this chapter is to examine the extent to which expert systems can provide a suitable solution to the problem of predicting toxicity, and to provide some pointers to those who want to try such a solution for themselves. We start with a description of two of the main approaches to building expert systems in Sec. 2. We then take a look in Sec. 3 at two particular expert systems, DEREK and STAR, that do this kind of risk prediction. The second of these systems works using a system of argumentation—that is, it builds up reasons for and against predictions in order to decide which is best. Because we believe that argumentation is a particularly good approach, we describe, in some detail in Sec. 4, both the general approach to argumentation used by STAR, and the directions in which we are developing the theory. Finally, in Sec. 5, we summarize the chapter.

2. EXPERT SYSTEMS

The key idea behind expert systems is that some problems are best solved by applying *knowledge* about the problem domain, knowledge that only people very familiar with the domain are likely to have. This naturally creates a need to represent that knowledge, and *knowledge representation* is a subject that has been widely researched. The knowledge needed to solve a problem rarely includes the exact answer to particular instance of the problem. Instead, the expert system has to take the knowledge that it has and infer new information from it that bears upon the exact problem it is solving. As a result, we are interested in how to perform this *reasoning* as well as how to represent knowledge. This section looks at two commonly used approaches to knowledge representation and their associated form of reasoning.

2.1. Rule-Based Systems

One of the earliest, and most successful approaches to knowledge representation is the use of *production rules* (7,8) similar to:

IF battery good AND battery charging
THEN battery ok

Such rules provide a very natural means of capturing the information of a domain expert—in this case an expert in the diagnosis of problems with the electrical system of a car. These rules also provide a relatively simple means of reasoning with this information, which we can briefly illustrate with the rules in Fig. 1.

If, for example, we are told that the battery is old and the alternator is broken, then we can reason as follows, "battery is old" can be used with R1 to learn that "battery is dodgy" is true, and "alternator is broken" can be used with R2 to learn that "battery is charging" is not true. Having established that these facts are true, they can then be used with R3 to learn that "battery is bad" is true, and so on. Finally, we can conclude that "radio is not working" and "lights are not working" are true. This kind of reasoning is known as *forward chaining*.

We can also use the rules in *backward chaining* to show the same thing. In this form of reasoning, we start from, for example, the desire to determine whether "radio is not working" and look for possible proofs of this fact. In this case, there is only one possibility, that presented by R4. To use this

R1	IF	battery is old
	THEN	battery is dodgy
R2	IF	alternator is broken
	THEN	NOT battery is charging
R3	IF	battery is dodgy AND NOT battery is charging
	THEN	battery is bad
R4	IF	battery is bad
	THEN	radio is not working
R5	IF	battery is bad
	THEN	lights are not working

Figure 1 An example of rule-based reasoning.

rule also requires that "battery is bad" be true, and again there is only one way to establish this fact—the use of R3. To apply R3, it must be the case that "battery is dodgy" is true and "battery is charging" is false, and these themselves can only be established by applying R1 and R2. To do this requires that "battery is old" and "alternator is broken" be true, and these, luckily, accord with what we were told to begin with.

What we have given here is a very simplified account of rule-based reasoning, but this is the essence of how it proceeds. There are additional complexities, such as how to do the necessary pattern matching—handled by algorithms like RETE (9) and TREAT (10)—but solutions have been found to these and are implemented in programming environments like OPS5 (11), CLIPS (12), and JESS (13). These environments allow one to write rules and then invoke forward and backward chaining on them, and so make it possible to simply construct the heart of an expert system.[c]

[c] And JESS, for example, by allowing arbitrary Java function calls from within rules, makes it possible to combine rules with conventional software.

Rules were used as the basis of many expert systems, including the early systems MYCIN (2,3), DENDRAL (4,5), and R1/XCON (6), as well as more recent systems, like DRACO, which helps astronomers sift through large amounts of data (14). MYCIN, for example, makes use of both forward and backward chaining when attempting to find a diagnosis. It starts using backward chaining to determine whether there is some organism (*significant organism* in the terminology of the system) that it should be treating, and then to determine what bacteria is most likely to be the cause. The first of these tasks is achieved by rules like:

IF organism-1 comes from a sterile site
THEN organism-1 is significant

and the second is achieved by rules like:

IF the identity of the organism is not known
with certainty,
AND the gram stain of the organism is gramneg,
AND the morphology of the organism is rod,
AND the aerobicity of the organism is aerobic
THEN there is strongly suggestive evidence that the
identity of the
organism is enterobactericeae.

Once the system has determined that there is a significant organism, and has a likely identity for that organism, it then forward chains to determine what therapy should be applied (in MYCIN, all therapies are courses of antibiotics). This forward chaining uses rules like:

IF the identity of the organism is bacteroides
THEN I recommend therapy chosen from among
the following drugs:
clindamycm
chloramphenicol

erythromycin
tetracycline
carbenecillin

One of the reasons that rules proved so popular as a mechanism for knowledge representation is that they are very *natural*. By this, we mean that it is relatively easy for anyone (including the domain expert) to understand them. It is relatively clear what rules mean, and it is relatively easy for the domain expert to learn how to write them down. However, there are problems with using rules. One such problem is that fact that rules lack a well-defined *semantics*. In other words while, because of their naturalness, it is clear *roughly* what rules mean, it is not clear *exactly* what they mean, and this lack of precision makes it hard to be sure exactly what an expert system is doing or how it will behave. In turn that can make it hard to trust for critical applications.

Another major problem with using rules as described so far as the basis of an expert system is that they are categorical. In our example above, we are only allowed to represent the fact that "battery is old" is true or is not true. There is no way to represent, for example, that we believe the battery to be old, but are not sure. This is a problem because so much information is not categorical. Indeed in most domains most of the information that an expert system must represent is *imperfect*, and this has prompted the development of a wide variety of mechanisms for representing and reasoning with these imperfect data (15).

Such mechanisms do not work well with rules. Although it is natural to try to associate some kind of a measure of belief with a rule,[d] producing something like:

IF battery good AND battery charging
THEN battery ok (0.8)

early attempts to do this were not very satisfactory. Indeed the best known system of attaching measures to rules, certainty factors (3,16), turned out to have some internal inconsistencies

---

[d] Indeed exactly this kind of approach was adopted in MYCIN.

(17,18). Because of these failings, it seemed that a better solution was to look for a different kind of knowledge representation.[e]

2.2. Bayesian Networks

Rather than thinking in terms of expert rules, let us consider describing a domain in terms of the important variables that it contains. For every variable X_i which captures some aspect of the current state of the domain, one way to express the imperfect nature of the information we have about X is to say that each possible value x_{i_j} of each X_i has some probability $\Pr(x_{i_j})$ of being the current value of X_i. Writing $\mathbf{x}$ for the set of all x_{i_j}, we have

$$\Pr : x \in \mathbf{x} \mapsto [0, 1]$$

and

$$\sum_j \Pr(x_{i_j}) = 1$$

In other words, the probability $\Pr(x_{i_j})$ is a number between 0 and 1 and the sum of the probabilities of all the possible values of X_i is 1. If X_i is known to have value x_{i_j} then $\Pr(x_{i_j}) = 1$ and if it is known not to have value x_{i_j} then $\Pr(x_{i_j}) = 0$.

Given two of these variables, X_1 and X_2, then the probabilities of the various values of X_1 and X_2 may be related to one another. If they are not related, a case we distinguish by referring to X_1 and X_2 as being *independent*, then for any two values x_{1_i} and x_{2_j}, we have

$$\Pr(x_{1_i} \wedge x_{2_j}) = \Pr(x_{1_i}) \Pr(x_{2_j})$$

---

[e] Here we are broadly tracing the historical development of these techniques. Initial work on rule-based systems made use of categorical rules. Later it became apparent that mechanisms for handling imperfect knowledge were required, and techniques like certainty factors were developed. When their failings became apparent, and other efforts such as Nilsson's probabilistic logic (19) were also found to be problematic, techniques like the Bayesian networks described in the next section were invented. Subsequently much more satisfactory combinations of probability and logic have been created (20,21), in turn overcoming limitations of Bayesian networks (which are inherently propositional).

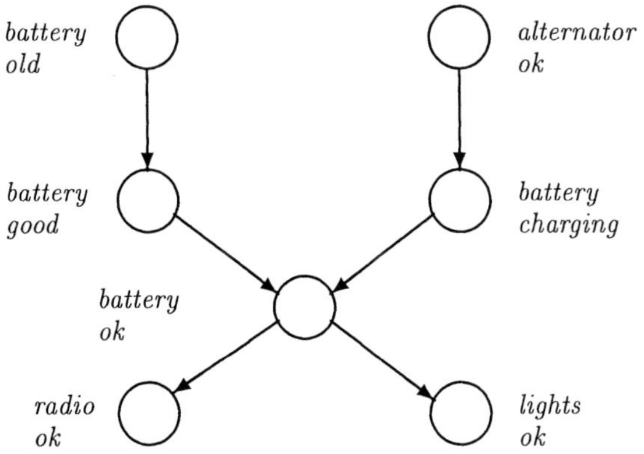

Figure 2 An example Bayesian network.

If the variables are not independent, then:

$$\Pr(x_{1_i} \wedge x_{2_j}) = \Pr(x_{1_i}|x_{2_j}) \Pr(x_{2_j})$$

where $\Pr(x_{1_i}|x_{2_j})$ is the probability of X_1 having value x_{1_i} given that X_2 is known to take value x_{2_j}. Such *conditional probabilities* capture the relationship between X_1 and X_2, representing, for instance, the fact that x_{1_i} (the value "wet," say, of the variable "state of clothes") becomes much more likely when x_{2_j} (the value "raining" of the variable "weather condition") is known to be true.

If we take the set of these X_i of which the agent is aware the set **X**—then for each pair of variables in **X** we can establish whether the pair is independent or not. We can then build up a graph in which each node corresponds to a variable in **X** and an arc joins two nodes if the variables represented by those nodes are not independent of each other. The resulting graph is known as a Bayesian network[f] (22), and provides a form of knowledge representation that is explicitly tailored to representing imperfect information.

Figure 2 is an example of a fragment of a Bayesian network for diagnosing faults in cars. It represents the fact

[f]The notion of independence captured in the arcs of a Bayesian network is somewhat more complex than that described here, but the difference is not relevant for the purposes of this chapter. For full details, see Ref. 22.

that the age of the battery (represented by the node *battery old*) has a probabilistic influence on how good the battery is, and that this in turn has an influence on whether the battery is operational (*battery ok*), the latter being affected also by whether the alternator is working and, as a result, whether the battery is recharged when the car moves. The operational state of the battery affects whether the radio and lights will work. In this network, it is expected that the observations that can be carried out are those relating to the lights and the radio (and possibly the age of the battery), and that the result of these observations can be propagated through the network to establish the probability of the alternator being okay and the battery being good. In this case, these latter variables are the ones that we are interested in since they relate to fixing the car.

As mentioned above, when building an expert system we are not only interested in how to represent knowledge, but also how to reason with it. It turns out that the graphical structure of a Bayesian network provides a convenient computational framework in which to calculate the probabilities of interest to the agent. In general, the expert system will have some set of variables whose values have been observed, and once these observations have been taken, we will want it to calculate the probabilities of the various values of some other set of variables. The details of how this may be achieved are rather complex (see Ref. 22 for details), but provide effective[g] algorithms that allow networks with several hundred variables to be solved in only a few seconds—a speed that is sufficient for all but the most exacting real-time domains. In brief these algorithms work by passing messages between the nodes in the graph. If we observe something that changes the probability of "battery old" in Fig. 2, this new value is sent to "battery good," which updates its probability, and sends a message to "battery ok." When "battery ok" updates in turn,

[g] It turns out that computing values of probabilities in Bayesian networks are not computationally efficient in general—the problem is NP-hard (23) even if the values are only computed approximately (24)—but in many practical cases, the computation can be performed in reasonable time.

it sends messages to "radio ok," "lights ok," and "battery charging," and so on through the network. The full range of algorithms for propagating probabilities through Bayesian networks is described in Refs. 25 and 26.

We should also note that there have been many successful applications of Bayesian networks. These include PATHFINDER (27), a system for diagnosis of diseases of the lymphatic system; MIDAS (28), a system for dealing with mildew in wheat; and a system for diagnosing faults in the space shuttle (29). These are all somewhat specialized systems, and ones that most of us will never come into contact with. However, there are expert systems based on Bayesian networks that we have all come into contact with at one time or another—these are the various systems employed by the Microsoft Windows® operating systems (from Windows 95 onward) for tasks such as troubleshooting (30).

2.3. Other Aspects of Expert Systems

Whatever form of knowledge representation is used by an expert system—whether rules, Bayesian networks, or other mechanisms like *frames* (31) and *semantic networks* (32)—there are a number of common features of any expert system.

These common features can be best illustrated by Fig. 3, which gives the general architecture of an expert system. Any such system has some form of *knowledge-base*, in which knowledge is stored in some knowledge representation. Associated with this is some form of *inference engine*, which

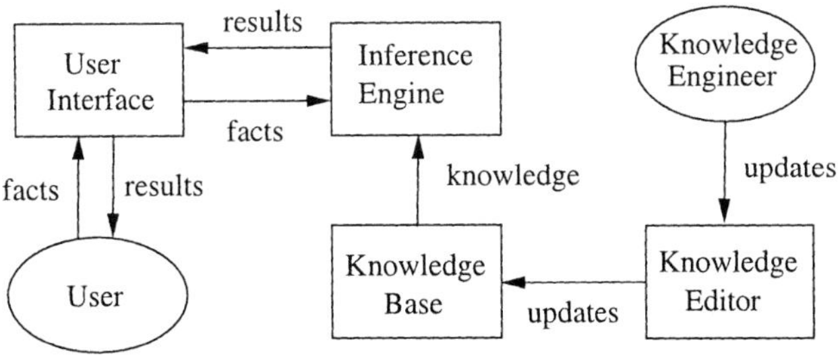

Figure 3 The general architecture of an expert system.

carries out the appropriate kind of reasoning on the knowledge-base. The user of the system interacts with it through some form of graphical user interface, and this interaction is typically to inform the system of some things the user knows to be true. This information sparks off some reasoning by the system, which then informs the user of the results of the inference. Finally, a *knowledge engineer* maintains and updates the knowledge-base by means of some kind of *knowledge editor*.

In addition to such common features, there are also a number of problems in developing expert systems whatever the underlying knowledge representation. Perhaps the most severe of these is the *knowledge acquisition bottleneck*. This refers to the fact that the step that most limits the speed of development of an expert system is the acquisition of the knowledge it uses. The traditional approach is to interview a domain expert and obtain their knowledge, but even if this is a suitable technique,[h] it generally proves to be very slow. As a result, there has been much work on trying to automate the process, including the development of techniques for *rule induction* (33,34), learning Bayesian networks from data (35), and *inductive logic programming* in which logical relations are inferred from data (36,37).

Another problem that it is worth remembering is the opposition which faced the adoption of many expert systems. When many organizations came to implement expert systems, they found that their employees objected to the idea of bringing in machines as experts. Understandably, those employees saw the use of expert systems as undermining their role—either replacing them in the job that they had previously done, or removing their chance to learn the expertise now encoded in the machine. This problem led to many expert systems (of the form we have been describing) being portrayed as *expert assistants* to some human operator—the role of the assistant being, for example, to remember and point out unlikely but plausible outcomes that the human operator should consider.

[h] There are many reasons why it might not be, including the fact that the expert in question may not wish to have their knowledge acquired for use in a computer system.

3. EXPERT SYSTEMS FOR RISK PREDICTION

A number of expert systems have been developed in the broad areas of toxicology and carcinogenicity risk prediction including TOPKAT (38), TOX-MATCH (39), CASETOX (40), HazardExpert (41), and MULTICASE (42). In this section, we look in some detail at DEREK, which predicts various forms of toxicity, and StAR, which predicts carcinogenicity.

3.1. Derek

DEREK is an expert system for the prediction of toxicology risk (43,44). It was developed by LHASA, U.K., a not-for-profit company based around the Department of Chemistry at the University of Leeds.[i] DEREK is a rule-based system, with many of the features of a classic expert system, the rules being developed by the expert toxicologists who work for LHASA and the various organizations that use DEREK.

A consultation with DEREK begins with the user entering the structure of the chemical in question. The system then compares the structure with rules such as (46):

IF a substance contains a carbamate group
bearing a small *N*-alkyl substituents approximately 5.2 angstroms from a nitrogen, oxygen or sulfur atom,
bearing a small- to medium-sized lipophilic substituent or substituents and ideally carrying a positive charge at biological pH
AND it is not too large to fit into the enzyme cavity
AND it has a $\log P = -0.5$ to $+3.0$
THEN the substance is likely to be insectidal.

[i] An early version of DEREK was developed by Schering Agrochemicals and donated to LHASA, which is now responsible for the development of the system (45).

The structural information for such rules are written in the language PATRAN (47,48), and additional information is recorded in the language CHMTRN.

These rules and the structural–activity relationships they encode, are used to identify any structural fragments, or *toxicophores*, that are suspected of being the cause of any toxicity in the chemical. These toxicophores are then displayed, along with a description of the toxicity they are suspected of causing—the kinds of toxicity covered by the system are mutagenicity, carcinogenicity, and skin sensitization. The performance of the system is typical of expert systems. Tested on 250 chemicals from the National Toxicology Program salmonella mutagenicity database, DEREK correctly predicted the genotoxicity of 98% of the 112 Ames positive compounds, and the non-genotoxicity of 70% of the Ames negative compounds (49). An analysis of this performance is contained in Ref. 50.

As mentioned above, the rule-base used by DEREK is under constant development, and some of the techniques used for this development have been published. A description of the development and validation of the part of the DEREK rule-base that relates to skin sensitization can be found in Ref. 57, while Ref. 46 explains how the REX system (52) can be used to automatically generate new rules for DEREK.

3.2. StAR

The StAR project, a collaboration between LHASA and the Imperial Cancer Research Fund[j] developed software for identifying the risk of carcinogenicity associated with chemical compounds (53,54), extending the work in DEREK.

In the carcinogenicity prediction domain, environmental and epidemiological impact statistics are often unavailable, so an approach known as *argumentation* is adopted. In this approach, the expert system builds arguments, based on whatever information is available, for or against the

[j] The ICRF has now become Cancer Research, U.K. The project also involved Logic Programming Associates and City University, London.

carcinogenicity of the chemical in question, and uses the interaction between these arguments to estimate the gravity of the risk. Thus, if there is one argument that a chemical might be carcinogenic (because it contains some functional group which is known to cause cancer in rats) then there is a risk that the chemical might cause cancer in humans. However, if there is a second argument which defeats the first (by, for instance, pointing out that the cancer-causing mechanism in rats involves an enzyme which is not present in humans) then the risk is considered to be lower. A British Government report on microbiological risk assessment identifies StAR as a major new approach to this important problem (55).

The demonstrator system produced by the StAR project is a prototype for a computer-based assistant for the prediction of the potential carcinogenic risk due to novel chemical compounds. A notion of hazard identification is taken as a preliminary stage in the assessment of risk, and the hazard identification used in StAR draws heavily on the approach taken in DEREK. As described above, DEREK is able to detect chemical substructures within molecules, known as structural alerts, and relate these to a rule-base linking them with likely types of toxicity. StAR builds on DEREK's ability to identify structural alerts, but uses a different set of alerts. In particular, the alerts used by StAR were taken from a U.S. FDA report identifying substructures associated with various forms of carcinogenic activity (56).

The user of the carcinogenicity risk adviser presents the system with the chemical structure of the compound to be assessed, together with any additional information which may be thought relevant (such as possible exposure routes, or species of animal that will be exposed to the chemical). The chemical structure may be presented using a graphical interface. The database of structural alerts is then searched for matches against the entered structure. If a match is found, a theorem prover tries to construct arguments for or against the hazard being manifest in the context under consideration. Having constructed all the relevant arguments, a report is generated on the basis of the available evidence, and the user can take appropriate action. Thus the StAR system is an expert assistant rather than a system that is intended to take action directly.

For a better understanding of how STAR works, let us look at some examples. For ease of presentation, these examples use a simplified database, and some of the following assessments may be chemically or biologically naive. The argumentation mechanism, however, is accurately described.

The first example is shown in Fig. 4. Here, the user has entered a relatively simple structure based on an aromatic ring. The system has identified that it contains an alert for epoxides (the triangular structure to the top right). While constructing arguments, the system has recognized that the Log *P* value is relevant in this case, and so queries the user for this information (loosely, the value of Log *P* gives a measure of how easily the substance will be absorbed into tissue). The functional group for epoxides is indicative of a direct acting carcinogen, and the value of Log *P* supplied by the user is supportive of the substance being readily absorbed into tissue. Hazard recognition plus supportive evidence, with no arguments countering potential carcinogenic activity, yields the classification of a "probable human carcinogen" (the result

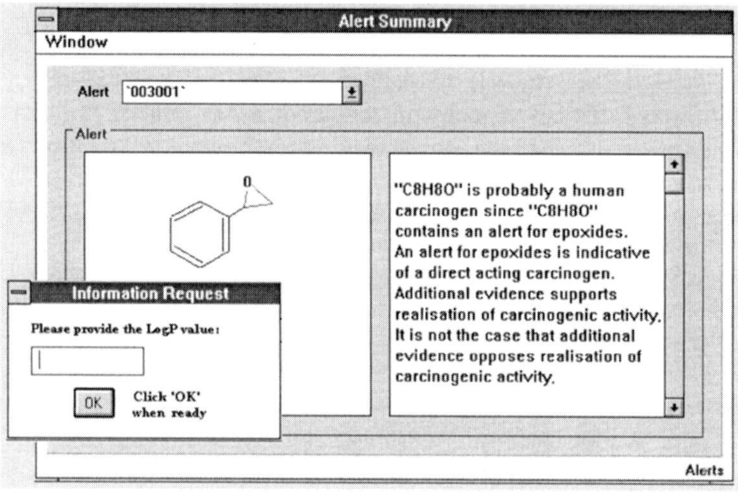

Figure 4 The STAR demonstrator: example 1.

might be different for different animals). Figure 4 shows the summary report. The query box is illustrated in this screen image, although it would normally have been closed by this stage.

The second example is shown in Fig. 5. This involves a structure which contains an alert for peroxisome proliferators. The top-most screen contains a simple non-judgmental statement to this effect. The lower screen contains the summary of the argumentation stage of analysis. Here, evidence is equivocal because there is evidence both for and against the carcinogenicity of the compound.

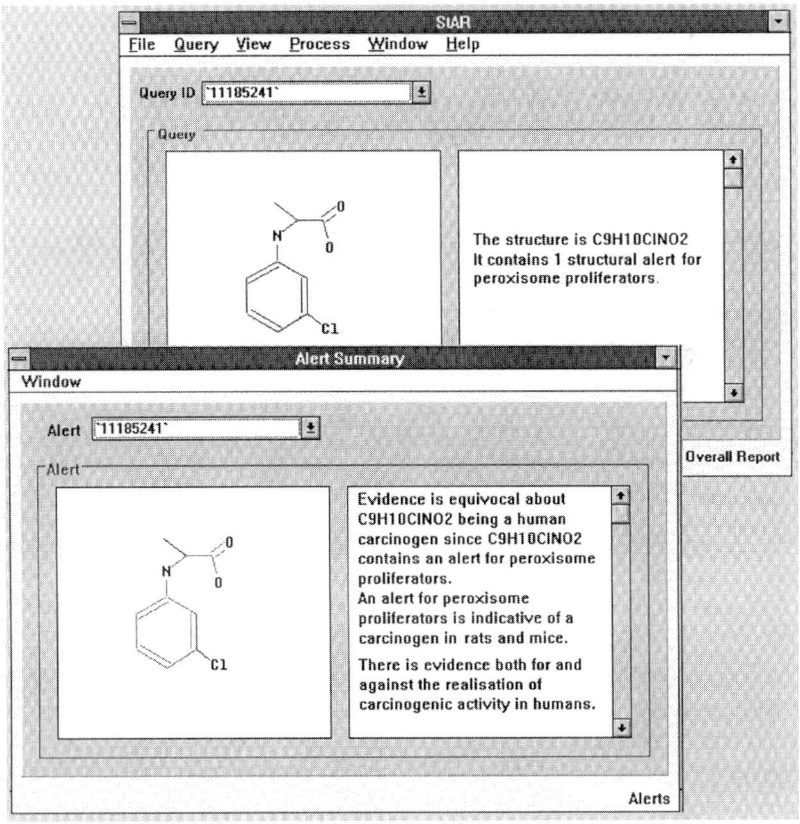

Figure 5 The StAR demonstrator: example 2.

The third example shows how StAR handles equivocal evidence in more detail. Figure 6 shows the reports generated for another compound for which there are arguments for and against carcinogenicity, and so no overall conclusion can be reached. The argument for carcinogenicity is that the structure contains the same peroxisome proliferators alert as in the previous example, and this is indicative of carcinogenic activity in rats and mice. Set against this is the argument that extrapolating from the result in rodents to carcinogenicity in humans is questionable since large doses were required to cause cancer in the test subjects. As indicated in Fig. 6, the StAR system can produce further levels of explanation—in this case explanation related to the "high doses required to obtain results in rats and mice."

The use of argumentation in StAR is discussed in more detail in Refs. 57,58, while the representation of chemical structure—part of the specialized knowledge representation required for the toxicology domain—is described in Ref. 59,

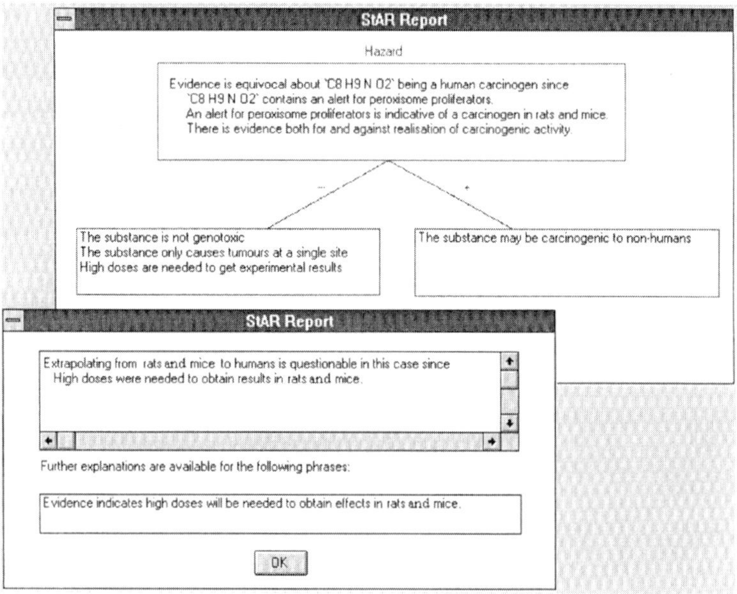

Figure 6 The StAR demonstrator: example 3.

and the main reference for the system of argumentation that underpins StAR is Ref. 60. It is also worth noting that, as part of the StAR project, experiments were carried out to test people's intuitive understanding of the terms used to express the results of the argumentation process. These results are explored in Refs. 61,62. The experiments on how people interpret arguments are useful in the context of StAR because argumentation is being used not only as a mechanism for reaching a decision about carcinogenicity, but also as a means of explaining it. In this sense, argumentation can be considered an extension of approaches like that of CASETOX (40) where properties are inferred on the basis of structural similarity. However, as we will discuss in the next section, argumentation can go far beyond this.

Before we pass on to this discussion, it is worth noting that the results from the StAR project have been incorporated in the successor to DEREK, DEREK for Windows. The model of argumentation used in DEREK for Windows is described in Ref. 63, and its use in DEREK for Windows is elaborated in Ref. 64. Finally, Button et al. (65) describe how an extended version of the argumentation system is applied in the METEOR system for predicting potential metabolic pathways for xenobiotics.

4. SYSTEMS OF ARGUMENTATION

Having given an example of the kind of system that can be built using an argumentation system, we turn to examining in more detail the kind of reasoning that can be handled using this kind of approach.

4.1. An Overview of Argumentation

An argument for a claim may be considered as a tentative proof for the claim. The philosopher Toulmin (66) proposed a generic framework for the structure of arguments which has been influential in the design of intelligent systems which use argumentation (67,58,68). Our analysis, informed by

Toulmin's structure, considers an argument to have the form of a proof, without necessarily its force.

Suppose ϕ is a statement that a certain chemical is carcinogenic at a specified level of exposure. Then an argument for ϕ is a finite, ordered sequence of inferences $G_\phi = (\phi_0, \phi_1, \phi_2, \ldots, \phi_{n\text{-}1})$. Each subclaim ϕ_i is related to one or more preceding subclaims ϕ_j, $j < i$, in the sequence as result of the application of an inference rule, R_i, to those subclaims. The rules

$$\bigcup_i \{R_i\}$$

underwrite the reason why ϕ is a reasonable conclusion, and they correspond to *warrants* in Toulmin's schema and are called *step-warrants* in Verheij's legal argumentation system (68). Note that R_i and R_j may be the same rule for different i and j.

We may present the sequence for a very simple argument graphically as follows:

$$\phi_0 \xrightarrow{R_1} \phi_1 \xrightarrow{R_2} \phi_2 \longrightarrow \cdots \longrightarrow \phi_{n-1} \xrightarrow{R_n} \phi$$

If any of these rules were rules of inference generally considered valid in deductive logic (modus ponens, say), then we would be confident that truth would be preserved by use of the rule. In other words, using a valid rule of Inference at step i means that whenever ϕ_{i-1} is true, so too is ϕ_i. If all the rules of inference are valid in this sense, then the argument G_ϕ constitutes a deductive proof of ϕ. As an example, consider the logical formulae in Fig. 7—a reformulation of the rules from Fig. 1—and the rules of inference in Fig. 8. The first of these rules says that if you can prove two things separately, then you can prove their conjunction. The second is just modus ponens. Now, if we are told that the battery is old and the alternator is broken, then we can give the following argument for $\neg$*lights_working*:

$$(battery\_old,\ battery\_dodgy,\ alternator\_broken,\\ \neg battery\_charging,\ battery\_bad,\ \neg lights\_working)$$

$r1$ $battery\_old \rightarrow battery\_dodgy$
$r2$ $alternator\_broken \rightarrow \neg battery\_charging$
$r3$ $battery\_dodgy \wedge \neg battery\_charging \rightarrow battery\_bad$
$r4$ $battery\_bad \rightarrow \neg radio\_working$
$r5$ $battery\_bad \rightarrow \neg lights\_working$

Figure 7 An example set of formulae.

$$\wedge\text{-I}\frac{\vdash \varphi \quad \vdash \psi}{\vdash \varphi \wedge \psi}$$

$$\rightarrow\text{-E}\frac{\vdash \varphi \quad \vdash \varphi \rightarrow \psi}{\vdash \psi}$$

Figure 8 Two rules for natural deduction.

Note that to represent even this straightforward example graphically requires a branching structure like that given in Fig. 9 (which covers the first part of the argument only, but enough to show the idea).[k]

This machinery would be sufficient if we were only interested in inferences that were valid in the sense of preserving truth. However, the situations of interest to us in toxicology (as indeed is the case in many domains) are when some or all of the inference rules are not valid.

In pure mathematics in general, once a theorem has been proven true, further proofs do not add to its truth, nor to the extent to which we are willing to believe the theorem to be true. However, even pure mathematicians may have variable belief in an assertion depending upon the means used to prove it. For example, constructivist

[k] A simple linear argument would be

$$p \wedge q \wedge r \xrightarrow{\wedge-E} p \wedge q \xrightarrow{\wedge-E} p$$

where $\wedge$-E is:

$$\frac{\vdash \varphi \wedge \phi}{\varphi}$$

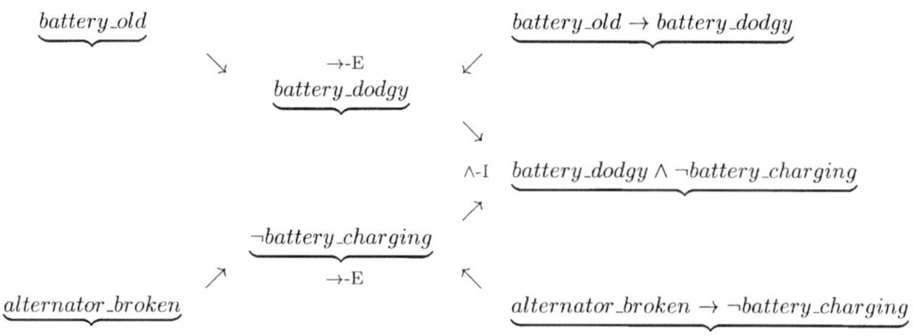

Figure 9 A branching argument structure.

mathematicians (e.g., Refs. 69,70) do not accept inference based on proof techniques which purport to demonstrate the existence of a mathematical object without also constructing it. Typically, such proofs use a *reductio ad absurdum* argument, showing that an assumption of non-existence of the object leads to a contradiction. Thus, constructivist mathematicians will seek an alternative proof for an assertion which a non-constructivist mathematician would accept as already proven.

Although originally a contentious notion within pure mathematics, constructivist mathematics has obvious applications to computing, and has recently been proposed as a medium for the foundations of quantum physics (71). Likewise, in another example, not all mathematicians accept the use of computers in proofs, or may do so only for some proofs. Computers have been used, for instance, to prove the Four Color Map Theorem (72) and to demonstrate the non-existence of projective planes of order 10 (73). For an interesting deconstruction of mathematical proofs as "objectively existing real things" see Appendix D of Ref. 74.

Argumentation extends this idea of different kinds of proof being more or less convincing. In general, all alternative arguments are of great interest, and the greater the number of independent arguments that exist for a claim, the stronger is the case for it, and the stronger may be our belief in its truth. However, in arriving at a considered view as to our belief in the truth of a claim ϕ we also need to consider the arguments against it, the arguments in favour of its negation $\neg\phi$ (which may not be the same thing), and any arguments which attack its supporting subclaims, ϕ_i.

Given these different arguments and counter-arguments, it is possible to define a symbolic calculus, called a Logic of Argumentation, which enables the combination ("flattening") of arguments for and against a proposition (76). Since an argument is a tentative proof of a claim, our degree of belief in the claim will likely depend upon the argument advanced for it. Thus, for each pair (ϕ,G_ϕ) consisting of a claim and an argument for it, we can associate a measure α_ϕ of our strength of belief in ϕ given G_ϕ. We represent this as a triple $(\phi,G_\phi, \alpha_\phi)$, which we call an *assessed argument.*[1] The belief-indicator may be a quantitative measure, such as a probability, or an element from a qualitative dictionary, such as $\{$*Likely, Unlikely*$\}$. In either case, we can define algebraic operations on the set of belief-indicators (the "denotation dictionary for belief") enabling us to generate the degree of belief in a combined argument, when we know the degrees of belief of the subsidiary arguments. In addition to belief-indicators, one can also define other labels for claim–argument pairs, such as the values of world-states and the consequences of actions arising from the claim (76).

The fact that we can attach measures to arguments based upon their relationship to other arguments (a relationship that can be based upon the contents of the grounds or the subclaims) is very powerful. It gives us a way of defining alternative *meta-theories* for argumentation systems, and of using these meta-theories for reasoning about the arguments themselves. This is the key feature of argumentation, and one that distinguishes it from other approaches to reasoning about toxicology.

---

[1]The use of "assessment" here is analogous to the concept of valuation in mathematical logic (75)

4.2. Argumentation Applied to Prediction of Carcinogenicity

The STAR system described above is one example of how argumentation can be applied to risk prediction, and the approach has been applied by workers at Cancer Research, U.K. to other risk prediction problems (see Ref. 77 for another example). There are other ways of using argumentation, however, and in this section, we describe one direction in which we are moving (78). We start with the question:

On what basis do scientists claim that a chemical substance is carcinogenic?

Such claims can be based upon evidence from a number of sources (adapted from Refs. 79,80):

- Using chemical theoretical reasoning, on the basis of the chemical structure of the substance and the known carcinogenicity of chemicals with congeneric structures.
- From mutagenicity tests, applying the substance to tissue-cultures in laboratory experiments.
- From experiments involving the application of the chemical to human or animal cadavers.
- From bioassays, applying the substance to animals in a laboratory experiment.
- From epidemiological studies of humans, either case–control studies (where a case group of people exposed to the substance is matched with a control group not so exposed, and their relative incidences of cancer compared), or cohort studies (where the incidence of the cancer among people exposed to the substance is compared with that in the general population, while controlling for other potential causal and interacting factors).
- From elucidation of theoretically sound bio-medical causal pathways.[m]

[m] These are E-theories in Pera's [(81), p. 154] typology of scientific theories.

Now, elucidation of causal pathways is generally not undertaken until evidence of an empirical nature is observed. Hence, we focus on the other categories of evidence. There are a number of comments one can make on the relative value of these different approaches. Reasoning from chemical structure is still an imprecise and immature science for most substances; indeed, automated prediction of carcinogenicity and other properties of chemicals on the basis of their structure is an active area of Artificial Intelligence research (82,83). Mutagenic tests may demonstrate carcinogenicity in principle, but do not reveal what will happen in a whole, living organism (with, for instance, viral defences), nor in an environment similar to that of people exposed to the substance. Experiments with cadavers have similar difficulties. Moreover, because the incidence rates of many cancers are very small, epidemiological studies may require large sample sizes, and so can be quite expensive. Also, the time-lag between exposure to typical environmental doses and the onset of a cancer can be very long (in the order of decades), so these studies can take years to complete. For these reasons and others, the most common form of assessment of potential carcinogenicity is the bioassay.

We therefore turn our attention to animal bioassays. Because of the difficulties in inferring conclusions about humans on the basis of evidence about animal species, most cautious scientists and policy makers would not *assert* carcinogenicity to humans from a bioassay: they would, at best, only claim that there is a (perhaps high) probability of human carcinogenicity.[n] However, although it is perhaps the most contentious, the animal-to-human inference is not the only inference being deployed in concluding such a probability. It is also not the only inference deployed when quantifying the extent of risk. It therefore behooves us to examine all

[n] Indeed, the USA Environmental Protection Agency guidelines (79) permit one to claim probable human carcinogenicity from (sufficiently strong) animal evidence alone. Although such a claim would be classed in the second of two categories of "probable", it is still above "possible" human carcinogenicity.

the modes of inference used. In doing so, we have abstracted from a number of descriptions and critiques of carcinogenic risk assessment processes (78,79,84–95), both ideal and actual.

For the purposes of exposition, we therefore suppose an archetypal animal bioassay for a chemical substance χ is undertaken. This will involve the administration of specific doses of χ to selected animal subjects, usually repeatedly, in a laboratory environment. Typically, two or three non-zero dose-levels are applied to the subject animals, along with a zero-dose to the control group. The rates at which cancers of a specific nature develop are then observed in each are group until a pre-determined time-point (usually the natural life-span of the animal). Those animals still alive at that time are then killed, and a statistical analysis of the hypotheses that exposure to the substance χ results in increased incidence of cancer is then undertaken. Suppose that, based on this animal bioassay, a claim is then made that χ is carcinogenic to humans at a specified dose. For ease of expression we will notate this claim by ϕ. In asserting ϕ from the evidence of the bioassay, a number of subsidiary inferences need to be made. We have expressed these in the form of *"FROM antecedent TO consequent."* This is short-hand for saying that an act of inference is undertaken whenever one assumes that the consequent is true (or takes a particular value) upon the antecedent being true (or, respectively, having taken a corresponding value).

The list of subsidiary inferences is as follows:

1. *FROM Administered dose TO Delivered dose*. Animal bodies defend themselves against foreign substances. Their ability to do this may be impacted by the amount of the foreign substance ingested or to which the animal is exposed. For example, chemicals applied to nasal tissues are initially repelled by defences in the tissues themselves. Larger doses may destroy this first line of defence, thereby permitting proportionately more of the chemical to enter the body's circulatory pathways than

would occur for smaller doses. In other words, the dose delivered to the target tissue or organ of the body may not be proportionate to the dose administered to the animal by the experimenter.

2. *FROM A sample of animals TO A population of the same species*. Reasoning from a sample to a population from which the sample is drawn in known as statistical inference.
3. *FROM A genetically uniform animal population TO A genetically more diverse population*. Animal subjects used in laboratory experiments are often closely related genetically, both in order to control for the impact of genetic diversity on responses and because, for reasons of convenience, subjects are used from readily available sources. Consequently, the animal subjects used in bioassays are often not as diverse genetically as would be a wild population of the same species.
4. *FROM An animal population TO The human population*. This is perhaps the most contentious inference-step in carcinogenicity claims from bioassays. Animals differ from humans in their physiology and in their body chemistry, so it is not surprising that they also differ from us in reactions to potential carcinogens. Indeed, they differ from each other. According to Graham et al. [(88), p. 18], writing more than a decade ago, "Several hundred chemicals are known to be carcinogenic to laboratory animals, but direct evidence of their human carcinogenicity is either insufficient or nonexistent." Formaldehyde, for instance, was found to cause significant nasal cancers in rats but not in mice (88), while epidemiological studies of humans whose professions exposed them to high levels of the chemical found no significant increases in such cancers. Conversely—and perversely—epidemiological studies did reveal significant increases in brain cancers and leukemias, for which there was no biologically plausible explanation (88).

5. *FROM A site specificity in bioassay animals TO A possibly different site specificity in humans.* Most chemicals are precarcinogens which must be altered by the body's metabolic processes into an actively carcinogenic form. This happens differently in different species, because the body-chemistries are different or because the physiology or relative sizes of organs are different. Hence, a chemical may cause liver cancer in one animal species, but not in another species, or act elsewhere in another.
6. *FROM Localized exposure TO Broader exposure.* Bioassays administer a chemical to a specific site in a specific way to the subject animals, as for example, in bioassays of formaldehyde applied to nasal passages to test for nasal cancer. In contrast, humans exposed to it may receive the chemical in a variety of ways. Morticians exposed to formaldehyde may receive it via breathing and by direct application to their skin, for example.
7. *FROM Large doses TO Small doses.* At typical levels of exposure, the incidences of most individual cancers in the general population are quite small, of the orders of a few percent or much less. At equivalent dose levels, then, bioassays will require very large sample sizes to detect statistically significant increases in cancer incidence. This would be prohibitively expensive, and so most bioassays administer doses considerably greater than the equivalent doses received (allowing for the relative sizes of the animal and human species) in the environment. In order to assert carcinogenicity, then, a conversion model—a dose–response curve—is required to extrapolate back from large to small dose levels.

While one might expect the dose–response curve to slope upwards with increasing dose levels, this is not always the case. For example, high doses of a chemical may kill cells before they can become cancerous; or a chemical may be so potent that even low doses initiate cancer in all cells able to be so initiated, and thus higher doses have no further or a lesser effect. Indeed, if the chemical is believed to be mutagenic as well as carcinogenic, then even a single molecule of the chemical should cause an effect. The issue of whether or not a threshold level for dose exists (below which no response would be observed) is a contentious one in most cases. Fuelling controversy is the fact that claims of carcinogenicity can be very sensitive to the dose–response model used. Two theoretically supported models for the risks associated with aflatoxin peanuts, for example, show human risk likelihood differing by a factor of 40,000 (41). Similarly, the Chief Government Medical Officer of Great Britain recently admitted that the number of people eventually contracting CJD in Britain as a result of eating contaminated beef may be anywhere between a few hundred and several million (96). For this reason, this inference is probably the most controversial aspect of carcinogenicity claims, after that of animal-to-human inference (Inference-Mode No. 4 above).

8. *FROM An animal dose-level TO A human equivalent*. The discussion of Inference-Mode No. 7 used the phrase "allowing for the relative sizes of the animal and human species." But how is this to be done? Is the dose extrapolated according to relative body weights of the two species (animal and human); or skin surface area (which may be appropriate for chemicals absorbed through the skin); or relative size of the organ affected? What is appropriate if different organs are affected in different species?
9. *FROM Administered doses TO Environmental exposure*. In order to expedite response times, bioassays may administer the chemical in a manner different to that likely to be experienced by humans exposed to it in their environment. For example, the chemical may be fed via a tube directly into

the stomach of the animal subject, which is unlikely to be the case naturally.

10. *FROM A limited number of doses TO Cumulative exposure.* Some chemicals may only produce adverse health effects after a lifetime of accumulated exposure. Body chemistry can be very subtle, and a small number of large doses of a chemical may have a very different impact from a much larger number of smaller doses, even when the total dose received is the same in each case.
11. *FROM A pure chemical substance TO A chemical compound.* Most chemicals to which people are exposed are compounds of several chemicals, not pure substances. Bioassay experiments, however, need to be undertaken with pure substances, so as to eliminate any spurious causal effects. Consequently, a bioassay will not be able to assess any effects due to interactions between substances which occur in a real environment, including any transformations which take place inside the human body.
12. *FROM The human population TO Individual humans.* Individuals vary in their reactions to chemical stimuli, due to factors such as their genetic profiles, lifestyles, and personalities. Risks of carcinogenicity may be much higher or much lower than claimed for specific groups or individuals.

These forms of inference could correspond to the inference rules R_i discussed in the previous section.

To claim human carcinogenicity on the basis of evidence from a bioassay thus depends on a number of different modes of inference, each, of which must be valid for the claim to stand. We could write:

The chemical χ is carcinogenic to humans at dose d based on a bioassay of animal species a if:

- *There is a relationship between administered dose and delivered dose in the bioassay, AND*

- *The sample of animals used for the experiment was selected in a representative manner from the population of animals, AND*
- *The animal population from which the sample was drawn is as genetically diverse as the animal population as a whole, AND*
- *The specific animal physiology and chemistry relevant to the activity of χ is sufficiently similar to human physiology and chemistry,*

⋮

and so on, through the remaining eight inference steps.

It is important to note that even if all modes of inference were valid in a particular case, our assertion could, strictly speaking, only be that the chemical χ is associated with an increase in incidence of the particular cancer. The assertion ϕ does not articulate, nor could a bioassay or epidemiological study prove, a causal pathway from one to the other. There may, for example, be other causal factors leading both to the presence of the chemical in the particular environment and to the observed carcinogenicity.

For the archetypal analysis above, we began with the assumption of just one bioassay being used as evidence to assert a claim for carcinogenicity. In reality, however, there is often evidence from more than one experiment and, if so, statistical meta-analysis may be appropriate (97). This may involve pooling of results across different animal species, or across both animal and human species.[o] None of these tasks are straightforward, and will generally involve further modes of inference, which we have yet to explore. The situation is further complicated by the fact that most chemical substances which adversely impact the body cause a number of effects—cell mutation, malignant tumors, benign tumors, toxicity to cells, cell death, cell replication, suppression of the immune system, endocrine disturbances, and so on.

[o] The U.S.A. Environmental Protection Agency Guidelines (79) deal, at a high level, with the second issue.

Some of these clearly interact—dead cells cannot then become cancerous, for instance—and the extent of interaction may be a non-linear function of the dose levels delivered. Simple claims about carcinogenicity often ignore these other effects and their interactions with the growth of malignant tumors ("carcinogenicity"). We do not deal with this issue here.

It is possible that working biomedical scientists and scientific risk assessors would consider the list above to be an example of extreme pedantry, and that many of these modes of inferences are no more than assumptions made in order to derive usable results. We have treated them as inference-modes so as to be quite clear about the reasoning processes involved. Our purpose in doing so is to make possible the automation of these processes, which we believe we can do using an argumentation formalism as described above.

For now this still remains to be done. However, we have begun to take some steps in this direction. In Ref. 98 we describe a formal system of argumentation that includes in the grounds of an argument the inference rules used (in effect replicating the full chain of reasoning in the grounds). This makes it possible to attack an argument not only in terms of the formulae used in its construction, but also the mode of inference. This, in turn, paves the way for argumentation based upon different logics, logics that can capture the different inference-modes described above.

We have also (99) investigated how argumentation may be used to support the process of scientific enquiry. This work shows how a system of argumentation may be used to keep track of different claims, about the toxicity of a chemical for instance, and to summarize the overall belief in the claim at a particular time. In addition, we have shown that this form of argumentation eventually converges on the right prediction about toxicity given the evidence—showing that the system exhibits the necessary soundness of reasoning. Future work is to combine these two pieces of work, allowing the different claims in the latter to be based on the different kinds of reasoning of the former.

5. SUMMARY

The aims of this chapter were to examine the extent to which expert systems can provide a suitable solution to the problem of predicting toxicity, and to provide some pointers to those who want to try such a solution for themselves. These aims were achieved in the following way.

First, we gave a description of the kind of knowledge representation and reasoning possible with production rules and Bayesian networks, two of the main approaches to building expert systems. Then we looked in some detail at two particular expert systems, DEREK and StAR, that predict toxicology risk. The heart of the StAR system is provided by a mechanism for argumentation—a system of reasoning that builds up reasons for and against predictions in order to decide which is best. This kind of reasoning is described in detail, along with the directions in which we are developing the theory. This, we feel, gives a good survey of the way in which expert systems techniques can be applied to toxicology risk prediction. The second aim was achieved by the provision of copious references throughout the chapter.

In summary, our view is that expert systems techniques are a good foundation from which to attack the problem of predicting toxicology risk. We feel that the most promising of these techniques is that of argumentation, a position supported by the U.K. Health and Safety Commission (55), although argumentation needs to be extended, along the lines described above, before its full value will be realized.

REFERENCES

1. Hamblin CL. Translation to and from Polish notation. Comput J 1962; 5:210–213.
2. Buchanan BG, Shortliffe EH. Rule-Based Expert Systems:. The MYCIN Experiments of the Stanford Heuristic Programming Project. MA: Reading, Addison-Wesley, 1984.
3. Shortliffe EH. Computer-Based Medical Consultations: MYCIN. Amsterdam, The Netherlands: Elsevier, 1976.

4. Buchanan BG, Sutherland GL, Feigenbaum EA. Heuristic DENDRAL: a program for generating explanatory hypotheses in organic chemistry. In: Meltzer B, Michie D, Swann M, eds. Machine Intelligence. Vol. 4. Edinburgh, Scotland: Edinburgh University Press, 1969:209–254.
5. Feigenbaum EA, Buchanan BG, Lederberg J. On generality and problem solving: a case study using the DENDRAL program. In: Meltzer B, Michie D, eds. Machine Intelligence. Edinburgh, Scotland: Edinburgh University Press Vol. 6. 1969:165–190.
6. McDermott J. R1: A rule-based configurer of computer systems. Artif Intell 1982; 19(1):41–72.
7. Davis R, Buchanan BG, Shortliffe EH. Production rules as a representation for a knowledge-based consultation system. Artif Intell 1977; 8:15–45.
8. Davis R, King JJ. An overview of production systems. In: Cock EWE, Michie D, eds. Machine Intelligence. Wiley 1969:8:300–334.
9. Forgy CL. Rete: a fast algorithm for the many pattern/many object pattern match problem. Artif Intell 1982; 19:17–37.
10. Miranker DP. Treat: a better match algorithm for AI production system matching. In: Proceedings of the National Conference on Artificial Intelligece. 1987:42–47.
11. Forgy CL. OPS5 users's guide. Technical Report CMU-CS-81-135. Pittsburgh: Carnegie-Mellon University, 1981.
12. http: //www.ghg.net/clips/CLIPS.html.
13. http://herzberg.ca.sandia.gov/jess/.
14. Miller G. The data reduction expert assistant. In: Heck A, Murtagh F, eds. Astronomy from Large Databases II. France: Hagenau, 1992.
15. Parsons S. Reasoning with imperfect information. In: Leondes CT, ed. Expert Systems. Vol. I. New York, NY, U.S.A.: Academic Press, 2002:I:79–117.
16. Shortliffe EH, Buchanan BG. A model of inexact reasoning in medicine. Math Biosci 1975; 23:351–379.

17. Heckerman DE. Probabilistic interpretation of MYCLN's certainty factors. In: Kanal LN, Lemmer JF, eds. Uncertainty in Artificial Intelligence. Amsterdam: Elesevier, 1986: 167–196.

18. Horvitz EJ, Heckerman DE. The inconsistent use of measures of certainty in artificial intelligence research. Kanal LN, Lemmer JF, eds. Uncertainty in Artificial Intelligence:. Amsterdam: Elsevier, 1986:137–151.

19. Nilsson NJ. Probabilistic logic. Artif Intell 1986; 28:71–87.

20. De Raedt L, Kersting K. Probabilistic logic learning. SIGKDD Explorations 2003; 5(1):31–48.

21. Friedman N, Getoor L, Koller D, Pfeffer A. Learning probabilistic relational models. In: Proceedings of the 16th International Joint Conference on Artificial Intelligence. Stockholm: Sweden, 1999:1300–1309.

22. Pearl J. Probabilistic Reasoning in Intelligent Systems; Networks of Plausible Inference. San Mateo, CA: Morgan Kaufmann, 1998.

23. Cooper GF. The computational complexity of probabilistic inference using belief network. Artif Intell 1990; 42:393–405.

24. Dagum P, Luby M. Approximating probabilistic inference in Bayesian belief network is NP-hard. Arti Intell 1993; 60:141–153.

25. Castillo E, Gutiérrez JM, Hadi AS. Expert System and Probabilistic Netowrk Models. Berlin, Germany: Springer Verlag, 1997.

26. Cowell RG, Dawid AP, Lauritzen SL, Spiegelhalter DJ. Probabilistic Networks and Expert Systems. Berlin, Germany: Springer Verlag, 1999.

27. Heckerman DE. Probabilistic Similarity Networks. MIT Press, 1991.

28. Jensen AL, Jensen FV. Midas: An influence diagram for management of mildew in winter wheat. In: Horvitz E, Jensen FV, eds. Proceedings of the 12th Conference on Uncertainty in Artificial Intelligence. San Francisco, CA: Morgan Kaufmann, 1996:349–356.

29. Horvitz EJ, Ruokangas C, Srinivas S, Barry M. A decision-theoretic approach to the.display of information for time-critical decisions: Project Vista. Technical Memorandum 96, Rockwell International Science Center. Palo Alto Laboratory, 1992.

30. Heckerman DE, Breese JS, Rommelse K. Decision-theoretic troubleshooting. Commun ACM 1995; 38:49–57.

31. Minsky M. A framework for representing knowledge. In: Winston PH, ed. The Psychology of Computer Vision. New York: McGraw-Hill, 1975:211–277.

32. Quillian MR. A design for an understanding machine. In: Colloqioum on Semantic Problems in Natural Language. Cambridge: Kings College, 1961.

33. Provost FJ, Kolluri V. A survey of methods for scaling up inductive algorithms. Data Mining Knowledge Discovery 1999; 3(2):131–169.

34. Quinlan JR. Generating production rules from decision trees. In: Proceedings of the 10th International Joint Conference on Artificial Intelligence. Los Altos, CA: Morgan Kaufmann, 1997:304–307.

35. Heckerman DE. A tutorial on learning in Bayesian networks. Jordan MI, ed. Learning in Graphical Models. Dordrecht, The Netherlands: Kluwer, 1998.

36. Bratko I, Muggleton SH. Applications of inductive logic programming. Commun ACM 1995; 38(11):65–70.

37. Muggleton SH, De Raedt L. Inductive logic programming: theory and methods. J Logic Program 1994; 19(20): 629–679.

38. Enslein K, Gombar VK, Blake BW. Use of SAR in prediction of carcinogenicity and mutagenicity of chemicals by the TOPKAT program. Mutat Res 1994; 305:47–61.

39. Kaufman JJ. Strategy for computer-generated theoretical and quantum chemical prediction of toxicity and toxicology (and pharmacology in general). Int J Quantum Chem 1981; 8:419–439.

40. Klopman G. Predicting toxicity through a computer automated structure evaluation program. Environ Health Perspect 1985; 61:269–274.

41. Smithing MP, Darvas F. HazardExpert, an expert system for predicting chemical toxicity. Finley JW, Robinson SF, Armstrong DJ, eds. Food Safety Assessment. Washington, DC: American Chemical Society, 1992:191–200.

42. Klopman G, Rosenkrantz HS. Approaches to SAR in carcinogenesis and utagenesis: prediction of carcinogenicity/mutagenicity using MULTICASE. Mutat Res 1994; 305:33–46.

43. Ridings JE, Barratt MD, Cary R, Earnshaw CG, Eggington CE, Ellis MK, Judson PN, Langowski JJ, Marchant CA, Payne MP, Watson WP, Yith TD. Computer prediction of possible toxic action from chemical structure; an update on the DEREK system. Toxicology 1996; 106:267–279.

44. Sanderson DM, Earnshaw CG. Computer prediction of possible toxic action from chemical structure; the DEREK system. Hum Exp Toxicol 1991; 10:261–273.

45. Judson PN, Coombes RD. Artificial intelligence systems for predicting toxicity. Pesticide Outlook 1996; 7(4):11–15.

46. Judson PN. Rule induction for systems predicting biological activity. J Chem Inf Comput Sci 1994; 34:148–153.

47. Marshall C. Computer Assisted Design of Organic Synthesis. Ph.D. Thesis, University of Leeds, 1984.

48. Myatt GJ. Computer Aided Estimation of Synthetic Accessibility. Ph.D. Thesis, University of Leeds, 1994.

49. Greene N. Computer software for risk assessment. J Chem Inf Comput Sci 1997; 37:148–150.

50. Greene N, Judson PN, Langowski JJ, Marchant CA. Knowledge based expert systems for toxicity and metabolism prediction: DEREK, acrostAR and meteor. SAR and QSAR in Environmental Research, 1999; 10:299–314.

51. Barratt MD, Langowski JJ. Validation and subsequent development of the derek skin sensitization rulebase by analysis of

the BgVV list of contact allergens. J Chem Inf Comput Sci 1999; 39:294–298.

52. Judson PN. QSAR and expert systems in the prediction of biological activity. Pestcide Sci 1992; 36:155–160.

53. Fox J. Will it happen? Can it happen? Sci Public Aff 1997; Winter:45–48.

54. Krause P, Judson P, Patel M. Qualitative risk assessment fulfills a need. In: Hunter A, Parsons S, eds. Applications of Uncertainty Formalisms. Berlin: Springer Verlag, 1998.

55. Health and Safety Commission. Advisory Committee on Dangerous Pathogens (UK), Microbiological Risk Assessment, Interim Report. HMSO, 1996.

56. Federal Drug Administration. General priniciples for evaluating the safety of compounds used in food-producing animals: appendix 1. Carcinogen Structure Guide. FDA, 1986.

57. Judson PN, Fox J, Krause PJ. Using new reasoning technology in chemical information systems. J Chem Inf Comput Sci 1996; 36:621–624.

58. Krause P, Fox J, Judson P. An argumentation based approach to risk assessment. IMA J Math Appl Bus Ind 1994; 5:249–263.

59. Tonnelier, CAG, Fox J, Judson PN, Krause PJ, Pappas N, Patel M. Representation of chemical structures in knowledge-based systems: The StAR system. J Chem Inf Comput Sci 1997; 37:117–123.

60. Krause P, Ambler S, Elvang-Gøransson M, Fox J. A logic of argumentation for reasoning under uncertainty. Computat Intell 1995; 11:113–131.

61. Ayton P, Pascoe E. Bias in human judgement under uncertainty? Knowledge Eng Rev 1995; 10:21–41.

62. Fox J, Hardman D, Krause P, Ayton P, Judson P. Risk assessment and communication: a cognitive engineering approach. Macintosh A, Cooper C, eds. Applications and Innovations in Expert Systems. Vol. III Cambridge University Press, , 1995.

63. Judson PN, Vessey JD. A comprehensive approach to argumentation. J Chem Inf Comput Sci 2003; 43:1356–1363.

64. Judson PN, Marchant CA, Vessey JD. Using argumentation for abssolute reasoning about the potential toxicity of chemicals. J Chem Inf Comput Sci 2003; 43:1364–1370.

65. Button WG, Judson PN, Long A, Vessey JD. Using absolute and relative reasoning in the prediction of the potential metabolism of xenobiotics. J Chem Inf Comput Sci 2003; 43:1371–1377.

66. Toulmin SE. The Uses of Argument. Cambridge, UK: Cambridge University Press, 1958.

67. Fox J, Krause P, Ambler S. Arguments, contradictions and practical reasoning. In: Proceedings of the European Conference on Artificial Intelligence 1992. Vienna, Austria 1992.

68. Verheij B. Automated argument assistance for lawyers. Proceedings of the 7th International Conference on Artificial Intelligence and Law, Oslo, Norway. New York, NY, U.S.A.: ACM, 1999:43–52.

69. Bishop E. Foundations of Constructive Analysis. New York City, NY, U.S.A.: McGraw-Hill, 1967.

70. Troelstra AS, van Dalen D. Constructivism in Mathematics: An Introduction (Two Volumes). Amsterdam, The Netherlands: North-Holland, 1988.

71. Bridges DS. Can constructive mathematics be applied in physics? J Philos Logic 1999; 28:439–453.

72. Appel K, Haken W. Every planar map is four colorable. Bull Am Math Soc 1976; 82:711–712.

73. Lam CWH, Swiercz S, Thiel L. The non-existence of finite projective planes of order 10. Can J Math 1989; 41:1117–1123.

74. Goguen J. An introduction to algebraic semiotics, with application to user interface design. In: Nehaniv CL, ed. Computation for Metaphors, Analogy, and Agents. Lecture Notes in Artificial Intelligence 1562. Berlin, Germany: Springer Verlag, 1999:242–291.

75. Popkorn S. First Steps in Modal Logic. Cambridge, UK: Cambridge University Press, 1994.

76. Fox J, Parsons S. Arguing about beliefs and actions. In: Hunter A, Parsons S, eds. Applications of Uncertainty Formalisms. Lecture Notes in Artificial Intelligence 1455. Berlin, Germany: Springer Verlag, 1998:266–302.

77. Parsons S, Fox J, Coulson A. Argumentation and risk assessment. In: Proceedings of the AAAI Spring Symposium on Predictive Toxicology. Stanford, March, 1999.

78. McBurney P, Parsons S. Dialectical argumentation for reasoning about chemical carcinogenicity. Logic J IGPL 2001; 9(2):191–203.

79. U.S.A. Environmental Protection Agency Guidelines for carcinogen risk assessment. U.S. Fed Register 51 33991–34003, 1986.

80. Graham JD, Rhomberg L. How risks are identified and assessed. Annals of the American Academy of Political and Social Science 1996; 545:15–24.

81. Pera M. The Discourses of Science. Chicago, IL, U.S.A.: University of Chicago Press, 1994.

82. Helma C, Kramer S, Pfahringer B. Carcinogenicity prediction for noncongeneric compounds: experiments with the machine learning program SRT and various sets of chemical descriptors. In: Proceedings of the 12th European Symposium on Quantitative Structure–Activity Relation-ships:. Copenhagen, Denmark: Molecular Modelling and Prediction of Bioactivity, 1998.

83. Srinivasan A, Muggleton SH, Sternberg MJE, King RD. Theories of mutagenicity: a study in first-order feature-based induction. Artif Intell 1995; 85:277–299.

84. Benford DJ, Tennant DR. Food chemical risk assessment. In: Tennant DR, ed. Food Chemical Risk Analysis. London, UK: Blackie Academic and Professional, 1997:21–56.

85. Boyce CP. Comparison of approaches for developing distributions for carcinogenic slope factors. Hum Ecol Risk Assess 1998; 4(2):527–577.

86. Broadhead CL, Combes RD, Balls M. Risk assessment: alternatives to animal testing. In: Tennant DR, ed. Food Chemical Risk Analysis. London, UK: Blackie Academic and Professional, 1997.

87. Graham JD. Historical perspective on risk assessment in the Federal government. Toxicology 1995; 102:29–52.

88. Graham JD, Green LC, Roberts MJ. In Search of Safety: Chemicals and Cancer Risk. Cambridge, MA, USA: Harvard University Press, 1988.

89. Jamieson D. Scientific uncertainty and the political process. Ann of the Am Acad Political Soc Sci 1996; 545:35–43.

90. Lovell DP, Thomas G. Quantitative risk assessment. In: Tennant DR, ed. Food Chemical Risk Analysis. London, U.K.: Blackie Academic and Professional, 1997:57–86.

91. Moolgavkar SH. Stochastic models of carcinogenesis. In: Rao CR, Chakraborty R, eds. Handbook of Statistics. Vol. 8. Statistical Methods in Biological and Medical Sciences. Amsterdam, The Netherlands: North-Holland, 1991:373–393.

92. Page T. A generic view of toxic chemicals and similar risks. Ecol Law Q 1978; 7(2):207–244.

93. Pollak RA. Government risk regulation. Ann Am Acad Political Soc Sci 1996; 545:25–34.

94. Rhomberg LR. A survey of methods for chemical health risk assessment among Federal regulatory agencies. Hum Ecol Risk Assess 1997; 3(6):1029–1196.

95. Shere ME. The myth of meaningful environmental risk assessment. Harvard Environ Law Rev 1995; 19(2):409–492.

96. Watt N. Millions still at risk from CJD. The Guardian, London, UK, 22 September 1999: p. 1.

97. Petitti DB. Meta-Analysis, Decision Analysis and Cost-Effectiveness Analysis: Methods for Quantitative Synthesis in Medicine. Oxford, U.K.: Oxford University Press, 1994.

98. McBurney P, Parsons S. Tenacious tortoises: a formalism for argument over rules of inference. In: Vreeswijk G, ed. Workshop on Computational Dialectics, Fourteenth European Conference on Artificial Intelligence (ECAI2000). Berlin, Germany: ECAI, 2000.

99. McBurney P, Parsons S. Representing epistemic uncertainty by means of dialectical argumentation. Ann Math Artif Intell 2001; 32(1–4):125–169.

6

Regression- and Projection-Based Approaches in Predictive Toxicology

LENNART ERIKSSON and ERIK JOHANSSON
Umetrics AB, Umeå, Sweden

TORBJÖRN LUNDSTEDT
Acurepharma AB, Uppsala, Sweden and BMC, Uppsala, Sweden

OVERVIEW

This chapter outlines regression- and projection-based approaches useful for QSAR analysis in predictive toxicology. The methods discussed and exemplified are: multiple linear regression (MLR), principal component analysis (PCA), principal component regression (PCR), and partial least squares projections to latent structures (PLS). Two QSAR data sets, drawn from the fields of environmental toxicology and drug design, are worked out in detail, showing the benefits of these methods. PCA is useful when overviewing a data set and exploring relationships among compounds and relationships among variables. MLR, PCR, and PLS are used for establishing the QSARs. Additionally, the concept of statistical molecular design is considered, which is an essential ingredient for selecting an informative training set of compounds for QSAR calibration.

1. INTRODUCTION

Much of today's activities in medicinal chemistry, molecular biology, predictive toxicology, and drug design are centered around exploring the relationships between X = chemical structure and Y = measured properties of compounds, such as toxicity, solubility, acidity, enzyme binding, and membrane penetration. For almost any series of compounds, dependencies between chemistry and biology are usually very complex, particularly when addressing in vivo biological data. To investigate, understand, and use such relationships, we need a sound description ("characterization") of the variation in chemical structure of relevant molecules and biological targets, reliable biological and pharmacological data, and possibilities of fabricating new compounds deemed to be of interest. In addition, we need good mathematical tools to establish and express the relationships, as well as informationally optimal strategies to select compounds for closer scrutiny, so that the resulting model is indeed informative and relevant for the stated purposes.

Mathematical analysis of the relationships between chemical structure and biological properties of compounds is often called quantitative structure–activity relationship (QSAR) modeling (1,2). Thus, QSARs link biological properties of a chemical to its molecular structure. Consequently, a hypothesis can often be proposed to identify which physical, chemical, or structural (conformational) features are crucial for the biological response(s) elicited. In this chapter, we will discuss two aspects of the QSAR problem, two parts which are intimately linked. The first deals with how to select informative and relevant compounds to make the model as good as possible (Sec. 2). The second involves methods to capture the structure–activity relationships (Sec. 3).

2. CHARACTERIZATION AND SELECTION OF COMPOUNDS: STATISTICAL MOLECULAR DESIGN

2.1. Characterization

A key issue in QSAR is the characterization of the compounds investigated, both concerning chemical and biological properties. This description of chemical and biological features may well be done multivariately, i.e., by using a wide set of chemical descriptors and biological responses (3). The use of multivariate chemical and biological data is becoming increasingly widespread in QSAR, both regarding drug design and environmental sciences. A multitude of chemical descriptors will stabilize the description of the chemical properties of the compounds, facilitate the detection of groups (classes) of compounds with markedly different properties, and help unravel chemical outliers. A multivariate description of the biological properties is highly recommended as well. This leads to statistically beneficial properties of the QSAR and improved possibilities of exploring the biological similarity of the studied substances. The absence of outliers in multivariate biological data is a very valuable indication of homogeneity of the biological response profiles among the compounds.

This rapidly developing emphasis on the use of many X-descriptors and Y-responses is at some contrast to the traditional way of QSAR-conduct, where single parameters are usually used to account for chemical properties, parameters that are often derived from measurements in chemical model systems (1). However, with the advancement of computers, quantum chemical theories, and dedicated QSAR software, it is becoming increasingly common to be confronted with a wide set of molecular descriptors of different kinds (4). An advantage of theoretical descriptors is that they are calculable for not yet synthesized chemicals.

Descriptors that are found useful in QSAR often mirror fundamental physico-chemical factors that in some way relate to the biological endpoint(s) under study. Examples of such molecular properties are hydrophobicity, steric and electronic properties, molecular weight, p*K*a, etc. These descriptors provide valuable insight into plausible mechanistic properties. It is also desirable for the chemical description to be reversible, so that the model interpretation leads forward to an understanding of how to modify chemical structure to possibly influence biological activity. (a deeper account of tools and descriptors used for representation of chemicals is provided elsewhere in this text.)

Furthermore, knowledge about the biological data is essential in QSAR. To quote Cronin and Schultz (5): "Reliable data are required to build reliable predictive models. In terms of biological activities, such data should ideally be measured by a single protocol, ideally even the same laboratory and by the same workers. High quality biological data will have lower experimental error associated with them. Biological data should ideally be from well standardized assays, with a clear and unambiguous endpoint." This article also discusses in depth the importance of appreciation of biological data quality, and that it is important to know the uncertainty with which the biological data were measured. (Issues related to representation of biological data are discussed elsewhere in this book.)

2.2. Selection of Representative Compounds

A second key issue in QSAR concerns the selection of molecules on which the QSAR model is to be based. This phase may perhaps also involve consideration of a second subset of compounds, which is used for validation purposes. Unfortunately, the selection of relevant compounds is an often overlooked issue in QSAR. Without the use of a formal selection strategy the result is often a poor and unbalanced coverage of the available structural (S-) space (Fig. 1, top). In contrast, statistical molecular design (SMD) (1–4) is an efficient tool resulting in the selection of a diverse set of

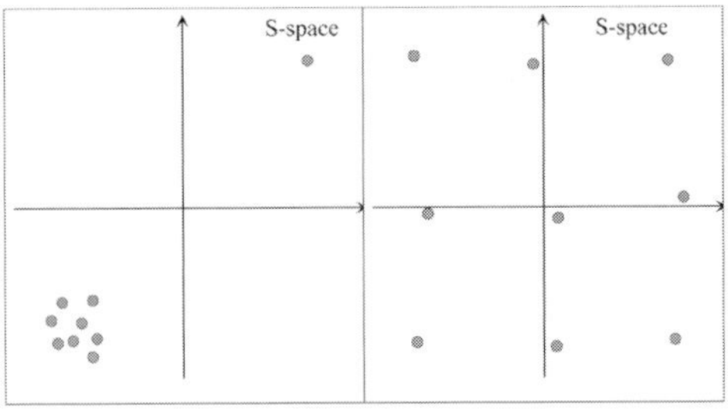

Figure 1 (Top) A set of nine compounds uniformly distributed in the structural space (S-space) of a series of compounds. The axes correspond to appropriate structural features, e.g., lipophilicity, size, polarizability, chemical reactivity, etc. The information content of the selected set of compounds is closely linked to how well the set is spread in the given S-space. In the given example, the selected compounds represent a good coverage of the S-space. (Bottom) The same number of compounds but distributed in an uninformative manner. The information provided by this set of compounds corresponds approximately to the information obtained from two compounds, the remote one plus one drawn from the eight-membered main cluster.

compounds (Fig. 1, bottom). One of the early proponents of SMD, was Austel (6), who introduced formal design on the QSAR arena.

The basic idea in SMD is to first describe thoroughly the available compounds using several chemical and structural descriptor variables. These variables may be measurable in chemical model systems, calculable using, e.g., quantum-chemical orbital theory, or simply based on atom- and/or fragment counts.

The collected chemical descriptors make up the matrix **X**. Principal component analysis (PCA) is then used to condense the information of the original variables into a set of "new" variables, the principal component scores (1–4). These score vectors are linear combinations of the original variables, and reflect the major chemical properties of the compounds. Because they are few and mathematically independent (orthogonal) of one another they are often used in a statistical experimental design protocols. This process is called SMD. Design protocols commonly used in SMD are drawn from the factorial and D-optimal design families (1–4).

3. DATA ANALYTICAL TECHNIQUES

In this section, we will be concerned with four regression- and projection-based methods, which are frequently used in QSAR. The first method we discuss is multiple linear regression, (MLR), which is a workhorse used extensively in QSAR (7). Next, we introduce three projection-based approaches, the methods of PCA, principal component regression (PCR), and projections to latent structures (PLS). These methods are particularly apt at handling the situation when the number of variables equals or exceeds the number of compounds (1–4). This is because projections to latent variables in multivariate space tend to become more distinct and stable the more variables are involved (3).

Geometrically, PCA, PCR, PLS, and similar methods can be seen as the projection of the observation points (compounds) in variable-space down on an A-dimensional hyper-plane. The positions of the observation points on this hyper-plane are given by the *scores* and the orientation of the plane in relation to the original variables is indicated by the *loadings*.

3.1. Multiple Linear Regression (MLR)

The method of MLR represents the classical approach to statistical analysis in QSAR (7,8). Multiple linear regression is usually used to fit the regression model (1), which models a single response variable, $\boldsymbol{y}$, as a linear combination of the X-variables, with the coefficients $\boldsymbol{b}$. The deviations between the data ($\boldsymbol{y}$) and the model ($\boldsymbol{Xb}$) are called residuals, and are denoted by $\boldsymbol{e}$

$$\boldsymbol{y} = \boldsymbol{Xb} + \boldsymbol{e} \tag{1}$$

Multiple linear regression assumes the predictor variables, normally called $\boldsymbol{X}$, to be mathematically independent ("orthogonal"). Mathematical independence means that the rank of $\boldsymbol{X}$ is K (i.e., equals the number of X-variables). Hence,

MLR does not work well with correlated descriptors. One practical work-around is long and lean data matrices—matrices where the number of compounds substantially exceeds the number of chemical descriptors—where inter-relatedness among variables usually drops. It has been suggested to preserve the ratio of compounds to variables above five (9). We note that one way to introduce orthogonality or near-orthogonality among the X-variables is through SMD (see Sec. 2.2).

For many response variables (columns in the response matrix **Y**), regression normally forms one model for each of the M Y-variables, i.e., M separate models. Another key feature of MLR is that it exhausts the **X**-matrix, i.e., uses all (100%) of its variance (i.e., there will be no **X**-matrix error term in the regression model). Hence, it is assumed that the X-variables are exact and completely (100%) relevant for the modelling of **Y**.

3.2. Principal Component Analysis (PCA)

Principal component analysis forms the basis for multivariate data analysis (10–13). This is an exploratory and summary tool, not a regression method. As shown by Fig. 2, the starting point for PCA is a matrix of data with N rows (observations) and K columns (variables), here denoted by **X**. In QSAR, the observations are the compounds and the variables are the descriptors used to characterize them.

PCA goes back to Cauchy, but was first formulated in statistics by Pearson, who described the analysis as finding "lines and planes of closest fit to systems of points in space" (10). The most important use of PCA is indeed to represent a multivariate data table as a low-dimensional plane, usually consisting of 2–5 dimensions, such that an overview of the data is obtained (Fig. 3). This overview may reveal groups of observations (in QSAR: compounds), trends, and outliers. This overview also uncovers the relationships between observations and variables, and among the variables themselves.

Statistically, PCA finds lines, planes, and hyper-planes in the K-dimensional space that approximate the data as well

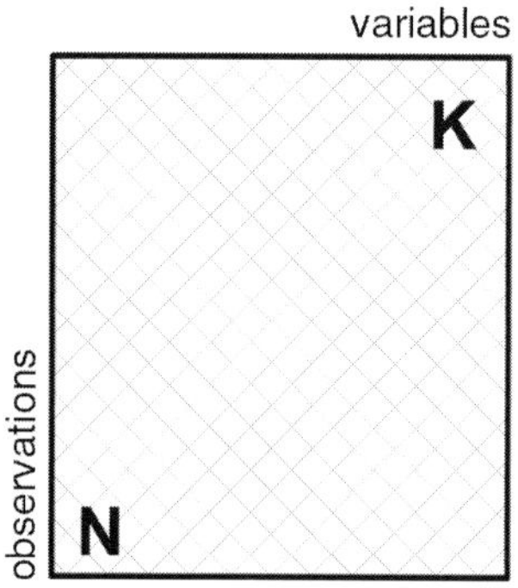

Figure 2 Notation used in PCA. The observations (rows) can be analytical samples, chemical compounds or reactions, process time points of a continuous process, batches from a batch process, biological individuals, trials of a DOE-protocol, and so on. The variables (columns) might be of spectral origin, of chromatographic origin, or be measurements from sensors and instruments in a process. (From Ref. 3.)

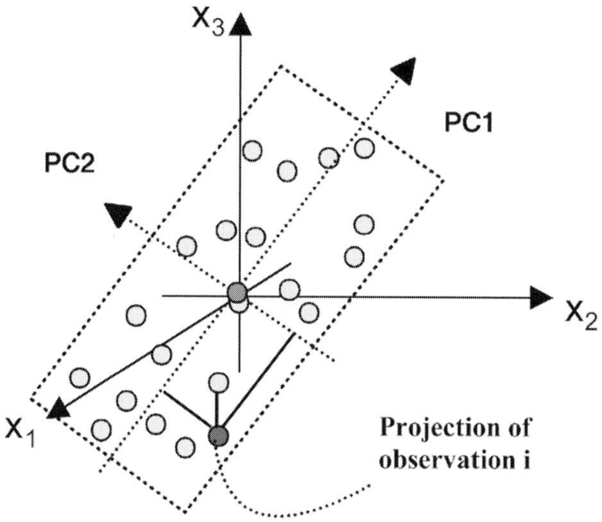

Figure 3 Two PCs form a plane. This plane is a window into the multidimensional space, which can be visualized graphically. Each observation may be projected onto this giving a score for each. The scores give the location of the points on the plane. The loadings give the orientation of the plane. (From Ref. 3.)

as possible in the least squares sense. It is easy to see that a line or a plane that is the least squares approximation of a set of data points makes the variance of the coordinates on the line or plane as large as possible (Fig. 4).

By using PCA a data table **X** is modeled as

$$\mathbf{X} = \mathbf{1}^{*}\mathbf{x}' + \mathbf{T}^{*}\mathbf{P}' + \mathbf{E} \tag{2}$$

In the expression above, the first term, $\mathbf{1}^{*}\bar{\mathbf{x}}'$, represents the variable averages and originates from the preprocessing step. The second term, the matrix product $\mathbf{T}^{*}\mathbf{P}'$, models the structure, and the third term, the residual matrix **E**, contains the noise (Fig. 5).

The principal component scores of the first, second, third, ..., components (t_1, t_2, t_3, ...) are columns of the score matrix **T**. These scores are the coordinates of the observations in the

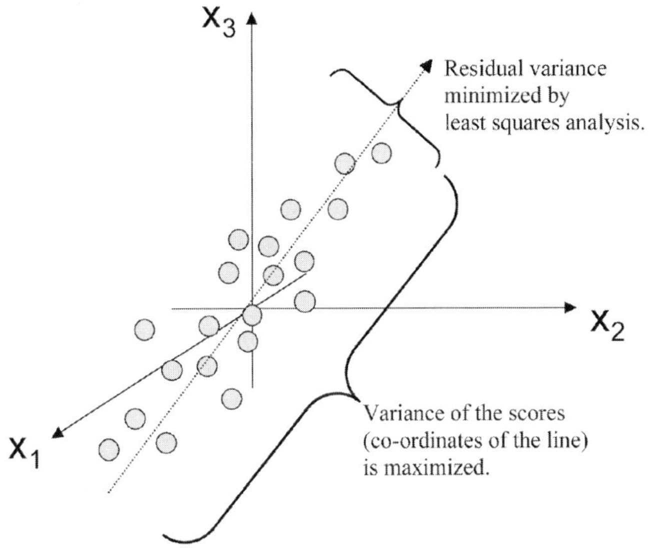

Figure 4 Principal component analysis derives a model that fits the data as well as possible in the least squares sense. Alternatively, PCA may be understood as maximizing the variance of the projection coordinates. (From Ref. 3.)

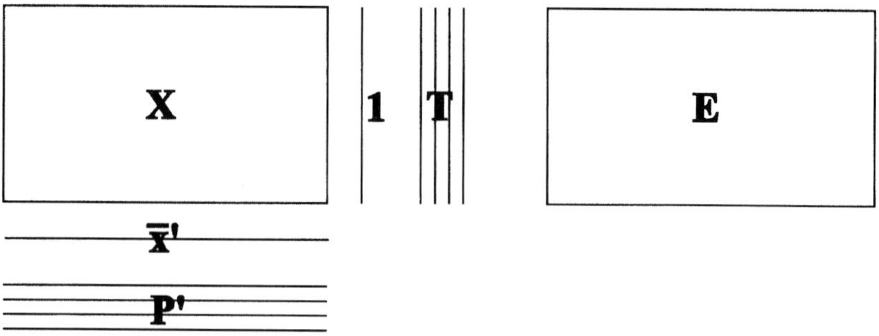

Figure 5 A matrix representation of how a data table X is modeled by PCA. (From Ref. 3.)

model (hyper-)plane. Alternatively, these scores may be seen as new variables which summarize the old ones. In their derivation, the scores are sorted in descending importance (t_1 explains more variation than t_2, t_2 explains more variation than t_3, and so on).

The meaning of the scores is given by the loadings. The loadings of the first, second, third, ..., components (p_1, p_2, p_3, ...) build up the loading matrix **P** (Fig. 5). Note that in Fig. 5, a prime has been used with **P** to denote its transpose.

3.3. Principal Component Regression (PCR)

Principal component regression can be understood as a hyphenation of PCA and MLR. In the first step, PCA is applied to the original set of descriptor variables. In the second step, the output of PCA, the score vectors (**t** in Fig. 2), are used as input in the MLR model to estimate Eq. (1).

Thus, PCR uses PCA as a means to summarize the original X-variables as orthogonal score vectors and hence the collinearity problem is circumvented. However, as pointed out by Jolliffe (13) and others, there is a risk that numerically small structures in the X-data which explain Y may disappear in the PC-modeling of X. This will then give bad predictions of Y from the X-score vectors (**T**). Hence, to begin with, a subset selection among the score vectors might be necessary prior to the MLR-step.

3.4. Partial Least Squares Projections (PLS) to Latent Structures

The PLS method (1,4,10–12) has properties that alleviate some of the difficulties noted for MLR and PCR. The matrix **T**—a projection of **X**—is calculated to fulfill two objectives, i.e., (i) to well approximate **X** and **Y**, and (ii) to maximize the squared covariance between **T** and **Y** (14). Hence, as opposed to MLR, PLS can handle correlated variables, which are noisy and possibly also incomplete (i.e., containing missing data elements). The PLS method has the additional advantage to handle also the case with several *Y*-variables. This is accomplished by using a separate model for the *Y*-data and computing a projection **U** (see below), which is modeled and predicted well by **T**, and which is a good description of **Y**.

The PLS regression method estimates the relationship between a matrix of predictors (**X**) of size N^*K and a matrix of responses (**Y**) of size N^*M (Fig. 6). This is accomplished by making the bilinear projections

$$\mathbf{X} = \mathbf{TP}' + \mathbf{E} \tag{3}$$

$$\mathbf{Y} = \mathbf{UC}' + \mathbf{F} \tag{4}$$

and connecting **X** and **Y** through the inner relation

$$\mathbf{U} = \mathbf{T} + \mathbf{H} \quad (\text{giving } \mathbf{Y} = \mathbf{TC}' + \mathbf{G}) \tag{5}$$

Here, **E**–**H** are residual matrices, **T** is an N^*A matrix of *X*-scores, **P** is a K^*A matrix of *X*-loadings, **U** is an N^*A matrix of *Y*-scores and **C** is an M^*A matrix of *Y*-weights. In addition, an *X*-weight matrix, $\mathbf{W}^*$, of size K^*A, is calculated, though it is not displayed in any equation above. The $\mathbf{W}^*$ matrix expresses how the *X*-variables are combined to form **T** ($T = X\mathbf{W}^*$). It is useful for interpreting which *X*-variables are influential for modelling the *Y*-variables. Finally, *A* is the number of PLS components, usually estimated by cross-validation (see below).

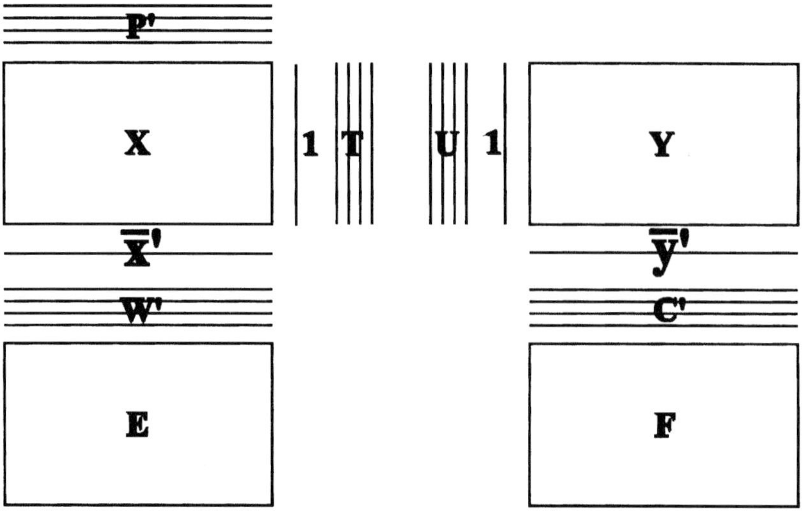

Figure 6 The matrix relationships in PLS. Here, $\mathbf{1}^*\mathbf{x}'$ and $\mathbf{1}^*\mathbf{y}'$ represent the variable averages and originate from the preprocessing step. The PLS scores comprise **T** and **U**, the X-loadings $\mathbf{P}'$, the X-weights $\mathbf{W}'$, and the Y-weights $\mathbf{C}'$ The variation in the data that was left out of the modeling forms the **E** and **F** residual matrices. (From Ref. 3.)

One way to understand PLS is that it simultaneously projects the X- and Y-variables onto the same subspace, **T**, in such a way that there is a good relationship between the predictor and response data. Another way to see PLS is that it forms "new" X-variables, **t**, as linear combinations of the old ones, and subsequently uses these new ts as predictors of **Y**. Only as many new ts are formed as are found significant by cross-validation. A graphical overview of PLS is provided in Fig. 7.

3.5. Model Performance Indicators

The performance of a regression model—and hence also a QSAR model—is usually described using R^2, the explained variation (or goodness of fit). It is defined as

$$R^2Y = 1 - \mathrm{RSS}/\mathrm{SS}_y \tag{6}$$

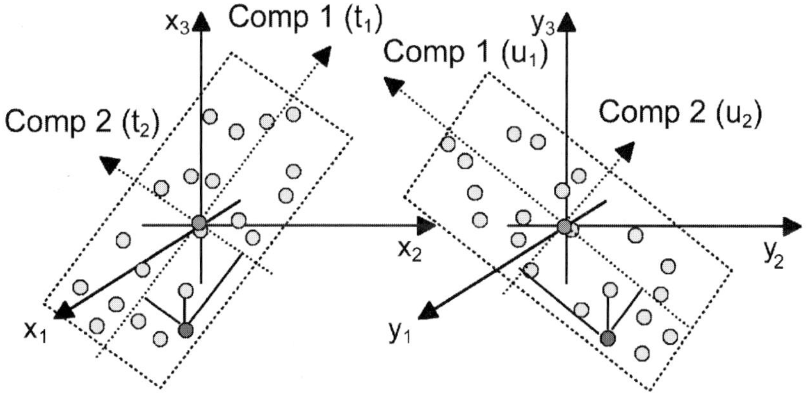

Figure 7 Two PLS components correspond to the insertion of model planes in the X- and Y-spaces. Upon projecting the observations onto these planes, the PLS score vectors of the first model dimension, $\mathbf{t}_1$ and $\mathbf{u}_1$, and the second model dimension, $\mathbf{t}_2$ and $\mathbf{u}_2$, are generated. (From Ref. 3.)

Here RSS is the sum of squares of the Y-residuals, and SS_y the initial sum of squares of the mean-centered response data. The predicted variation (or goodness of prediction), Q^2Y, is also becoming increasingly used in this context. The value Q^2Y is defined as

$$Q^2Y = 1 - \text{PRESS}/\text{SS}_y \tag{7}$$

Here PRESS is the sum of squares of the predictive residuals of the responses. The difference between R^2Y and Q^2Y is thus that the former is based on the *fitted* residuals

$$e_i = y_i - \sum b_k x_{ik} \tag{8}$$

while PRESS is based on the *predictive* residual ($e_{(i)}$) calculated as

$$e_{(i)} = y_i - \sum b^*_{xk} x_{ik} \tag{9}$$

Here b_k^* are the model coefficients calculated from data *without* observation i.

The size of Q^2Y is often estimated via cross-validation (15,16). During cross-validation some of the data points are kept out, and are then predicted by the model and compared with the measured values. This procedure is repeated until each data point has been eliminated once and only once. Then PRESS is formed and the number of PLS components resulting in the lowest PRESS-value is considered optimal.

Cross-validation is also used in the context of PCA, the difference being that PRESS and related statistics refer to X-data, and not Y-data (as in PLS). Hence, we here use R^2X/Q^2X or R^2Y/Q^2Y-notations to distinguish between methods used.

4. RESULTS FOR THE FIRST EXAMPLE—MODELING AND PREDICTING IN VITRO TOXICITY OF SMALL HALOALKANES

4.1. Introduction and Background

In the first example, the aim is to contrast the methods introduced in Sec. 3. For this purpose, we shall deal with a series of halogenated aliphatic hydrocarbons and their in vitro genotoxicity and cytotoxicity. We call this dataset CELLTEST. The complete CELLTEST data set consists of $N=58$ observations (compounds), $K=6$ descriptors (X-variables) and $M=2$ responses (Y-variables) (17,18).

The objective of the study was to set up a QSAR model enabling large-scale prediction of the two endpoints for very many similar untested compounds (17,18). In order to accomplish this, SMD was used to encode a diverse subset of 16 compounds. These compounds were tested for their genotoxic and cytotoxic potencies (17,18). Ten out of the 16 tested compounds were defined as the training set (17,18) and will here be used as a basis to calculate QSARs. The remaining six compounds will be used for assessing the predictive ability of the developed QSARs. Further details are found in the original publications (17,18). The identity of the compounds, and the nature of the X- and the Y-variables is seen in the legends to Figs. 8–10.

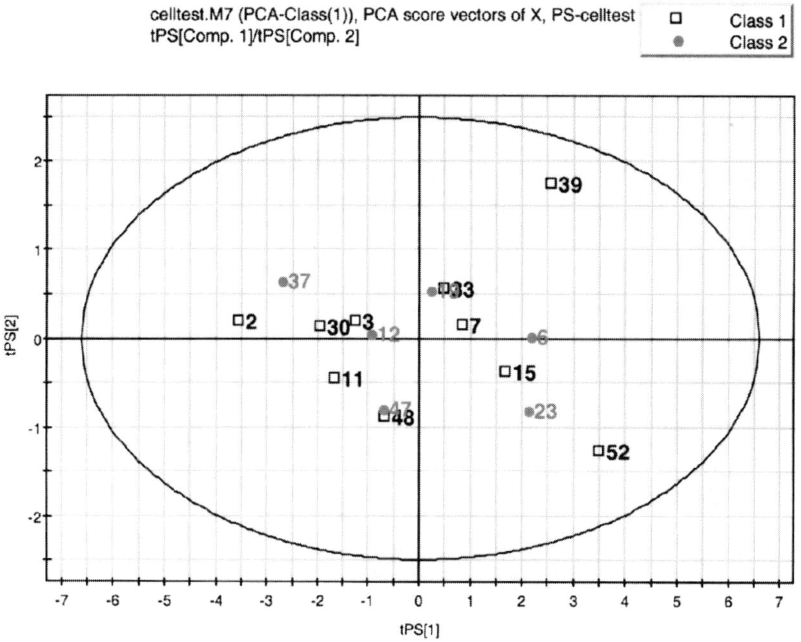

Figure 8 The PCA t_1/t_2 score plot of model for CELLTEST training data set. Each plot mark represents one aliphatic hydrocarbon. Open squares designate the training set; solid dots represent the prediction set (as classified into the model of the training set). The compounds are: (2) dichloromethane, (3) trichloromethane, (6) tetrachloromethane, (7) fluorotrichloromethane, (11) 1,2-dichloroethane, (12) 1-bromo-2- chloroethane, (15) 1,1,2,2-tetrachloroethane, (19) 1,2-dibromoethane, (23) 1,2,3-trichloropropane, (30) 1-bromoethane, (33) 1,1-dibromoethane, (37) bromochloromethane, (39) fluorotribromomethane, (47) 1-chloropropane, (48) 2-chloropropane, (52) 1-bromobutane.

4.2. Obtaining an Overview: PCA Modeling

The PCA-modeling of the six X-variables of the training set yielded a three-component model with $R^2X = 0.98$ ("explained variation") and $Q^2X = 0.93$ ("predicted variation"). There are no signs of deviating compounds (Fig. 8). The loading plot of the first two model dimensions (Fig. 9) indicates that the two HPLC retention indices (LCI and LC2)

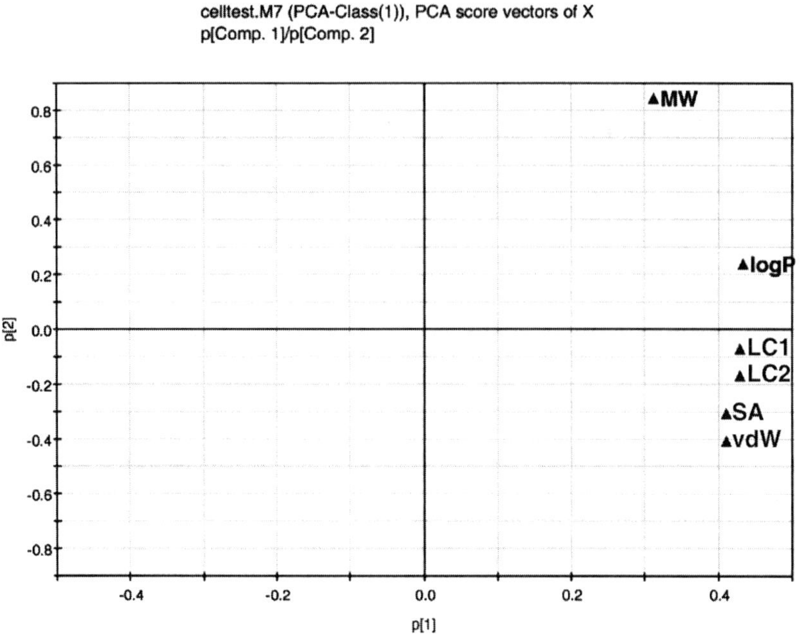

Figure 9 The PCA p_1/p_2 loading plot of model for CELLTEST data set. In this plot, one considers the distance to the plot origin. The closer to the origin a variable point lies, the less informative it is for the model. Variables that are close to each other provide similar information (= are correlated). Variable description: Mw = molecular weight; vdW = van der Waals volume; log P = logalogarithm of octanol/water partition coefficient; SA = accessible molecular surface area; LC1 = log HPLC retention times for Supelcosil LC-08 column; LC2 = log HPLC retention times for Nucleosil 18 column.

are very correlated, as are surface area (SA) and van der Waals volume (vdW). Log P is also correlated with these four descriptors, whereas Mw partly encodes unique information.

The conclusions of the PCA-model are as follows:

- There are no outliers. Therefore, there is no need to delete any compound prior to regression (QSAR) modeling.

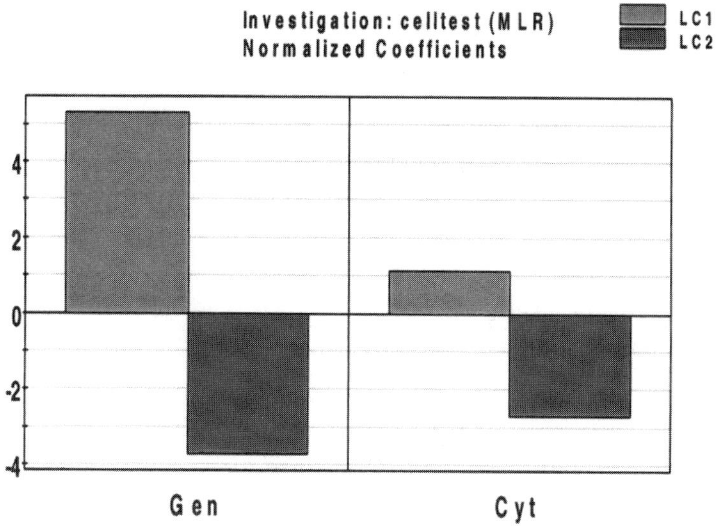

Figure 10 Overview plot of regression coefficients of model M12 for both responses. Genotox = log slope of concentration–response curve from the DNA-precipitation assay using Chinese hamster V79 cells; Cytotox = log inhibitory concentration (in mM) decreasing cell viability by 50% in the MTT-assay using human HeLa cells.

- The first PC mainly describes lipophilicity, the second predominantly molecular weight, and the third chiefly variation in surface area and volume not accounted for in the first component.

4.3. One- and Two-Parameter QSARs for Single *Y*-Variables: MLR Modelling

Table 1 shows the correlation matrix of the training set. There are strong correlations among the six chemical descriptors, and the absolute value of the correlation coefficients invariably lie between 0.5 and 1.0. Clearly, as MLR cannot cope with strongly correlated descriptor variables, care must be exercised in the regression modeling. We deployed MLR and for each of the two responses six one-parameter QSARs were calculated. The regression results are summarized by Table 2 (see models M1–M6).

Table 1 Correlation Matrix of Example 1

	1	2	3	4	5	6	7	8	9
1		HW	vdw	log *P*	SA	LC1	LC2	Gen	Cyt
2	MW	1	0.505474	0.753596	0.495771	0.584111	0.6258	0.731489	−0.654851
3	vdw	0.505474	1	0.77058	0.99441	0.804914	0.811741	0.669773	−0.645101
4	log *P*	0.753596	0.77058	1	0.771746	0.915963	0.923825	0.931572	−0.949996
5	SA	0.495771	0.99441	0.771746	1	0.799627	0.807417	0.663032	0.635862
6	LC1	0.534111	0.804914	0.915963	0.799627	1	0.998609	0.919258	−0.929651
7	LC2	0.5258	0.811741	0.923825	0.807417	0.998609	1	0.911918	−0.932788
8	Gen	0.731489	0.669773	0.931572	0.663032	0.919258	0.911918	1	−0.961786
9	Cyt	−0.654851	−0.645101	−0.949996	−0.635862	−0.929651	−0.932788	−0.961786	1

Table 2 Summary of Regression Modeling Results of Example 1

			Gentox	Gentox	Gentox	Gentox	Gentox	Cytotox	Cvtotax	Cvtotox	Cytotox	Cytotox
Model	Method	Parameter (s)	R2X	R2Y	R2Yadj	Q2Yint	Q2Yext	R2X	R2Y	RZYadj	Q2Yint	Q2Yext
1G, 1C	MLR	Mw	1.00	0.54	0.48	0.39	0.13	1.00	0.43	0.36	0.25	0.12
2G, 2C	MLR	vdW	1.00	0.45	0.38	0.18	0.57	1.00	0.42	0.34	0.14	0.23
3G, 3C	MLR	log *P*	1.00	0.87	0.85	0.80	0.88	1.00	0.90	0.89	0.85	0.81
4G, 4C	MLR	SA	1.00	0.44	0.37	0.15	0.53	1.00	0.40	0.33	0.12	0.19
5G, 5C	MLR	LC1	1.00	0.85	0.83	0.72	0.93	1.00	0.86	0.85	0.77	0.97
6G, 6C	MLR	LC2	1.00	0.83	0.81	0.71	0.93	1.00	0.87	0.85	0.78	0.96
7G, 7C	MLR	Mw, vdW	1.00	0.66	0.56	0.15	0.49	1.00	0.56	0.44	−0.02	0.27
8G, 8C	MLR	Mw, log *P*	1.00	0.87	0.83	0.75	0.90	1.00	0.91	0.89	0.79	0.77
9G, 9C	MLR	Mw, SA	1.00	0.65	0.56	0.20	0.51	1.00	0.56	0.43	0.04	0.26
10G, 10C	MLR	Mw, LC1	1.00	0.93	0.90	0.78	0.91	1.00	0.90	0.87	0.70	0.96
11G, 11C	MLR	Mw, LC2	1.00	0.92	0.90	0.77	0.91	1.00	0.91	0.88	0.72	0.96
12G, 12C	MLR	LC1, LC2	1.00	0.86	0.82	0.72	0.94	1.00	0.87	0.83	0.74	0.95
13G, 13C	PCR	*t*1, *t*2, *t*3	1.00	0.94	0.91	0.80	0.85	1.00	0.97	0.95	0.88	0.98
14G, 14C	PCR	*t*1, *t*2	1.00	0.86	0.82	0.68	0.92	1.00	0.82	0.76	0.64	0.78
15G, 15C	PCR	*t*1, *t*3	1.00	0.89	0.85	0.75	0.90	1.00	0.94	0.92	0.88	0.96
16G, 16C	PCR	*t*2, *t*3	1.00	0.14	0.11	−1.41	0.02	1.00	0.19	−0.05	−1.41	0.34
17G, 17C	PCR	*t*1	1.00	0.80	0.78	0.69	0.93	1.00	0.78	0.76	0.68	0.72
18G, 18C	PCR	*t*2	1.00	0.06	−0.06	−1.34	0.13	1.00	0.03	−0.09	−1.33	0.02
19G, 19C	PCR	*t*3	1.00	0.08	−0.03	−0.24	0.13	1.00	0.15	0.05	−0.26	0.44
20	PLS	All	0.90	0.94	0.93	0.87	0.95	0.90	0.95	0.93	0.89	0.99

Explanatory note: For each of the models M1–M19 actually, two models have been derived, one for each response variable. Hence, model 1G means first model for genotoxicity response, model 1C first model for cytotoxicity response, and so on.

Table 2 is divided into two parts along the vertical direction. The five left-most columns with numerical entries relate to the first biological response (Genotox) and the five right-most columns to the second response (Cytotox). For each response, the following five model performance indicators are listed:

- R^2X = amount of explained sum of squares of X (i.e., how much variation is used to model the biological response);
- R^2Y = amount of explained sum of squares of Y;
- R^2Y_{adj} = amount of explained variance of Y;
- Q^2Y_{int} = amount of predicted variance of Y based on internal prediction (↔ cross-validation) (15,16);
- Q^2Y_{ext} = amount of predicted variance of Y based on external prediction of the six compounds in the prediction set.

As shown by Table 2, the one-parameter QSARs based either on log P, LC1, or LC2 are the successful ones from a predictive point of view. Interpretation of these three models shows (no plots provided) that an increasing value of the chemical descriptor is coupled to an increasing toxicity for both biological responses.

In this context, it might be tempting to use more than one descriptor variable to see whether predictions can be sharpened. We note, however, that due to the strong correlations among the six X-variables, there is a substantial risk that the interpretation of the model(s) might give misleading hints about structure–activity relationships.

We decided to use molecular weight as foundation for two-parameter correlations, as this is the descriptor that is least correlated with the other chemical descriptors. Models M7–M11 of Table 2 represent the five two-parameter combinations used. Again log P, LC1 and LC2 work best, but compared with models M1–M6, there is practically no gain in predictive power.

Finally, in order to illustrate how model interpretation may break down with correlated descriptors, we calculated model M12 for each response, in which the two most correlated descriptors LC1 and LC2 were employed. Figure 10 shows a plot of the regression coefficients of the QSAR model for each of the two responses. Since LC1 and LC2 are positively correlated with a correlation coefficient of >0.99, we would expect their regression coefficients to display the same sign. However, as seen from Fig. 10, LC1 and LC2 have *opposite* signs in both cases. Regarding the first response variable LC2 has the wrong sign, whereas the converse is true for the second endpoint.

In conclusion, by using MLR, very good one-parameter QSARs are obtainable using either $\log P$, LC1, or LC2, and this holds true for both responses. As shown by the last example (model 12), however, some caution is needed when using correlated descriptors. Here, the risk is that the *interpretation* may break down, because the coefficients do not get the right numerical size and they may even get the wrong sign (see Refs. 19 and 20 for a more elaborate theoretical account of this phenomenon). Note, however, that the *predictive ability* is not influenced in any way. On the contrary, Q^2s for model M12 are in every aspect as good as for the previous best MLR models (models M3, M5, M6, M8, M9, and M11).

4.4. Obtaining an Orthogonal Representation of the *X*-Descriptors and Regressing this Against Single *Y*-Variables: PCR Modeling

In order to circumvent the collinearity problem, the three PCA score vectors derived in Sec. 4.2 were used to replace the original descriptor variables. The three score vectors, henceforth denoted t_1, t_2, and t_3, can be thought of as new variables optimally summarizing the original X-variables. Their modelling ability of the six original variables is excellent as R^2X exceeds 0.98 (see Sec. 4.2).

We developed seven different PCR models for each response variable (see models M13–M20 in Table 2). Apart from the model in which all three score vectors are present (model M13), the best model from a predictive point of view is Ml5 in which only t_1 and t_3 are active (i.e., t_2 is omitted). Figure 11

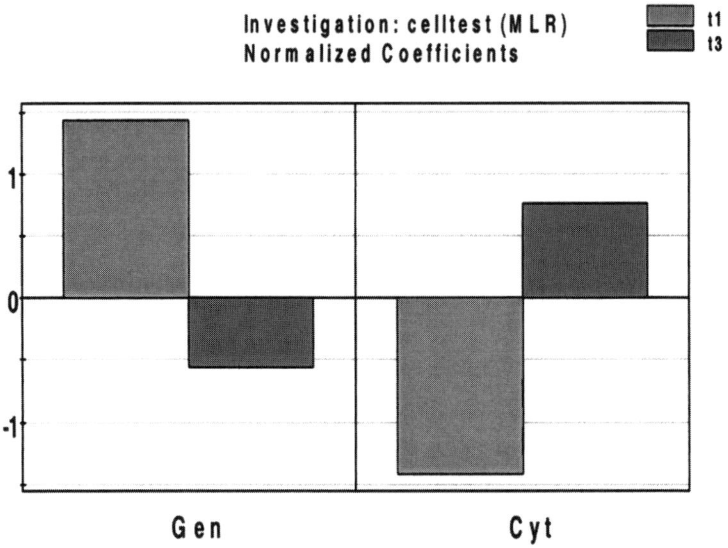

Figure 11 Overview plot of regression coefficients of model M13 for both responses.

displays the regression coefficients of M15. We recall from Sec. 4.2 that t_1 models lipophilicity, t_2 variation in molecular weight, and t_3 is heavily influenced by SA and vdW volume.

The modeling results of the remaining PCR models (M14 and M16–M19) indicate that including t_2 in the set of *X*-variables is not favorable as far as predictive power is concerned. Thus, it appears that variation in molecular weight is not critically linked to the variation in biological activity among the studied chemicals.

4.5. Using All *X*- and *Y*-Variables in One Shot: PLS Modeling

In contrast to MLR, PCR and the like, PLS can handle many *X*- and *Y*-variables in one single model. This is possible since PLS is based on the assumption of correlated variables, which may be noisy and incomplete (3,4,10). Hence, PLS allows correlations among the *X*-variables, among the *Y*-variables, and between the *X*- and the *Y*-variables, to be explored. The model performance statistics of the here derived two-component PLS model is given in Table 2 (as model M20).

A popular way of expressing PLS model information is through a loading plot. Figure 12 shows such a plot of model M20. We can discern three features. Firstly, the two responses are strongly inversely correlated. Secondly, the six X-variables form two distinct subgroups. Thirdly, the X-variables Mw, log P, LCI and LC2 are most important, but the contributions from SA and vdW are not negligible.

The PLS model uses 90% ($R^2X = 0.90$) of X to explain and predict the Y-data, not 100% as MLR. These X-residuals are of diagnostic interest. They can be used to calculate the typical distance to the model in the X-data (here abbreviated DModX) for a compound (Fig. 13). Fig. 14 shows DModX for each compound in the training set and the prediction set. We can also see the critical distance corresponding to the 0.05 probability level. This critical distance indicates the "tolerance volume"

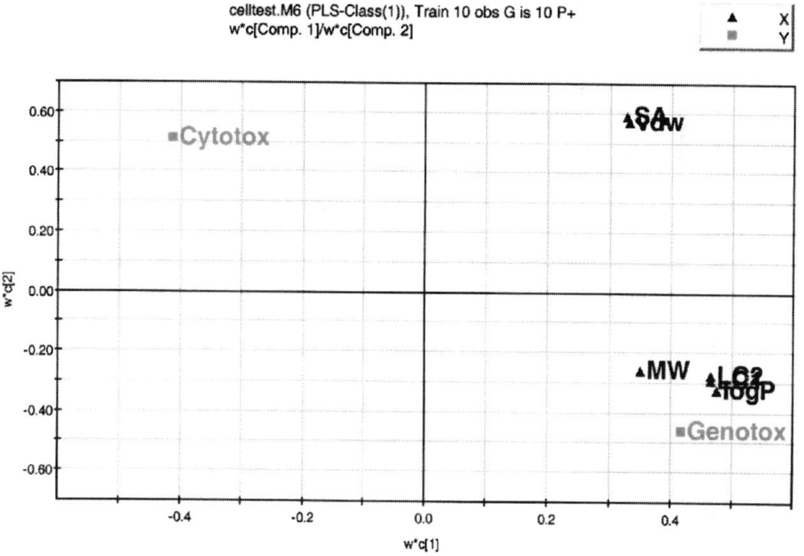

Figure 12 The PLS loading plot of model M20. Interpretation in accordance with Fig. 9.

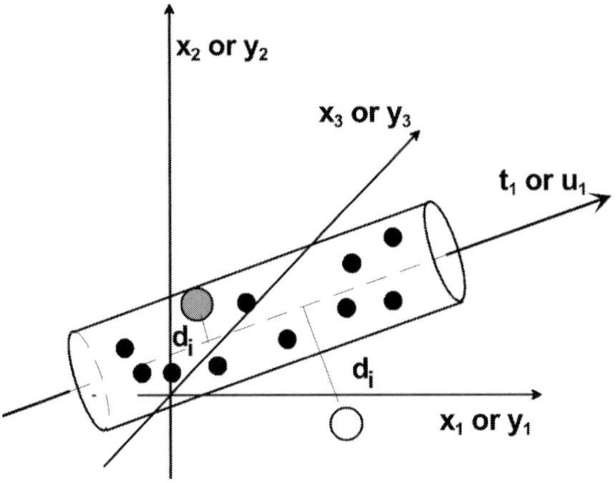

Figure 13 A geometrical interpretation of a compound's distance to the model (DModX). A value for DModX can be calculated for each compound and these values may be plotted together with a typical deviation distance (Dcrit) in order to reveal moderate outliers. Dcrit can be interpreted as the radius of the "beer-can" inserted around the compounds. (From Ref. 3.)

around the model, i.e., the range of the model in the *X*-data (3,4). Apparently, all compounds—also those featuring in the prediction set—are positioned inside the range of the model, i.e., they fit the model well. Hence, predictions for the prediction set compounds can be considered as realistic.

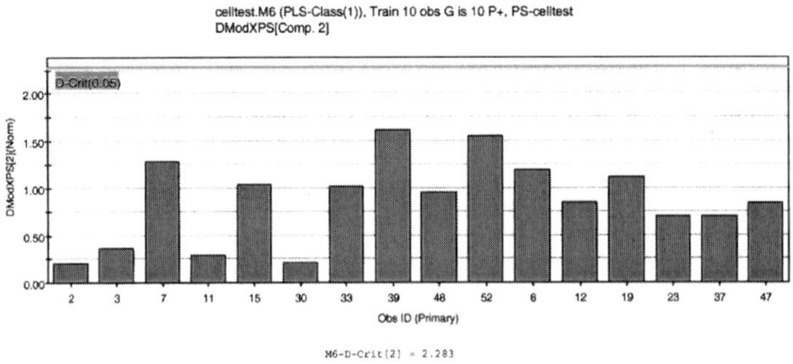

Figure 14 DModX plot of model M20. There are no outliers in the CELLTEST data set.

The final plot related to this example shows the relationship between observed and predicted cytotoxicity values (Fig. 15). The soundness and the excellent predictive power of the established QSAR cannot be mistaken. A similar result is the case for the other response variable (genotoxicity), but no such plot is provided.

4.6. Discussion of First Example

The CELLTEST data set is a small data set, here chosen merely because it allows a simple illustration of MLR, PCR, and PLS. It can be seen from the regression results of Table 2 that this is a "well-behaving" data set. There are no influential outliers and the predictive ability—as quantified by Q^2_{int} and Q^2_{ext}—is basically the same regardless of choice of regression method.

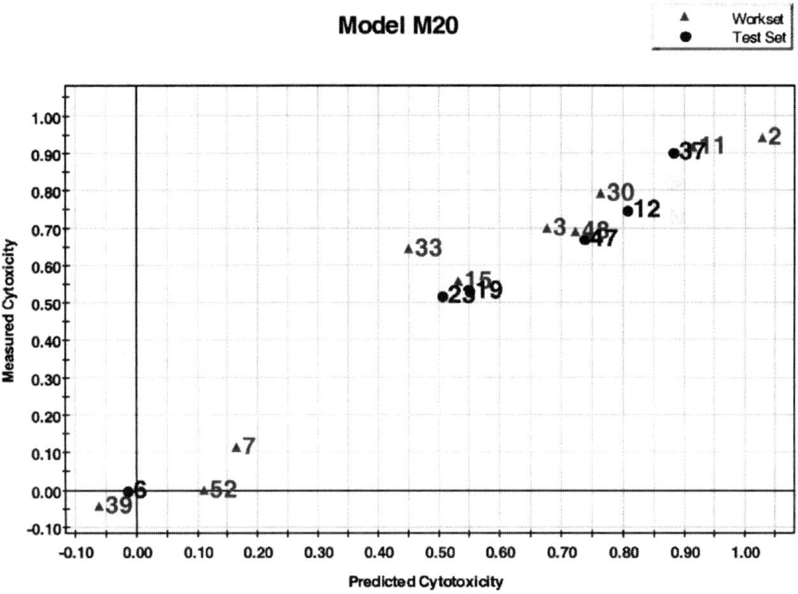

Figure 15 Observed and predicted cytotoxicities. Prediction set compounds are depicted by dots and training set compounds by triangles. Notation as in Fig. 8.

For a data set like CELLTEST, with strongly correlated *X*-variables, variable selection is needed in order for MLR to work properly. As an example of how NOT to do things we developed model M12, based on the two descriptors LC1 and LC2. Figure 10 shows the regression coefficients, which are totally misleading. According to these, LC1 and LC2 are positively and negatively correlated with the two responses, although their correlation coefficient surpasses 0.99. Thus, the problem lies in the interpretation of this model, which is terribly misleading, whereas the predictive power is unaffected.

Alternatively, to get around the problem of correlated *X*-descriptors, one may choose to first establish a PCA model of the *X*-block. The output of PCA is a set of new variables, score vectors, that are orthogonal and limited in number, and that may well be used in MLR. This combination of PCA and MLR makes up what is known as PCR.

As is evident from Table 2, the first and the third score vectors are most important for modeling and predicting the *Y*-data. This also shows that the order of emergence of PCA score vectors does not necessarily parallel their importance in QSAR work. The root for this behavior is that PCA is a maximum variance model of *X*, and these maximum variance directions are not always optimally correlated with *Y*.

Subsequently, a PLS model was established in which all *X*- and *Y*-variables were modeled at the same time. We note that it is not a drawback to use several variables accounting for lipophilieity (i.e., log *P*, LCI and LC2). On the contrary, being weighted averages of the original descriptors, the score vectors of the PLS model will become more stable and appear more distinctly the more appropriate variables are used as their basis.

Moreover, because the PLS model does not exhaust the *X*-matrix, i.e., uses less than 100% of the *X*-variance in the modeling of *Y*, it is possible to construct a diagnostic tool reflecting whether a compound conforms with the model. We here call this diagnostic tool DModX, and it can be thought as defining the validity range of the QSAR model. Both MLR and PCR lack a corresponding *X*-residual-based range-defining diagnostic statistic.

5. RESULTS FOR THE SECOND EXAMPLE—LEAD FINDING AND QSAR-DIRECTED VIRTUAL SCREENING OF HEXAPEPTIDES

5.1. Introduction and Background

We shall here study a data set named HEXAPEP. This data set originates from Torbjörn Lundstedt and Bernt Thelin. In the HEXAPEP data set, 16 hexapeptides were synthesized according to a statistical molecular design. Two biological activities, here coded as BA1 and BA2, were measured on each peptide. The experimental objective was to get a high BA1 and a low BA2.

The objective of the multivariate data analysis was to elucidate structure–activity relationships of the hexapeptides, and to identify the most promising lead compound for further exploration. To encode the structural variability among the hexapeptides, three amino acid descriptor scales developed by Hellberg, the so called z-scales, were used (21,22). This resulted in 18 X-variables, distributed as three scales for each of the six amino acid positions (cf. Fig. 16).

Due to the large number of X-variables (18), we shall here use the PLS method to accomplish the QSAR analysis. However, before we discuss the QSAR analysis, a short recollection of the z-scales is warranted.

5.2. Review of the *z*-Scales

In a QSAR study, the structural and chemical variation in a series of compounds needs to be translated into a set of descriptor variables. One way to accomplish this in peptide QSAR is to quantitatively characterize the properties of the amino acids, i.e., the peptide building blocks. It has been shown that such relevant amino acid descriptor scales may be derived using PCA on multiproperty matrices containing several appropriate physico-chemical variables. With such descriptor scales, the variation in amino acid sequence within a series of peptide analogs is quantified by using these scales for each amino acid in the peptide sequence.

	Z_1	Z_2	Z_3
Ala	0.07	-1.73	0.09
Val	-2.69	-2.53	-1.29
Leu	-4.18	-1.03	-0.96
Ile	-4.44	-1.68	-1.03
His	2.41	1.74	1.11
Gly	2.23	-5.36	0.31

Z- Scales → Insert Z-scale according to sequence → PLS

	Z_1	Z_2	Z_3	Z_1	Z_2	Z_3	LogBA
Ala-Ala	0.07	-1.73	0.09	0.07	-1.73	0.09	2.12
Leu-Val	-4.18	-1.03	-0.96	-2.69	-2.53	-1.29	1.48
Leu-Gly	-4.18	-1.03	-0.96	2.23	-5.36	0.31	0.66
His-His	2.41	1.74	1.11	2.41	1.74	1.11	4.32
Gly-His	2.23	-5.36	0.31	2.41	1.74	1.11	0.81
Val-Ile	-2.69	-2.53	-1.29	-4.44	-1.68	-1.03	3.90

Figure 16 The three z-scales (to the left) of the 20 coded amino acids are three principal component scores resulting from a PC-analysis of a 20×29 multiproperty matrix. In QSAR modeling, these three scales can be used to describe the properties of a peptide, simply by arranging the triplets of numbers according to the amino acid sequence. In this way, an amino acid sequence is translated into a series of numbers corresponding to triplets of z-scales. Subsequently, this kind of quantitative description of peptide sequences may be used in PLS-modeling to obtain QSAR models. (From Ref. 3.)

Some fifteen years ago, Hellberg et al. (21,22) presented three quantitative descriptive scales for the 20 coded amino acids, which were called z_1, z_2 and z_3, or simply the "z-scales." These scales were derived by PCA of a data matrix consisting of a multitude of physico-chemical variables, such as molecular weight, pKa's, ^{13}C NMR-shifts, etc. These z-scales reflect the most important properties of amino acids and are, therefore, often referred to as the "principal properties" of amino acids.

With the three z-scales it is possible to numerically quantify the structural variation within a series of related peptides, by arranging the z-scales according to the amino acid sequence (Fig. 16). Moreover, this parameterization has been demonstrated to be useful for the selection of training sets of representative peptides by means of SMD (see Ref. 3 and references therein).

Because the z-scales are simple and may be naturally interpreted as reflecting hydrophobicity, size, and electronic properties, they can be used to translate sequences of amino acids (i.e., peptides) into an understandable parameterization. The use of the original z-scales in peptide QSAR is discussed in the next section.

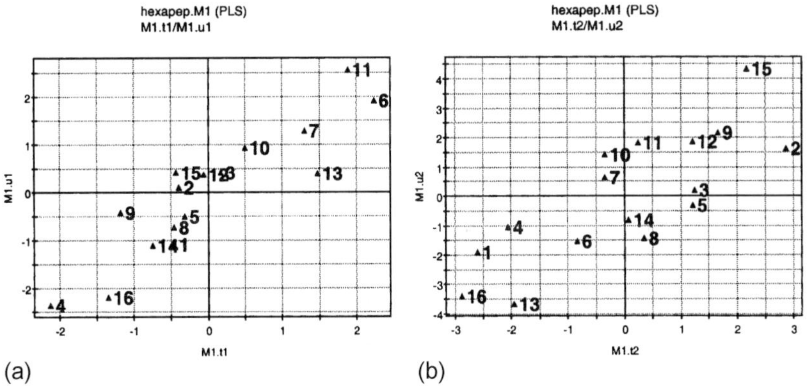

Figure 17 (a) The PLS t_1/u_1 score plot of HEXAPEP QSAR-model. (b) The PLS t_2/u_2 score plot of HEXAPEP QSAR-model.

5.3. Initial PLS-Modeling

To achieve a QSAR-model for the hexapeptides we used PLS. The PLS expression was a two-component model with $R^2Y = 0.86$ and $Q^2Y = 0.20$, i.e., a weakly predictive model. As manifested by the PLS t_1/u_1 score plot in Fig. 17a, there is a strong correlation between X (the z-scales) and Y (the two responses). No outlier is seen. Also in the second PLS component, the correlation band is relatively narrow (Fig. 17b).

The question that arises is why Q^2Y (=0.20) is so low? The regression coefficients of the BA1 and BA2 responses are plotted in Fig. 18a and b, respectively. These coefficients are similar in profile, albeit with an inverse relationship, because the two biological activities are negatively correlated ($r = -0.80$). The regression coefficients indicate that amino acid positions 1–3 affect the biological activities of the hexapeptides, whereas positions 4–6 do not. Hence, it was decided to refit the QSAR using only the three first positions.

5.4. Refined PLS-Modeling

Besides the removal of nine X-variables (three z-scales in position 4, 5, and 6), one additional change was made, namely

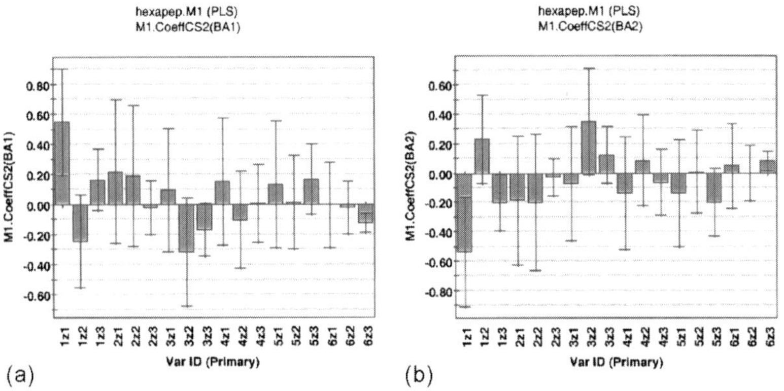

Figure 18 (a) The PLS regression coefficients of BA1. (b) The PLS regression coefficients of BA2.

to alter the number of cross-validation groups from the software default of 7 groups to 16 groups. This was done because cross-validation may often run into problems with designed data. Hence, by leaving out only one observation (hexapeptide), cross-validation might be stabilized. It is stressed, however, that in general leaving several observations out is a preferable and more realistic approach.

When fitting the refined PLS-model, only one component was obtained. This model had $R^2Y = 0.80$ and $Q^2Y = 0.39$, i.e., some improvement in predictive power in comparison with the original model. As shown by Fig. 19a, there are some modeling differences between the two responses, and BA2 is handled better by the model. The PLS t_1/u_1 score plot rendered in Fig. 19b indicates a strong correlation structure between the z-scales of positions 1–3 and the two biological activities. Hence, there is no doubt that changes in peptide sequences can be associated with changes in biological performance.

We can see from the regression coefficients (Fig. 20a and b) that a hydrophilic amino acid (high z_1) in position 1 and a small-sized amino acid (low $z2$) in position 3 are beneficial for increasing BA1. Because the two biological activities are strongly inversely related, the pattern of the regression coefficients of BA2 is inverted. However, basically, the same model interpretation holds true.

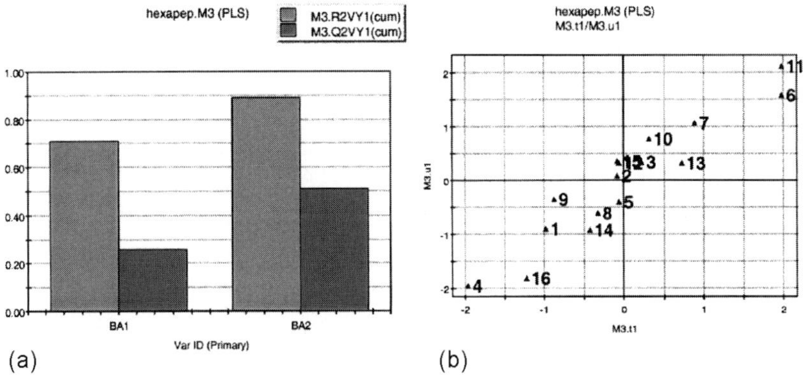

Figure 19 (a) Overview of modelling performance of refined HEXAPEP QSAR. For BA1 R^2Y amounts to 0.71 and Q^2Y to 0.26. The corresponding values for BA2 are 0.89 and 0.51. It is mainly a weakly deviating behavior of peptide 15 (no plot given) that degrades the predictive power with regard to BA1. (b) PLS t_1/u_1 score plot of refined HEXAPEP QSAR. This plot reveals the strong association between X and Y.

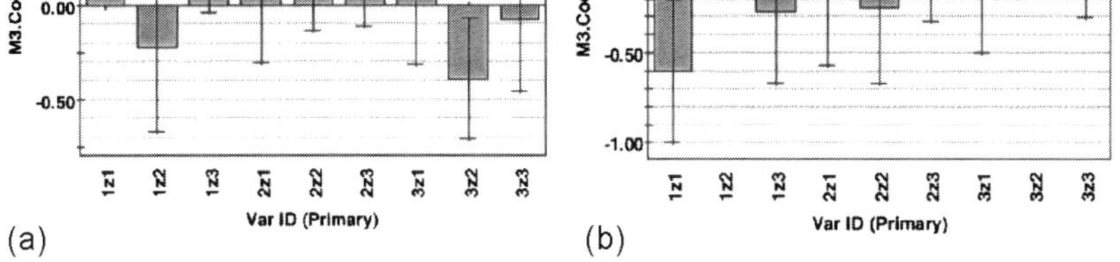

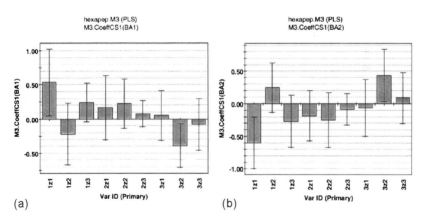

Figure 20 (a) The PLS regression coefficients of BA1 of refined HEXAPEP QSAR. The coefficients display the modelling impact of the z-scales of positions 1–3. (b) The PLS regression coefficients of BA2 of refined HEXAPEP QSAR. The inverse relationship is due to the strong negative correlation between BA1 and BA2 ($r = -0.80$).

5.5. Molecular Design of New Hexapeptides

In order to propose new and interesting hexapeptides, we have to consider the following premises:

- Postion 1: high z_1, low z_2, high z_3. Matching amino acids are: serines, glycine (G), and aspartate (D).
- Position 2: high z_1–z_3. Matching amino acids are: histidine (H), aspartate (D), and asparagine (N).
- Position 3: high z_1, low z_2, low z_3. Matching amino acids are: glycine (G), serine (S), and threonine (T).
- Position 4–6: for convenience fixed as alanine–alanine–alanine (AAA).

Thus, in the proposition of new hexapeptides, the general sequence X–X–X–alanine–alanine–alanine seems useful. In all, there are 27 sequence combinations (of XXXAAA) using three candidate amino acids for each of the three varied positions. This is a rather small number compared with the 20^6 (= 64 millions) theoretically possible compounds.

5.6. Virtual Screening of New Hexapeptides

The established peptide QSAR was used to predict the biological activities of the 27 proposed hexapeptides. These results are summarized in Fig. 21. The best tested hexapeptide of the training set, DYGDWG (no. 11), has BA1 = 89 and BA2 = 12 as measured values. The model used predicts BA1 = 90 and BA2 = 18. In the light of this information, the predictions for the hexapeptides GDGAAA (no. 32), DDGAAA (no. 41), GNGAAA (no. 29), and GHGAAA (no. 26) appear very encouraging. One of these compounds might tentatively be chosen as a new lead.

5.7. Discussion of Second Example

In the example reviewed, we have seen how peptide QSAR may be used for lead finding. The underlying training set of 16 peptides was constructed through the use of SMD. By fitting a multivariate peptide QSAR-model, it was then possible

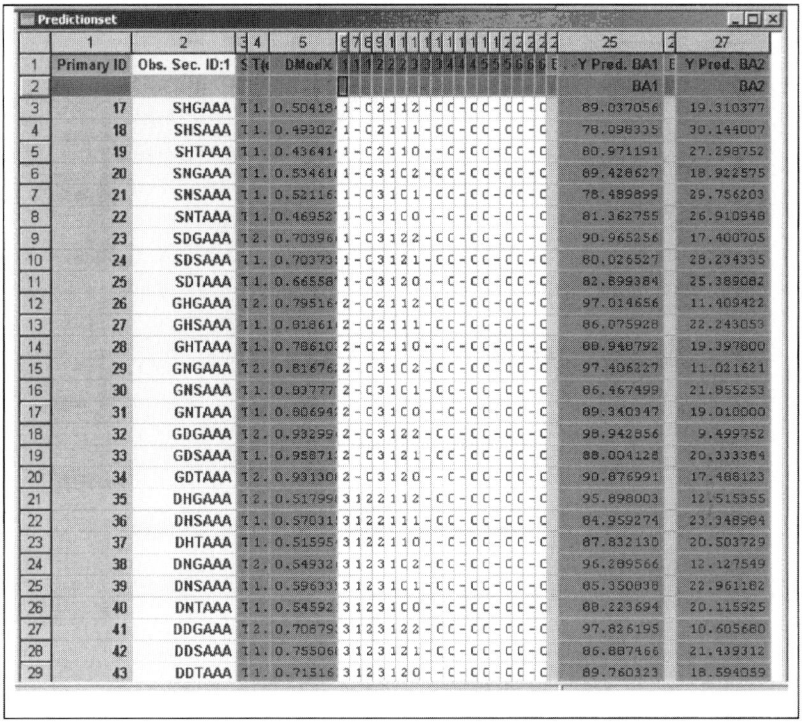

	1	2	3	4	5	6–24	25	27
1	Primary ID	Obs. Sec. ID:1	S	T(	DModX	1 1 1 2 2 2 3 3 3 4 4 4 5 5 5 6 6 6 E	Y Pred. BA1	Y Pred. BA2
2							BA1	BA2
3	17	SHGAAA	T	1.	0.50418	1 - C 2 1 1 2 - C C - C C - C C - C	89.037056	19.310377
4	18	SHSAAA	T	1.	0.49302	1 - C 2 1 1 1 - C C - C C - C C - C	78.098335	30.144007
5	19	SHTAAA	T	1.	0.43641	1 - C 2 1 1 0 - - C - C C - C C - C	80.971191	27.298752
6	20	SNGAAA	T	1.	0.53461	1 - C 3 1 C 2 - C C - C C - C C - C	89.428627	18.922575
7	21	SNSAAA	T	1.	0.52116	1 - C 3 1 C 1 - C C - C C - C C - C	78.489899	29.756203
8	22	SNTAAA	T	1.	0.46952	1 - C 3 1 C 0 - - C - C C - C C - C	81.362755	26.910948
9	23	SDGAAA	T	2.	0.70396	1 - C 3 1 2 2 - C C - C C - C C - C	90.965256	17.400705
10	24	SDSAAA	T	1.	0.70373	1 - C 3 1 2 1 - C C - C C - C C - C	80.026527	28.234335
11	25	SDTAAA	T	1.	0.66558	1 - C 3 1 2 0 - - C - C C - C C - C	82.899384	25.389082
12	26	GHGAAA	T	2.	0.79516	2 - C 2 1 1 2 - C C - C C - C C - C	97.014656	11.409422
13	27	GHSAAA	T	1.	0.81861	2 - C 2 1 1 1 - C C - C C - C C - C	86.075928	22.243053
14	28	GHTAAA	T	1.	0.78610	2 - C 2 1 1 0 - - C - C C - C C - C	88.948792	19.397800
15	29	GNGAAA	T	2.	0.81676	2 - C 3 1 C 2 - C C - C C - C C - C	97.406227	11.021621
16	30	GNSAAA	T	1.	0.83777	2 - C 3 1 C 1 - C C - C C - C C - C	86.467499	21.855253
17	31	GNTAAA	T	1.	0.80694	2 - C 3 1 C 0 - - C - C C - C C - C	89.340347	19.010000
18	32	GDGAAA	T	2.	0.93299	2 - C 3 1 2 2 - C C - C C - C C - C	98.942856	9.499752
19	33	GDSAAA	T	1.	0.95871	2 - C 3 1 2 1 - C C - C C - C C - C	88.004128	20.333384
20	34	GDTAAA	T	2.	0.93130	2 - C 3 1 2 0 - - C - C C - C C - C	90.876991	17.488123
21	35	DHGAAA	T	2.	0.51799	3 1 2 2 1 1 2 - C C - C C - C C - C	95.898003	12.515355
22	36	DHSAAA	T	1.	0.57031	3 1 2 2 1 1 1 - C C - C C - C C - C	84.959274	23.348984
23	37	DHTAAA	T	1.	0.51595	3 1 2 2 1 1 0 - - C - C C - C C - C	87.832130	20.503729
24	38	DNGAAA	T	2.	0.54932	3 1 2 3 1 C 2 - C C - C C - C C - C	96.289566	12.127549
25	39	DNSAAA	T	1.	0.59633	3 1 2 3 1 C 1 - C C - C C - C C - C	85.350838	22.961182
26	40	DNTAAA	T	1.	0.54592	3 1 2 3 1 C 0 - - C - C C - C C - C	88.223694	20.115925
27	41	DDGAAA	T	2.	0.70879	3 1 2 3 1 2 2 - C C - C C - C C - C	97.826195	10.605680
28	42	DDSAAA	T	1.	0.75506	3 1 2 3 1 2 1 - C C - C C - C C - C	86.887466	21.439312
29	43	DDTAAA	T	1.	0.71516	3 1 2 3 1 2 0 - - C - C C - C C - C	89.760323	18.594059

Figure 21 Predicted biological activities of the 27 new hexapeptides. Peptides 26, 29, 32, and 41 are best and have better predictions than DYGDWG, the best tested hexapeptide.

to understand which positions in the peptide sequence most effectively elicited the biological activity. In this case, positions 1–3 were found most influential. In order to arrive at a new peptide lead, these three positions were modified by inserting new amino acids according to a pattern dictated by a new SMD.

It might be argued that QSAR-modeling is not really necessary in an application of the HEXAPEP-type. A simple visual inspection of the BA-data would do the job equally well, and enable an identification of the most promising hexapeptide. This argument is, however, not true. This is because such an eye-balling exercise would not reveal the fundamental knowledge regarding (1) which positions correlate with biological activity, and (2) by which amino acids one should modify the peptide sequence.

In general, the use of QSAR modeling for the final evaluation of many test data stabilizes the interpretation and points to important chemical properties. Also, in case of several responses, the use of all data decreases the risk of false negatives (i.e., an actually active compound that is classified as inactive) and false positives (an inactive compound classified as active). This is because random variations of individual pharmacological tests are averaged out in large data sets. Moreover, with PLS-modeling, rather than visual inspection, outliers and erroneous test results are revealed by model diagnostics, which further strengthens the credibility of the results.

Once a lead has been put forward, the next step is lead optimization. Lead optimization is usually less complicated than lead finding. Often one can identify some kind of scaffold, on which small to moderate changes can be inflicted by manipulating substituents and other small molecular features. In the HEXAPEP data set, the scaffold corresponded to the chain of five peptide bonds, and the structural variation occurred in the amino acid side chains.

Regardless of the type of chemical backbone (scaffold) and the structural features varied, lead optimization is achieved smoothly and efficiently by coupling SMD with QSAR and possibly also virtual screening exercises. The principles of SMD were introduced in Sec. 2 and are not reiterated here. In fact, SMD is an indispensable tool in lead optimization. It makes it possible to explore the principal property space around a lead in an informationally optimal manner, thereby increasing the likelihood of finding an optimal candidate drug.

In the HEXAPEP data set, SMD laid the ground for a strong peptide QSAR-model. By interpreting the regression coefficients of this model, the most important amino acid positions were identified. Hence, it was possible to focus on these positions in the virtual screening.

6. DISCUSSION

6.1. SMD, Projections, and QSAR—A Framework for Predictive Toxicology

Whenever one wishes to model one or several biological or environmental response variables, Y, by a linear model based on a set of correlated X-variables, the PLS method is an interesting choice. The graphical representation of the PLS model parameters and residuals facilitates its use and interpretation. The rather natural assumptions underlying the method, namely that the chemical and biological variables are correlated and possibly also noisy and incomplete, is more in line with reality than those of classical regression (MLR and the like).

In PLS, the variable correlations are modeled as arising from a small set of latent variables, where all the measured variables are modelled as linear combinations of these latent variables. In QSAR, such latent variables are often interpretable in terms of the size, lipophilicity, polarity, polarizability, and hydrogen bonding properties of molecules and their substituents. Hopefully, such an interpretation will also help us understand which factors are regulating the biological response mechanisms.

Multiple linear regression is also a method that deserves a place in the QSAR-modeler's toolbox, as long as the assumptions of MLR are met. Since MLR is sensitive to the existence of correlated variables, care must be exercised to verify that the chemical descriptors used are not too strongly interrelated. As a rule of thumb, in MLR, one should avoid using descriptors having a correlation surpassing 0.3 (in absolute value). Ways to increase the applicability of MLR include SMD, or the use of PCA to obtain an orthogonal summary representation of the original X-variables. This latter representation may then serve as the entry point for PCR.

Besides the regression tools (MLR, PCR, and PLS), PCA is very useful in QSAR, but unfortunately its applicability in predictive toxicology is often underestimated. Many important conclusions may be drawn using PCA prior to regression, rendering the regression analysis smoother and less time-consuming. Principal component analysis is especially advantageous in the situation when the response data (Y) comprises more than around five to six responses. Then the correlation structure among the response variables may be explored, and also the inherent dimensionality of the Y-matrix be determined. With few (2–4) Y-variables scatter plots of raw data are useful to understand the properties of the responses. The discoveries made in a PCA loading plot have a decisive impact on the subsequent regression analysis. In addition, PCA informs about any subgroupings among the compounds, possibly motivating more than one QSAR to be formulated.

Apart from describing the regression- and projection-based methodologies used in predictive toxicology, we have here also placed a lot of emphasis on the concept of SMD. The purpose of using SMD is to compose a training set consisting of relevant, informative and diverse compounds. Unfortunately, this step is often neglected in QSAR research. It is of crucial importance for any QSAR model, irrespective of its origin and future use, that the series of substances used to develop the model displays a well-balanced distribution in the chemical space, e.g., a PC score plot, and accommodates representative compounds. The accomplishment of this is greatly facilitated by an SMD, whereby all the major structural and chemical features of the compounds are spanned systematically and simultaneously. This is much more efficient than varying just one variable or property at a time.

Projection methods, SMD, and QSAR may be seen as forming a framework for sound modelling in predictive toxicology. To introduce this framework, we have in detail described analyses of two QSAR examples, of varying nature and sophistication. We have tried to highlight the richness of plots of model parameters and residuals, and other diagnostics, which are very useful in the quest for developing predictively valid and interpretationally sound QSARs.

6.2. Outlier Detection and Model Interpretation

Plots of PCA, PCR, and PLS scores are invaluable for overviewing relationships among the compounds, e.g., for finding outliers in the X- or Y-data, or in the relation between X and Y. In a similar manner, plots of residuals, for example, in terms of the DModX and DModY bar charts, are essential tools for identifying moderate outliers in the data. To support decision making, some statistical tools are available. For score plots, a parameter known as Hotelling T^2 may be used to define a 95% or 99% tolerance region, and for the DModX bar chart, referring to a critical distance corresponding to the 0.05 level is a common practice (3). Compounds outside the tolerance volume defined by Hotelling T^2 deviate from normality, and compounds exceeding the critical distance in DModX do not fit the model well.

It is here appropriate to make a sharp distinction between *strong* and *moderate* outliers. Strong outliers, which are found in score plots, *conform* with the overall correlation structure (although they may have an extreme character), whereas moderate ones, which are found in residual plots, *break* the general correlation structure. A moderate outlier does not have the same profound effect on the QSAR model building as does a strong outlier. But for QSAR diagnostics, it is important to unravel outliers in residuals, because their presence indicates lack of homogeneity in the X-and/or$Y$-data.

For the interpretation of QSAR models, one may consider the loadings, p, in PCA/PCR, the weights, w^*c, in PLS, or in PLS resort to the regression coefficients, b, derived from the underlying latent variable model. In MLR the regression coefficients are utilized in model interpretation. All such loadings, weights, and coefficients reflect the relationships among the X- and the Y-variables. Thus, one obtains information on what X gives Y, or, how to "set" X to get a desired Y. This implies that with a good QSAR, it is not only possible to understand the mechanism of biological performance, but also to extract clues of how to modify the chemical structure to get enhanced biological performance.

6.3. Model Validation Tools

Furthermore, in predictive toxicology, it is important to consider how to accomplish a predictively valid QSAR. In principle, four tools of model validation are discernible (3), i.e., (i) external validation, (ii) cross-validation, (iii) response permutation, and (iv) plots of model parameters and residuals, and the underlying data:

- External validation is rare in QSAR, but was exemplified in the first example, where six compounds were used as an external prediction set. This is the most demanding way of testing the predictive validity.
- Cross-validation is used to assess the predictive significance using only the training set data. Hence, this procedure is sometimes called internal validation. Cross-validation has been used in the derivation of all models referred to in this chapter.
- Response permutation is based on fitting PLS models to random numbers (permuted Y-data) and obtaining reference distributions of R^2Y and Q^2Y, which, in turn, may be used to assess the statistical significance of the "real" R^2Y and Q^2Y. The combination of cross-validation and response permutation offers a powerful way of exploring the validity of a QSAR. Further discussion in provided in Ref. 3.
- Finally, plots of model parameters and residuals, and of the underlying data, are necessary for model understanding and acceptance. Such plots have been used, in different combinations, throughout this chapter.

6.4. Many Variables and Many Compounds

An attractive property of projection methods, such as PCA and PLS, lies in their ability to cope with almost any type of data matrix, i.e., a matrix which may have many variables (columns), many compounds (rows), or both. For instance, the precision and reliability of the PCA and PLS parameters related to the observations (compounds) is enhanced by increasing the number of relevant variables. This property is readily understood by realizing that the "new variables," the scores (t_a in PCA, t_a/u_a in PLS) are estimated as weighted averages of the X- and Y-variables, since PCA and PLS operate as projection methods. Any (weighted) average becomes more precise and distinct the more numerical values are used as its basis. Hence, PCA and PLS work well with "short and wide" matrices, i.e., matrices with many more columns than rows.

Analogously, the PCA and PLS parameters supplying information related to the variables, for example, loadings, weights, R^2, Q^2, etc., get more reliable the larger the number of observations. This is because the loadings, weights, etc., are linear combinations, i.e., weighted averages, of the N observation vectors. Thus, PCA and PLS also work well with "long and lean" data structures, i.e., data tables with many more rows than columns.

Obviously, PCA and PLS are also capable of handling "square" matrices, i.e., situations in which the number of rows and columns is approximately equal.

6.5. Related Methods

Another tool used for regression-like modelling in QSAR with many and collinear descriptor variables is ridge regression (RR). RR uses another approach to cope with the near singularities of the **X**-matrix in the regression problem. Here a small number, d, is added to all the diagonal elements of the variance–covariance matrix of **X**, i.e., **X′X**, before its inversion in the regression algorithm (10). This is closely related to the omission of all principal components with singular values smaller than d, and indeed RR and PCR demonstrate similar performance. Ridge regression can be seen as a mathematical circumvention of the multicollinearity problem rather than a solution to it (10,23).

In addition, other methods, such as, neural networks (NN) are often employed in QSAR analysis. The NNs are equivalent to a certain type of non-linear regression, and often use as input the output of a PCA-model. The NNs are detailed elsewhere in this text.

6.6. Extensions of Projection-Based Approaches

There also exist extensions of the projection methods aimed at addressing more complex problem types. From a QSAR point of view, the elaborate versions of most immediate interest are (i) non-linear PCA and PLS, (ii) PLS-discriminant analysis (PLS-DA), and (iii) hierarchical PCA and PLS (3,4,24).

A non-linear regression problem may be modelled with PLS by making the t_a/u_a inner-relation non-linear, or simply by expanding the **X**-matrix with the squares and cubes of its columns, and then using this augmented matrix to model **Y** (3,4,24). In predictive toxicology, non-linear modelling might be of relevance when the series of compounds shows large variation chemically and biologically, e.g., a biological response variable spanning several orders of magnitude.

The objective with PLS-DA is to develop a model that is capable of separating classes of observations on the basis of their X-variables. To encode in Y information of class memberships, a "dummy" **Y**-matrix is created that has G-1 columns (for G classes) with ones and zeroes, such that the gth column is 1 and the others 0 for observations of class g when $g < G - 1$. For the "last" class, G, all columns have the value of -1. A PLS model is fitted between X and the artificial Y, and in this way a discriminant plane is found in which the observations are well separated according to their class belongings (3,4,24). This approach works well provided that each class is "tight" and occupies only a small portion in the X-space. In predictive toxicology, PLS-DA might be of interest in the classification of molecular structures according to chemical or biological properties, say, separating acids from bases, or actives from inactives. PLS-DA was successfully used for classifying some 500+ environmental pollutants into four classes based on expected mechanism of biological action (25).

Hierarchical PCA and PLS are two emerging techniques (3,4,24,26) which have much to offer when working with very many variables. In such a situation, plots and lists of loadings, weights, and coefficients, tend to become messy and the results are often difficult to overview. Instead of reducing the number of variables, and thus endangering the validity of the model (3,4,24,26), a better alternative is often to divide the variables into conceptually meaningful blocks and apply hierarchical PCA or PLS. With hierarchical modeling, each block of variables is first summarized by a few scores. Next, these scores are collected and put together in a new matrix, consisting only of the scores. This new matrix is then modelled with PCA or PLS to get an overview of the data.

In predictive toxicology and QSAR, the blocks of variables may correspond to different regions of the compounds, and different categories of variables (lipophilicity, size, reactivity properties, etc.). Hierarchical block-relations may be specified within both the X- and the Y-data. Berglund et al. (27) recently used hierarchical PCA in a novel way to approach the alignment problem of rigid and flexible molecules at their receptor site in drug design studies.

6.7. Concluding Remarks

There is a strong trend in QSAR to use more and more variables to characterize molecules, fragments, reaction pathways, etc. This is particularly true in the pharmaceutical industry where new drugs are produced (27,28), but also within the environmental sciences (25,29). The reasons for this are obvious. First, we strongly believe that we know and understand more about our compounds when we measure many properties (variables) rather than a handful. Second, computers, electronics, and the rapid instrumental development of spectrophotometers, chromatographs, etc., provide us with a multitude of data for any scrutinized compound. This abundance of data makes multivariate projection methods, such as, PCA and PLS, of primary importance. Because of their flexibility, we do not any more need awkward variable selection schemes, but are able to use the information in our full set of measured and calculated variables. This is very appealing in QSAR and predictive toxicology.

REFERENCES

1. Dunn WJ III. Quantitative structure–activity relationships (QSAR). Chemometrics Intell Lab Syst 1989; 6:181–190.

2. Eriksson L, Johansson E. Multivariate design and modeling in QSAR. Chemometrics Intell Lab Syst 1996; 34:1–19.

3. Eriksson L, Johansson E, Kettaneh-Wold N, Wold S. Multi- and Megavariate Data Analysis—Principles and Applications, Umetrics AB, 2001, ISBN 91-973730-1-X.

4. Wold S, Sjöström M, Andersson PM, Linusson A, Edman M, Lundstedt T, Norden B, Sandberg M, Uppgård L. Multivariate design and modelling in QSAR, combinatorial chemistry, and bioinformatics. Klaus Gundertofte, Flemming Jörgensen, eds. Molecular Modelling and Prediction of Bioactivity. New York: Kluwer Academic/Plenum Publishers, 2000:27–44.

5. Cronin MTD, Schultz TW. Pitfalls in QSAR. J Mol Struct—THEOCHEM, 2003; 622:39–51.

6. Austel V. Eur J Med Chem 1982; 17:9–16.

7. Turner L, Choplin F, Dugard P, Hermens J, Jaeckh R, Marsmann M, Roberts D. Structure–activity relationships in toxicology and ecotoxicology: an assessment. Toxicol In Vitro 1987; 1:143–171.

8. Hermens JLM. Quantitative structure–activity relationships of environmental pollutants. Hutzinger O, ed. Handbook of Environmental Chemistry. Reactions and Processes. Berlin: Springer-Verlag1989:2E:111–162.

9. Topliss JG, Edwards RP. Chance factors in studies of quantitative structure-activity relationships. J med Chem 1979; 22:1238–1244.

10. Höskuldsson A. Prediction Methods in Science and Technology. Copenhagen, Denmark: Thor Publishing, 1996.

11. Wold S, Albano C, Dunn WJ III, Edlund U, Esbensen K, Geladi P, Hellberg S, Johansson E, Lindberg W, Sjöström M. Multivariate data analysis in chemistry. In: Kowalski BR, ed. Chemometrics—Mathematics and Statistics in Chemistry. Vol. D Dordrecht, The Netherlands: Reidel Publishing Company, 1984; 1–81.

12. Wold S, Sjostrom M, Eriksson L. PLS-regression: a basic tool of chemometrics. Chemometrics Intell Lab Syst 2001; 58:109–130.

13. Jolliffe IT. Principal Component Alanysis. Berlin: Springer Verlag, 1986.

14. Trygg J. O2-PLS for qualitative and quantitative analysis in multivariate calibration. J Chemometrics 2002; 16:283–293.

15. Wold S. Cross-validatory estimation of the number of components in factor and principal components models. Technometrics 1978; 20:397–405.

16. Osten DW. Selection of optimal regression models via cross-validation. J Chemometrics 1988; 2:39–48.

17. Eriksson L, Jonsson J, Hellberg S, Lindgren F, Sjöström M, Wold S, Sandström B, Svensson. A strategy for ranking environmentally occurring chemicals. Part V: the development of two genotoxicity QSARs for halogenated aliphatics. Environ Toxicol Chem 1991; 10:585–596.

18. Eriksson L, Sandström BE, Tysklind M, Wold S. Modelling the cytotoxicity of halogenated aliphatic hydrocarbons. Quantitative structure–activity relationships for the IC50 to human HeLa cells. Quant Struct Activ Rel 1993; 12:124–131.

19. Mullet GM. Why regression coefficients have the wrong sign. J Qual Technol 1976; 8:121–126.

20. Eriksson L, Hermens JLM, Johansson E, Verhaar HJM, Wold S. Multivariate analysis of aquatic toxicity data with PLS. Aquat Sci 1995; 57:217–241.

21. Hellberg S, Sjöström M, Wold S. The prediction of Bradykinin potentiating potency of pentapeptides. An example of a peptide quantitative structure–activity relationship. Acta Chem Scand 1986; B40:135–140.

22. Hellberg S, Eriksson L, Jonsson J, Lindgren F, Sjöström M, Skagerberg B, Wold S, Andrews P. Minimum analogue peptide sets (MAPS) for quantitative structure–activity relationships. Int J Peptide Protein Res 1991; 37:414–424.

23. Lindgren, F. Third generation PLS: Some Elements and Applications, Ph.D. Thesis, Umeå University, Umeå, Sweden, 1993.

24. Wold S, Trygg J, Berglund A, Antti H. Some recent developments in PLS modeling. Chemometrics Intell Lab Syst 2001; 58:131–150.

25. Nouwen J, Lindgren F, Hansen B, Karcher W, Verhaar HJM, Hermens JLM. Classification of environmentally

occurring chemicals using structural fragments and PLS discriminant analysis. Environ Sci Technol 1997; 31:2313–2318.

26. Eriksson L, Johansson E, Lindgren F, Sjöström M, Wold S. Megavariate analysis of hierarchical biological data. J Comput-Aided Mol Des 2001; 16:711–726.
27. Berglund A, De Rosa MC, Wold S. Alignment of flexible molecules at their receptor site using 3D descriptors and Hi-PCA 1997; 11:601–612.
28. Gabrielsson J, Lindberg NO, Lundstedt T. Multivariate methods in pharmaceutical applications. J Chemometrics 2002; 16:141–160.
29. Damborsky J. Quantitative structure–function relationships of the single-point mutants of haloalkane dehalogenase: a multivariate approach. Quant Struct Activ Rel 1997; 16:126–135.

INTRODUCTORY READING

1. Wold S, Dunn WJ III. Multivariate quantitative structure–activity relationships (QSAR): conditions for their applicability. J Chem Inf Comput Sci 1983; 23:6–13.

GLOSSARY

Calibration set: see Training set.

Cross-validation: Internal validation technique used to estimate predictive power.

Descriptor variables: see Predictor variables.

External validation: Predictions on external validation set used to estimate predictive power.

Explained variance, R^2: Shows how well a model can be made to fit the data.

Loadings: Shows orientation of low-dimensional PCA or PLS hyper-plane.

MLR: Multiple linear regression.

Observations: Denotes compounds in QSAR.

PCA: Principal component analysis.

PCR: Principal component regression.

PLS: Partial least squares projections to latent structures.

Predicted variance, Q^2: Shows how well a model can be made to predict the data.

Predictor variables: Variables used to describe chemical properties of compounds.

QSAR: Quantitative structure–activity relationships.

Residuals: Unexplained variance.

Response variables: Variables used to describe biological/toxicological properties of compounds.

Scores: Shows location of observations (compounds) on low-dimensional PCA or PLS hyper-plane.

SMD: Statistical molecular design.

Test set: see Validation set.

Training set: Group of compounds used to calculate the predictive model.

Variables: Denotes predictors or responses in QSAR.

Validation set: Group of compounds used to verify the predictive power of a model.

X-variables: see Predictor variables.

Y-variables: see Response variables.

7

Machine Learning and Data Mining

STEFAN KRAMER
Institut für Informatik, Technische Universität München, Garching, München, Germany

CHRISTOPH HELMA
Institute for Computer Science, Universität Freiburg, Georges Köhler Allee, Freiburg, Germany

1. INTRODUCTION

In this chapter, we will review basic techniques from knowledge discovery in databases (KDD), data mining (DM), and machine learning (ML) that are suited for applications in predictive toxicology. We will discuss primarily methods which are capable of providing new insights and theories. Methods, which work well for predictive purposes but do not return models that are easily interpretable in terms of toxicological knowledge (e.g., many connectionist and multivariate approaches), will not be discussed here, but are discussed elsewhere in this book.

Also not included in this chapter, yet important, are visualization techniques, which are valuable for giving first clues about regularities or errors in the data. The chapter will feature data analysis techniques originating from a variety of fields, such as artificial intelligence, databases, and statistics. From artificial intelligence, we know about the structure of search spaces for patterns and models, and how to search them efficiently. Database literature is a valuable source of information about efficient storage of and access to large volumes of data, provides abstractions of data management, and has contributed the concept of query languages to data mining. Statistics is of utmost importance to data mining and machine learning, since it provides answers to many important questions arising in data analysis. For instance, it is necessary to avoid flukes, that is, patterns or models that are due to chance and do not reflect structure inherent in the data. Also, the issue of prior knowledge has been studied to some extent in the statistical literature.

One of the most important lectures in data analysis is that one cannot be too cautious with respect to the conclusions to be drawn from the data. It is never a good idea to rely too much on automatic tools without checking the results for plausibility. Data analysis tools should never be applied naively—the prime directive is "know your data." Therefore, sanity checks, (statistical) quality control, configuration management, and versioning are a necessity. One should always be aware of the possible threats to validity.

Regarding the terminology in the paper, we will talk about instances, cases, examples, observations interchangeably. Instances are described in terms of attributes/features/variables (e.g., properties of the molecules, LD_{50} values), which also will be used as synonyms in the chapter. If we are considering prediction, we are aiming at the prediction of one (or a few) dependent variables (or target classes/variables, e.g., LD_{50} values) in terms of the independent variables (e.g., molecular properties).

In several cases, we will refer to the *computational complexity* of the respective methods. The time complexity of an algorithm gives us an asymptotic upper bound on the runtime of the algorithm as a function of the size of the input problem. Thus, it gives us the worst-case behavior of algorithms. It is written in the *O*() ("big O") notation, which in effect suppresses constants. If the input size of a dataset is measured in terms of the number of instances n, then O(n) means that the computation scales linearly with n. (Note that we are also interested in the scalability in the number of features, m.) Sometimes, we will refer to the space complexity, which makes statements about the worst-case memory usage of algorithms. Finally, we will assume basic knowledge of statistics and probability theory in the remainder of the chapter (see Ref. 2 for an introductory text).

This chapter consists of four main sections: The first part is an *introduction to data mining*. Among other things, it introduces the terminology used in the rest of the chapter. The second part focuses on so-called *descriptive data mining*, the third part on *predictive data mining*. Each class of techniques is described in terms of the inputs and outputs of the respective algorithms, sometimes including examples thereof. We also emphasize the typical usage and the advantages of the algorithms, as well as the typical pitfalls and disadvantages. The fourth part of the chapter is devoted to *references* to the relevant literature, available *tools*, and *implementations*.

1.1. Data Mining (DM) and Knowledge Discovery in Databases

This section shall provide a non-technical introduction to data mining (DM). The book *Data Mining* by Witten and Frank (2) provides an excellent introduction into this area and it is quite readable even for non-computer scientists. A recent review (3) covers DM applications in toxicology. Another recommended reading is *Advances in Knowledge Discovery and Data Mining* by Fayyad et al. (4).

First, we will have to clarify the meaning of DM and its relation to other terms frequently used in this area, namely knowledge discovery in databases (KDD) and machine learning (ML). Common definitions (2,5–8) are:

- *Knowledge discovery* (KDD) is the non-trivial process of identifying valid, novel, potentially useful, and ultimately understandable structure in data.

- *Data mining (DM)* is the actual data analysis step within this process. It consists of the application of statistics, machine learning, and database techniques to the dataset at hand.
- *Machine learning (ML)* is the study of computer algorithms that improve automatically through experience. One ML task of particular interest in DM is *classification*; that is, to classify new unseen instances on the basis of known *training* instances.

This means that *knowledge discovery* is the process of supporting humans in their enterprise to make sense of massive amounts of data; *data mining* is the application of techniques to achieve this goal; and *machine learning* is one of the techniques suitable for this task. Other DM techniques originate from diverse fields, such, as statistics, visualization, and database research. The focus in this chapter will be primarily on DM techniques based on machine learning.

In practice, many of these terms are not used in their strict sense. In this chapter, we will also use sometimes the popular term DM, when we mean KDD or ML.

Table 1 shows the typical KDD process as described by Fayyad et al. (6). In the following, we will sketch the adapted process for the task of extracting structure–activity

Table 1 The Knowledge Discovery (KDD) Process According to Fayyad et al.

1. Definition of the goals of the KDD process
2. Creation or selection of a data set
3. Data cleaning and preprocessing
4. Data reduction and projection
5. Selecting data mining methods
6. Exploratory analysis and model/hypothesis selection
7. Data mining (DM)
8. Interpretation/evaluation
9. Utilization

[a]From Ref. 6.

relationships (SARs) from experimental data. The steps closely resemble those from the generic process by Fayyad:

1. Definition of the goal of the project and the purpose of the SAR models (e.g., predictions for untested compounds, scientific insight into toxicological mechanisms).
2. Creation or selection of the dataset (e.g., by performing experiments, downloading data).
3. Checking the dataset for mistakes and inconsistencies, and perform corrections.
4. Selection of the features which are relevant to the project and transformation of the data into a format, which is readable by DM programs.
5. Selection of the DM technique.
6. Exploratory application and optimization of the DM tools to see if they provide useful results.
7. Application of the selected and optimized DM technique to the dataset.
8. Interpretation of the derived model and evaluation of its performance.
9. Application of the derived model, e.g., to predict the activity of untested compounds.

The typical KDD setting involves several iterations over these steps. Human intervention is an essential component of the KDD process. Although most research has been focused on the DM step of the process (and the present chapter will not make an exception), the other steps are at least equally important. In practical applications, the data cleaning and preprocessing step is the most laborious and time-consuming task in the KDD process (and therefore often neglected).

In the following sections, we will introduce a few general terms that are useful for describing and choosing DM systems on a general level. First, we will discuss the structure of the data, which can be used by DM programs. Then, we will have a closer look at DM as search or optimization in the space of patterns and models. Subsequently, we will distinguish between *descriptive* and *predictive* DM.

1.2. Data Representation

Before feeding the data into a DM program, we have to transform it into a computer-readable form. From a computer scientists point of view, there are two basic data representations relevant to DM, both will be illustrated with examples.

Table 2 shows a table with physico-chemical properties of chemical compounds. For every compound, there are a fixed number of parameters or features available, therefore it is possible to represent the data in a single table. In this table, each row represents an example and each column an attribute. We call this type of representation *propositional.*

Let us assume we want to represent chemical structures by identifying atoms and the connections (bonds) between them. It is obvious that this type of data does not fit into a single table, because each compound may have a different number of atoms and bonds. Instead we may write down the atoms and the relations (bonds) between them as in Fig. 1. This is called a *relational* representation. Other biologically relevant structures (e.g., genes, proteins) may be represented in a similar manner.

The majority of research on ML and DM has been devoted to propositional representations. However, there exists a substantial body of work on DM in relational representations. Work in this area is published under the heading of *inductive logic programming* and *(multi-)relational data mining.* As of this writing, only very few commercial products

Table 2 Example of a Propositional Representation of Chemical Compounds Using Molecular Properties

CAS	log P	HOMO	LUMO	...	MUTAGEN
100-01-6	1.47	−9.42622	−1.01020	...	1
100-40-3	3.73	−9.62028	1.08193	...	0
100-41-4	3.03	−9.51833	0.37790	...	0
...	...	...	...	...	...
99-59-2	1.55	−9.01864	−0.98169	...	1
999-81-5	−1.44	−9.11503	−4.57044	...	0

```
% Trichloromethane [67-66-3]

atom('67-66-3','67-66-3_1').
element('67-66-3_1',c).

atom('67-66-3','67-66-3_2').
element('67-66-3_2',cl).

atom('67-66-3','67-66-3_3').
element('67-66-3_3',cl).

atom('67-66-3','67-66-3_4').
element('67-66-3_4',cl).

atom('67-66-3','67-66-3_5').
element('67-66-3_5',h).

bond('67-66-3','67-66-3_1_2').
connected('67-66-3_1','67-66-3_2','67-66-3_1_2').
bond_type('67-66-3_1_2',single).

bond('67-66-3','67-66-3_1_3').
connected('67-66-3_1','67-66-3_3','67-66-3_1_3').
bond_type('67-66-3_1_3',single).

bond('67-66-3','67-66-3_1_4').
connected('67-66-3_1','67-66-3_4','67-66-3_1_4').
bond_type('67-66-3_1_4',single).

bond('67-66-3','67-66-3_1_5').
connected('67-66-3_1','67-66-3_5','67-66-3_1_5').
bond_type('67-66-3_1_5',single).
```

Figure 1

are explicitly dealing with relational representations. Available non-commercial software packages include ACE by the KU Leuven http://www.cs.kuleuven.ac.be/~ml/ACE/Doc/ and Aleph by Ashwin Srinivasan http://web.comlab.ox.ac.uk/oucl/research/areas/machlearn/Aleph/. One of the few commercial products is Safarii http://www.kiminkii.com/safarii.html by the Dutch company Kiminkii.

One of the implications of choosing a relational representation is that the complexity of the DM task grows substantially. This means that the runtimes of relational DM algorithms are usually larger than those of their propositional relatives. For the sake of brevity, we will not discuss relational DM algorithms in the remainder of the chapter.

1.3. DM as Search for Patterns and Models

The DM step in the KDD process can be viewed as the search for structure in the given data. In the most extreme case, we are interested in the probability distribution of all variables (i.e., the full joint probability distribution of the data). Knowing the joint probability distribution, we would be able to answer all conceivable questions regarding the data. If we only want to predict one variable given the other variables, we are dealing with a *classification* or *regression* task: *Classification* is the prediction of one of a finite number of discrete classes (e.g., carcinogens, non-carcinogens), *regression* is the prediction of a continuous, real-valued target variable (e.g., LD_{50} values). In prediction, we just have to model the dependent variable given the independent variables, which requires less data than estimating the full joint probability distribution. In all of the above cases, we are looking for *global* regularities, that is, *models* of the data. However, we might just as well be satisfied with *local* regularities in the data. Local regularities are often called *patterns* in the DM literature. Frequently occurring substructures in molecules fall into this category, for instance. Other examples for patterns are dependencies among variables (functional or multivalued dependencies) as known from the database literature (9). Again, looking for patterns is an easier task than predicting a target variable or modeling the joint probability distribution.

Most ML and DM approaches, at least conceptually, perform some kind of search for *patterns* or *models*. In many cases, we can distinguish between (a) the search for the *structure* of the pattern/model (e.g., a subgroup or a decision tree), and (b) the search for *parameters* (e.g., of a linear classifier or a Bayesian network). Almost always the goal is to optimize some scoring or loss function, be it simply the absolute or relative frequency, information-theoretic measures that evaluate the information content of a model, numerical error measures such as the root mean squared error, the degree to which the information in the data can be compressed, or the like. Sometimes, we do not explicitly perform search in the space of patterns or models, but, more directly, employ optimization techniques.

Given these preliminaries, we can summarize the elements of ML and DM as follows. First, we have to fix the representation of the data and the patterns or models. Then, we often have a partial order and a lattice over the patterns or models that allows an efficient search for patterns and models of interest. With these ingredients, data mining often boils down to search/optimization over the structure/parameters of patterns/models with respect to some scoring/loss function.

Finally, *descriptive DM* is the task to describe and characterize the data in some way, e.g., by finding frequently occurring *patterns* in the data. In contrast, the goal of *predictive DM* is to make predictions for yet unseen data. Predictive DM mostly involves the search for classification or regression models (see below). Please note that clustering should be categorized as descriptive DM, although some probabilistic variants thereof could be used indirectly for predictive purposes as well.

Given complex data, one popular approach is to perform descriptive DM first (i.e., to find interesting patterns to describe the data), and perform predictive DM as a second step (i.e., to use these patterns as descriptors in a predictive model). For instance, we might search for frequently occurring substructures in molecules and then use them as features in some statistical models.

2. DESCRIPTIVE DM

2.1. Tasks in Descriptive DM

In the subsequent sections, we will discuss two popular tasks in descriptive DM. First, we will sketch clustering, the task of finding groups of instances, such that the similarity within the groups is maximized and the similarity between the groups is minimized. Second, we will sketch frequent pattern discovery and its descendants, where the task is to find all patterns with a minimum number of occurrences in the data (the threshold being specified by the user).

2.2. Clustering

The task of clustering is to find groups of observations, such that the intragroup similarity is maximized and the intergroup similarity is minimized. There are tons of papers and books on clustering, and it is hard to tell the advantages and disadvantages of the respective methods. Part of the problem is that the evaluation and validation of clustering results is, to some degree, subjective. Clustering is unsupervised learning in the sense that there is no target value to be predicted.

The content of this section is complementary to that of Marchal et al. (10): We focus on the advantages and disadvantages of the respective techniques, their computational complexity and give references to recent literature. In the section on resources, several pointers to existing implementations will be given. In the following exposition, we will closely follow Witten and Frank (12).

Clustering algorithms can be categorized along several dimensions:

- *Categorical vs. probabilistic:* Are the observations assigned to clusters categorically or with some probability?
- *Exclusive vs. overlapping:* Does the algorithm allow for overlapping clusters, or is each instance assigned to exactly one cluster?
- *Hierarchical vs. flat:* Are the clusters ordered hierarchically (nested), or does the algorithm return a flat list of clusters?

Practically, clustering algorithms exhibit large differences in computational complexity (the worst-case runtime behavior as a function of the problem size). Methods depending on pair-wise distances of all instances (stored in a so-called proximity matrix) are at least quadratic in time and space, meaning that these methods are not suitable for very large datasets. Other methods like k-means are better suited for such problems (see below).

As stated above, the goal of clustering is to optimize the conflicting goals of maximal homogeneity and maximal separation. However, in general, the evaluation of clustering results is not trivial. Usually, the evaluation of clusters involves the inspection by human domain experts. There exist several approaches to evaluating and comparing methods:

- If clustering is viewed as density estimation, then the likelihood of the data given the clusters can be estimated. This method can be used to evaluate clustering results on fresh test data.
- Other measures can be applied, for instance, based on the mean (minimum, maximum) distance within a cluster and the mean (minimum, maximum) distance between instances coming from different clusters.
- Take a classification task and see whether the known class structure is rediscovered by the clustering algorithm. The idea is to hide the class labels, cluster the data, and check to which degree the clusters contain instances of the same class.
- Vice versa, one can turn the discovered clusters into a classification problem: Define one class per cluster, that is, assign the same class label to all instances within each cluster, then apply a classifier to the dataset and estimate how well the classes can be separated. In this way, we can also obtain an interpretation of the clusters found.

Example applications can be found in the area of microarray data (10). For instance, we might have the expression levels of several thousand genes for a group of patients. The user may be interested in two tasks: clustering genes and clustering patients. In the former case, we are interested in finding genes that behave similarly across all patients.

In the latter case, we are looking for subgroups of patients that share a common gene expression profile.

2.2.1. Hierarchical Clustering

In hierarchical clustering, the goal is to find a hierarchy of nested clusters. Most algorithms of this category are *agglomerative,* that is, they work bottom-up starting with single-instance clusters and merge the closest clusters until all data points are lying within the same cluster. Obviously, one of the design decisions is how to define the distance between two *clusters,* as opposed to the distance between two *instances.*

Hierarchical clustering algorithms can also work top-down, in which case they are called *divisive.* Divisive hierarchical clustering starts off with all instances in the same cluster. In iterations, one cluster is selected and split according to some criterion, until all clusters contain a single instance. There are many more agglomerative than divisive clustering algorithms in the literature, and usually divisive clustering is more time-consuming.

In general, hierarchical clustering is at least quadratic in the number of instances, which makes it impractical for very large datasets. Both agglomerative and divisive methods produce a so-called dendrogram, i.e., a graph showing at which "costs" two clusters are merged or divided. Dendrograms can readily be interpreted, but have to be handled with care. It is clear that hierarchical clustering algorithms by definition detect hierarchies of clusters in the data, whether they exist or not ("to a hammer everything looks like a nail"). A frequent mistake is to apply this type of clustering uncritically and to present the clustering results as the structure inherent in the data, as opposed to the result of an algorithm.

2.2.2. *k*-Means

k-Means clusters the data into *k* groups, where *k* is specified by the user. In the first step, *k* cluster centers are chosen (e.g., randomly). In the second step, the instances are assigned to the clusters based on their distance to the cluster centers determined in the first step. Third, the centroids of the clusters from step two are computed. These steps are repeated until convergence is reached.

The complexity of k-means (if run for a constant number of iterations—so far no results about the convergence behavior of this old and simple algorithm are known) is $O(I \times k \times n)$, where I is the number of iterations, k is the number of centroids, and n is the size of the dataset. The linearity in the number of instances makes the algorithm well suited for very large datasets.

Again, a word of caution is in order: The results can vary significantly based on initial choice of cluster centers. The algorithm is guaranteed to converge, but it converges only to a local optimum. Therefore, the standard approach is to run k-means several times with different random seeds ("random restarts"). Another disadvantage of k-means is that the number of clusters has to be specified beforehand. In the meantime, algorithms automatically choosing k, such as X-means (11), have been proposed.

2.2.3. Probabilistic/Model-Based Clustering

Ideally, clustering boils down to density estimation, that is, estimating the joint probability distribution of all our random variables of interest. The advantage of this view is that we can compare clustering objectively using the log-likelihood. From a probabilistic perspective, we want to find the clusters that give the best explanations of the data. Also, it is desirable that each instance is not assigned deterministically to a cluster, but only with a certain probability. One of the most prominent approaches to this task is *mixture modeling,* where the joint probability distribution is modeled as a weighted sum of some base probability distributions, such as Gaussians. In the latter case, we are speaking of Gaussian mixture models. In mixture modeling, each cluster is represented by one distribution, governing the probabilities of feature values in the corresponding cluster. Since we usually consider only a finite number of clusters, we are speaking of *finite mixture models* (12).

One of the most fundamental algorithms for finding finite mixture models is the EM (expectation-maximization) algorithm (13). EM can be viewed as a generalization of k-means as sketched above. EM also relies on random initializations and random restarts, and often converges in a few iterations to a *local* optimum. Still, probabilistic/model-based clustering can be computationally very costly and relies on reasonable assumptions about the distributions governing the data.

2.2.4. Other Relevant References

Another popular clustering algorithm is CLICK by Ron Shamir (http://www.cs.tau.ac.il/~rshamir), which is operating in two phases: the first phase is divisive, the second agglomerative (14). Implementations are available from the website of the university of Tel Aviv. A survey paper by the same author (15) compares clustering algorithms in the context of gene expression data. Another recent experimental comparison of several algorithms on gene expression data has been performed by Datta and Datta (16). Finally, fuzzy c-means (17), an algorithm based on k-means and fuzzy logic, appears to be popular in the bioinformatics literature as well.

2.3. Mining for Interesting Patterns

In this section, we introduce the task of mining for interesting patterns. Generally speaking, the task is defined as follows: We are given a language of patterns L, a database D, and an "interestingness predicate" q, which specifies which patterns in L are of interest to the user with respect to D. The task is then to find all patterns $p \in L$ such that $q(p, D)$ holds. Alternatively, we might have a numeric measure of interestingness that is to be maximized. For space reasons, we will only focus on the former problem definition here.

In its simplest form, interesting pattern discovery boils down to frequent pattern discovery. That is, we are interested in all patterns that occur with a minimum frequency in a given database. Why is frequency an interesting property?

First, if a pattern occurs only in a few cases, it might be due to chance. Second, frequent patterns may be useful when it comes to prediction. Infrequent patterns are not likely to be useful, when we have to generalize over several instances in order to make a reasonable prediction.

Frequent pattern discovery can be defined for many so-called *pattern domains*. Pattern domains are given by the types of data in D and the types of patterns in L that are searched in D. Obviously, L is to a large extent determined by the types of data in D. For instance, we might want to analyze databases of graphs (e.g., 2D structures of small molecules). The language L could then be defined as general subgraphs. Alternatively, we might look for free (that is, unrooted) trees or linear paths (e.g., linear fragments, see Ref. 18) in D.

The most common pattern domain is that of so-called *itemsets*. Consider a database of supermarket transactions consisting of items that are purchased together. Let I be the set of all possible products (so-called items). Then every purchase is a subset $X \subseteq I$ of these products (ignoring multiplicities). Sets of items X are commonly called *itemsets*. Thus, the transaction database D is a multiset of itemsets. The classic DM problem then consists of finding all itemsets occurring with a minimum frequency in the database. Note that the pattern language L is in this case the power set of I (i.e., $L = P(I)$). Therefore, the patterns $p \in L$ as well as the examples in the database $\mathbf{x} \in D$ are itemsets. A pattern p is contained in a transaction $\mathbf{x}$ if $p \subseteq \mathbf{x}$. The interestingness predicate q for frequent itemsets is defined as $q(p, D) \iff |\{\mathbf{x} \in D | p \subseteq \mathbf{x}\}| \geq$ *min_freq*. In other words, an itemset is of interest to the user if it occurs frequently enough in the transaction database.

Let us illustrate the concepts with an example. In the following, itemsets of the form $\{x_{i,1}, x_{i,2}, \ldots, x_{i,m}\}$ are written shortly as $\{x_{i,1}x_{i,2} \ldots x_{i,m}\}$. The database consists of six transactions: $D = \{x_1x_2x_3x_4, x_1x_3, x_3x_4, x_1x_3, x_1x_2x_3, x_1x_4\}$. If we are asking for all itemsets with a minimum absolute frequency of 3 in this database, we obtain the following set of solution patterns: $\{x_1, x_3, x_4, x_1x_3\}$.

Algorithms for frequent pattern discovery generally suffer from the combinatorial explosion that is due to the structure of the pattern space: the worst-case complexity is mostly exponential. The practical behavior of these algorithms, however, depends very much on properties of the data (e.g., how dense the transaction database is) and, conversely, on the minimum frequency threshold specified by the user. Often, the question is how low the minimum frequency can be set before the programs run out of memory. Another problem with most algorithms for frequent pattern discovery is that they usually return far too many patterns. Therefore, the relevant patterns have to be organized in a human-readable way or ranked, before they can be presented to the user. Finally, users should be aware that frequent patterns might only describe one part of the dataset and ignore the rest. In other words, all the frequent patterns might occur in the same part of the database, and the remaining part is not represented at all.

On the positive side, search strategies like levelwise search (19) are quite general, so that their implementation for new pattern domains, say, strings or XML data, is mostly straightforward, if necessary.

Examples for frequent pattern discovery are the search for frequently occurring molecular fragments in small molecules, the search for motifs in protein data, or the search for coexpressed genes in microarray experiments.

Continuing the example from above, we might have not only one database, but two: $D_1 = \{x_1x_2x_3x_4, x_1x_3, x_3x_4\}$ and $D_2 = \{x_1x_3, x_1x_2x_3, x_1x_4\}$. The user queries for all itemsets with a minimum absolute frequency of 2 in D_1 and a maximum absolute frequency of 0 in D_2. The set of solutions contains only one element: $\{x_3x_4\}$. This is the kind of query we would like to pose in differential data analysis. For instance, we might be looking for differentially expressed genes. The general idea is to build systems that are able to answer complex queries of this kind. The user is posing constraints on the patterns of interest, and the systems employ intelligent search techniques to come up with solutions satisfying the constraints. The use of constraints and query languages in DM is studied in the areas of *constraint-based mining* and *inductive databases*. Unfortunately, no industrial-strength implementations are available at the time of this writing.

3. PREDICTIVE DM

3.1. Tasks in Predictive DM

In toxicology, we can differentiate between two basic types of effects: those with a threshold (the majority of toxicological effects) and those without (e.g., carcinogenicity, mutagenicity). For the last category, it is possible to distinguish between active and inactive compounds, whereas for other endpoints a number (e.g., LD_{50}) is used to indicate the toxic potency.

If it is sufficient to distinguish between different classes of compounds (e.g., active and inactive), we have, in computer science terms, a *classification* problem. If we want to predict numerical values we need to use *regression* techniques. Most DM techniques focus on classification.

It is possible to transform a regression problem into a classification problem. We use categories of activities (e.g., low, medium, high) instead of the activities themselves. Although this looks very much like a workaround, discretization may even improve performance, especially in the presence of noise.

3.2. Benchmark Learners: *k*-Nearest Neighbor and Naive Bayes

Given a new dataset, it is useful to know the performance of a few simple benchmark classification algorithms. In the following, we will present two such algorithms: k-nearest neighbor and Naive Bayes. These are benchmark algorithms in the sense that the performance of more complicated learners should be compared against their performance. The question is whether any additional effort is justified with respect to the performance of these two algorithms.

k-Nearest neighbor is one of the simplest learning schemes. Consider we are given training instances ($\mathbf{x_i}$, $\mathbf{y_i}$), where

$\mathbf{x_i} \in \Re^m$ (i.e., are real values) and $\mathbf{y_i}$ is the class to be predicted. For two class problems, we assume that $\mathbf{y_i} \in \{+1, -1\}$, meaning that we have positive respectively negative examples. At training time, nothing happens—we just memorize all instances in the training set. At testing time, we are given one test instance $\mathbf{q}$ after the other. For each $\mathbf{q}$, we determine the k closest training instances according to some distance measure d, and predict their majority class for $\mathbf{q}$. For instance, we might use the Euclidean distance as a distance measure $d(\mathbf{x_i}, \mathbf{q})$, telling us how far test instance $\mathbf{q}$ is from a training instance $\mathbf{x_i}$. If we are dealing with a regression problem (i.e., if $\mathbf{y_i} \in \Re$), then we are not taking the majority class of the closest k training instances, but the mean of these instances.

Practically, the precise choice of k, say $k = 3$ or $k = 5$, often does not make much of a difference. Theoretically and practically, it *does* make a difference whether $k = 1$ or $k > 1$, where $k > 1$ is to be preferred. k-Nearest neighbor is a good choice if we have not too many, real-valued features. The basic algorithm from above is easily fooled by a large number of irrelevant features. For instance, if we had a dataset with 1 relevant and 100 irrelevant features, then the k-nearest neighbors would not tell us much about the true classification of the test instances. One possible solution is to assign weights to the features according to their relevance or, more radically, to select a subset of relevant features. However, the discussion of such techniques is outside the scope of this chapter. Note that the only way to use domain knowledge in k-nearest neighbor is to encode it into the distance measure. With the choice of an appropriate distance measure, classification may become trivial. Regarding the computational complexity, even the basic variant of k-nearest neighbor as described here is linear in the number of training instances. However, the computations may become costly in practical applications.

Another conceptually and computationally simple learning scheme is Naive Bayes. For simplicity, we focus on two-class problems again (e.g., with the classes carcinogen and non-carcinogen). Let us assume, for example, that we intend to estimate the probability that a compound is a carcinogen based on the presence or absence of predefined structural features (e.g., amino group, aromatic ring, etc.). For this purpose, we can estimate for each feature the probability that it indicates (non-)carcinogenicity, and obtain a final prediction by combining the values of all features.

In Naive Bayes, we attempt to estimate the probability $P(Y|X)$ (read: the probability of Y, given X; in our example the probability of carcinogenicity given the presence or absence of an amino group), where Y is the random variable of the class (e.g., carcinogen/non-carcinogen) and X is the random variable of the presence or absence of the structural features (e.g., one of them could be the presence or absence of an amino group).

If we had only one feature, say, the presence or absence of an *amino* group, then it would be easy to estimate $P(Y|X)$ from the training data. For instance, one could estimate $P(carcinogen = true|amino = present) = amino_{carcinogens}/amino_{all}$ (with $amino_{carcinogens}$: the number of carcinogens with an amino group, and $amino_{all}$: the number of all compounds in the training set with an amino group). Given a test instance with an *amino* group, we would predict that this compound is carcinogenic if $P(carcinogen = true|amino = present) > P(carcinogen = false|amino = present)$ according to our estimates.

If we want to consider several features for classification, we first apply Bayes theorem to rewrite $P(Y|X)$ as $P(Y|X) = P(X|Y)\ P(Y)/P(X)$. $P(X|Y)$ denotes the probability of X given Y, in our example the probability of carcinogenicity depending on the presence or absence of structural features. $P(Y)$ stands for the probability of Y, in our case, of carcinogenicity. $P(X)$ finally is in our case the probability of the presence or absence of an amino group in the complete dataset (P (*amino*) estimated by $amino_{all}/n_{all}$).

The estimation of probabilities for classifications based on multiple features would require a lot of data, because we would have to estimate the probability of all feature value *combinations* of X for each class Y [i.e., we would need an estimation, of $P(X|Y)$]. To simplify the task, we may assume the independence of the features $X_1, \ldots, X_m$. Then we can obtain the overall probabilities for each class by multiplying the contributions of all fragments and predict the activity of the class with the highest probability. More formally, we then use the decision function $f(\mathbf{x}) = argmax_{y \in \{+1,-1\}} \Pi_{i=1}^{m} P(Y=y|X_i=x_i) P(Y=y)$.

In other words, we only have to estimate the probability of feature values occurring in the respective classes, and the probability of the classes themselves, in order to make a prediction. Therefore, Naive Bayes requires only one scan through the database, where we are counting the occurrences of feature values in each class and the number of instances per class.

Since Naive Bayes works linearly in the number of instances and features, it is well suited for very large databases. It is not the method of choice if many strong variable dependencies are known. However, it has been observed to work well even if the independence assumption is violated, because the classification is more robust against this violation than the probabilities behind the classification (20). If we do not want to make such independence assumptions, we might want to model the full joint probability distribution of the data. This can be done using Bayesian networks (21), which model the joint probability distribution by explicitly stating known independencies. Due to space limitations, Bayesian networks are outside the scope of this chapter.

3.3. Inductive ML Rule Learning and Decision Tree Learning

The task in inductive ML is to find useful generalizations from observational data in order to make accurate predictions for unseen cases. Traditional techniques from inductive ML have been devised to find predictive models that are, at least in principle, comprehensible and human-readable. In other words, we are interested in models that are not black boxes making some predictions, but models that are amenable to interpretation by human domain experts. Of course, there exists a trade-off between predictive accuracy and comprehensibility: highly predictive models often are not comprehensible, and simple, comprehensible models often do not perform sufficiently. This trade-off is rarely ever addressed in ML research.

In this section, we will cover only methods that expect a *propositional representation* (i.e., data in tabular form; Table 2). Extensions for *relational data* (Fig. 1), exist for some of the algorithms, but their discussion is beyond the scope of this chapter. The majority of research on inductive learning deals with models in the form of rules and decision trees.

Rules consist of a series of preconditions (e.g., for the presence of certain substructures) and a conclusion (e.g., a classification carcinogen/non-carcinogen) as in Figure 2. Generally the preconditions are connected by a logical *AND*. Therefore, all tests of the preconditions have to succeed if a rule is to be applied. There are, however, extensions that allow general logical expressions for rule formulation.

Figure 2 depicts an example set of rules derived from PART, a rule learner implemented in the WEKA workbench (2). It is quite obvious how to read it:

If a compound fulfills the criteria of the first rule, it is classified as positive; if it fulfills the criteria of the second rule, it is classified as negative. Both rules test for the presence or absence of molecular substructures. The numbers in brackets summarize the correctly/incorrectly classified compounds in the training set.

```
c=c-n-o no AND
n=o yes AND
c-c=c-c=c no AND
o-c~c~c-n-o no AND
o-c-c-c-o no: 1 (73.0/4.0)

c-c-c-c yes AND
c-c-c-c-c-o-c-c-c no AND
o=c-n-c=o no AND
c-c~c~c-n no AND
c-c~c~c-c no: 0 (76.0/4.0)
```

Figure 2

A variety of algorithms have been developed for classification [e.g., CN2 (22), RIPPER/SLIPPER (23,24), C4.5 rules (25)] and regression (e.g., CUBIST, see the section on actual implementations) rule learning. The books by Mitchell (8) and Witten and Frank (2) provide more information about the algorithmic details, the next section describes some commercial and non-commercial implementations.

Trees are the result of a "divide-and-conquer" approach. At each node of the tree, a particular attribute is tested (e.g., the absence/presence for boolean attributes, comparison with a constant for real values). The leaf nodes give a classification for all instances that reach the leaf.

Figure 3 shows an example of a *decision tree* generated by C4.5 (25), one of most popular DM algorithms. It may require a little more explanation than rules, but is equally well understandable.

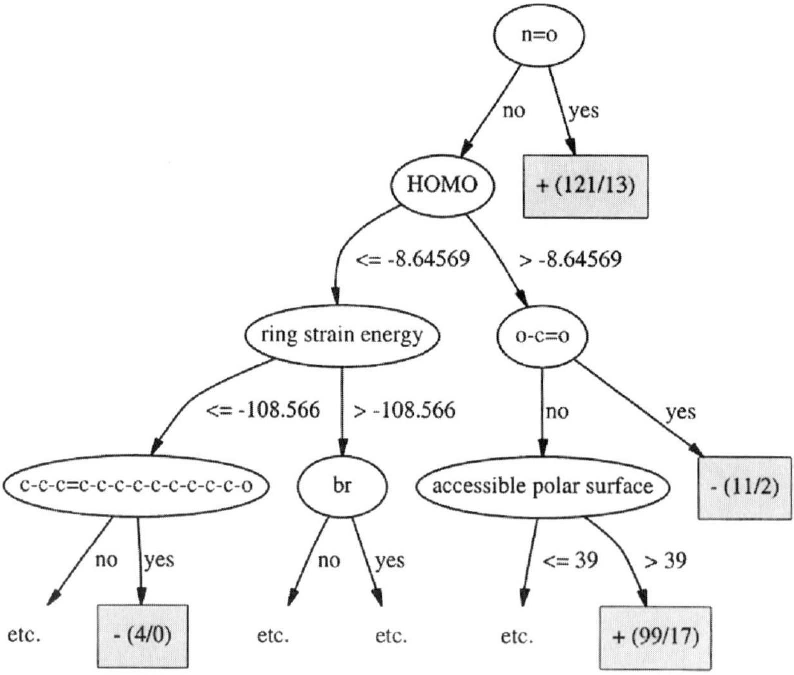

Figure 3

If a compound contains a N=O fragment (usually a nitro group) it is classified as active. One hundred twenty-one compounds were classified correctly, 13 incorrectly. If it does not contain such a fragment, it is checked if the HOMO is higher or lower than −8.6 eV, followed by the ring strain energy and the presence of O–C=O fragments.

Of course, there are methods available that can generate more sophisticated models involving for example regression to predict numerical values (2).

The most popular algorithm for decision trees is C4.5 (25) and its variants and reimplementations (C5.0 and J48 (2)). Numerical values can be predicted by regression (e.g., CART (26) and model (M5 (27)) trees. For details, we have to refer to textbooks (2,8) again, some reference implementations can be found in the next section.

3.4. Linear Classifiers and Support Vector Machines

Because an earlier chapter gives a thorough introduction to support vector machines (SVMs), we only want to touch upon the general topic of linear classifiers briefly. Linear classifiers are models separating the instance space by a hyperplane. In general, linear classifiers have a decision function $f(\mathbf{x}) = sgn(\beta^T\mathbf{x} + \beta_0)$, where $\beta = (\beta_{1,\ldots,}\beta_m) \in \Re^m$ and $\beta_o \in \Re$. If the instances are linearly separable, then there are (mostly infinitely) many possibilities to draw a hyperplane (i.e., to choose the parameters β_i) separating the instances from the two classes. Thus, the design decision arises: which hyperplane should we choose? In SVMs, the hyperplane with the maximum distance to the closest instances from the two classes is used. The resulting decision boundary is called the *maximum margin hyperplane*. As shown in the chapter on SVMs, the setting can be extended toward noisy data and non-linear separation. The latter is achieved by a mathematical trick, where the given input space is implicitly mapped into a so-called feature space of higher dimensionality. In the higher-dimensional feature space, we can find a linear separation that corresponds to a non-linear separation in the originally given input space.

Linear models and in particular SVMs can easily be tuned with respect to the noise level in the data. Another advantage is that they are, for good reasons, not so much haunted by the curse of dimensionality as many other learning algorithms. A disadvantage is that SVMs are usually at least quadratic in the number instances, which make them hardly applicable to very large datasets. Moreover, only linear SVMs (and perhaps quadratic SVMs) are interpretable; other types of SVMs are essentially black-box classifiers.

4. LITERATURE AND TOOLS/IMPLEMENTATIONS

In this section, we provide pointers to the relevant literature and implementations of the surveyed techniques.

4.1. Books on ML and DM

A general and comprehensible introduction to ML and DM is the book *Data Mining: Practical Machine Learning Tools and Techniques with JAVA Implementations* by Witten and Frank (2). The book explains complex concepts from data analysis in intuitive terms.

A landmark book on ML is the introduction *Machine Learning* by Mitchell (8). The most instructive and useful chapters for our purposes are those on concept learning (version spaces), Bayesian learning (Naive Bayes and a sketch of Bayesian networks), instance-based learning (k-nearest neighbor and the like), decision trees, and rule learning. Also included, but a bit outdated, are the chapters on analytical learning and hybrid analytical/empirical learning. The book also covers ML topics outside the data analysis context (reinforcement learning). Moreover, a chapter on computational learning theory sketches the basic theoretical concepts behind concept learning. Not covered are the following topics: descriptive DM (interesting pattern discovery and clustering), support vector machines, and statistical techniques such as hidden markov models.

The book *Principles of Data Mining* by Hand et al. (28) gives a systematic introduction to the concepts of DM. For instance, the book is divided into chapters on score functions (measures of interestingness and loss functions), search and optimization, and so forth. It also covers, to some extent, memory and database management issues in DM.

A rather technical account of many important ML techniques is given in *The Elements of Statistical Learning* by Hastie et al. (29). Apart from the authors' own achievements (the Lasso, MARS, generalized additive models, etc.), the book covers general topics from clustering to decision trees, self-organizing maps (SOMs), support vector machines, ensemble methods (e.g., bagging and boosting), and many more.

A good introduction to Bayesian networks is by Jensen (30). The book is out-of-print at this point in time, but still available via libraries. According to the author, the successor *Bayesian Networks and Decision Graphs* (21) is a proper substitute for this textbook.

4.2. Implementations and Products

Most of the techniques described in this chapter are implemented in the open-source DM workbench WEKA [(2), see also http://www.cs.waikato.ac.nz/ml/weka/], in packages of the open-source statistical data analysis system R [(31), see also http://cran.r-project.org], and in commercial products such as Clementine (http://www.spss.com/spssbi/clementine/).

The WEKA workbench is a open-source system implemented in JAVA. As of this writing, WEKA (version 3.4) includes:

- clustering methods: k-means, model-based clustering (EM) and a symbolic clustering method Cobweb
- methods for finding frequent itemsets (APriori, etc.)
- Bayesian learning: Naive Bayes, a basic implementation of Bayesian networks
- instance-based learning: k-nearest neighbor

- inductive learning methods: state-of-the-art rule learners such as PART and RIPPER, a reimplementation of the standard decision tree algorithm C4.5 called J48
- support vector machines
- logistic regression

The statistical workbench and programming language R [http://cran.r-project.org/ (31)], is the open-source variant of the commercial product S-Plus. It includes many *packages* http://spider.stat.umn.edu/R/doc/html/packages.html and http://www.bioconductor.org) implementing the most popular clustering and classification techniques:

- package mva includes: k-means, hierarchical clustering (function `hclust`, with single, complete and mean linkage)
- package `mclust`—model-based clustering (EM, mixture modeling)
- package `cluster` by Struyf et al. (32) includes implementations of popular clustering techniques such as Diana, Fanny, and Pam
- package `knn`: k-nearest neighbor
- package `rpart`: an implementation of the classical decision and regression tree algorithm CART (26)
- other relevant packages: `ipred` (bagging), `randomForest` (interfacing with an Fortran implementation of random forests), `LogitBoost` (a boosting variant), `e1071` function `svm` (support vector machines)

As an example of a commercial product, we briefly present Clementine (version 8.0) by SPSS (http://www.spss.com/spssbi/clementine/). Clementine 8.0 features, among other things:

- C5.0, C&RT (a CART variant (26)
- APriori and frequent itemsets
- k-means, self-organizing maps (SOMs)
- logistic regression

Clementine offers a graphical user interface to all DM activities and supports complex processes by means of prefabricated templates.

RuleQuest (http : //www.rulequest.com/), an Australian company, offers a few stand-alone implementations of valuable DM tools:

- C5.0, the commercial successor of the classic decision tree learner C4.5
- Cubist, a tool for learning regression rules
- Magnum Opus, a tool for frequent pattern discovery and association rule mining

All three products are easy to use and scale up nicely in the size of the databases.

Hugin (http : //www.hugin.com) and Netica (http : //www.norsys.com/) are two of the few industrial-strength implementations of Bayesian networks.

5. SUMMARY

This chapter started with a brief review of the knowledge discovery process and the role of DM. We distinguished between descriptive and predictive DM tasks and described the most important techniques that are suitable for predictive toxicology applications. Finally, we recommended some books for further reading and gave a survey of commercial and non-commercial implementations of DM algorithms.

REFERENCES

1. Frasconi P. Artificial Neural Networks and Kernel Machines in Predictive Toxicology. 2004. This volume.
2. Witten IH, Frank E. Data Mining. San francisco, CA: Morgan Kaufmann Publishers, 2000.
3. Helma C, Gottmann E, Kramer S. Knowledge discovery and DM in toxicology. Stat Methods Med Res 2000; 9:329–358.
4. Fayyad UM, Piatesky-Shaprio G, Smyth P, Uthurusamy R. Advances in Knowledge Discovery and Data Mining. AAAI/MIT Press, 1996.

5. Fayyad U, Piatetsky-Shapiro G, Smmyth P. From DM on knowledge discovery: an overview. In: Fayyad UM, Piatesky-Shaprio G, Smyth P, Uthurusamy R, eds. Advances in Knowledge Discovery and Data Mining. Menlo Park, CA: AAAI Press, 1996:1–30.

6. Fayyad U, Piatetsky-Shapiro G, Smmyth P. The KDD process for extracting useful knowledge from volumes of data. Commun ACM 1996; 39:27–34.

7. Fayyad U, Uthurusamy R. Data mining and knowledge discovery in databases. Commun ACM 1996; 39:24–26.

8. Mitchell TM. Machine Learning. The McGraw-Hill Companies, Inc., 1997.

9. O'Neil P, O'Neil E. Database: Principles Programming, and Performance. 2nd ed. Morgan Kaufmann, 2000.

10. Marchal K, De Smet F, Engelen K, De Moor B. Computational Biology and Toxicogenomics. 2004. This volume.

11. Pelleg D, Moore A. X-means: Extending K-means with efficient estimation of the number of clusters. In: Proceeding of the 17th International Conference on Machine Learning San Francisco, CA: Morgan Kaufmann, 2000; 727–734.

12. McLachlan GJ, Peel D. Finite Mixture Models. New York: John Wiley and Sons, 2000.

13. McLachlan GJ, Krishnan T. The EM Algorithm and Extensions. New York: John Wiley and Sons, 1997.

14. Sharan R, Eukon R, Shamir R. Cluster analysis and its applications to gene expression data. In: Ernst Schering Workshop on Bioinformatics and Genome Analysis, Berlin, Heidelberg, New York: Springer Verlag, 2002; 83–108.

15. Shamir R. A clustering algorithm based on graph connectivity. Inf Processing Lett 2000; 76:175–181.

16. Datta S, Datta S. Comparisons and validation of statistical clustering techniques for microarray gene expression data. Bioinformatics 2003; 19(4):459–466.

17. Gustafson DE, Kessel WC. Fuzzy clustering with a fuzzy covariance matrix. In: Proceedings of the IEEE Conference on Decision and Control. IEEE Press, 1979; 761–766.

18. Helma C. `lazar`: lazy Structure–activity relationships for toxity prediction. In: Helma C, ed. Predictive Toxicology. New York: Marcel Dekker, 2004.

19. Mannila H, Toivonen H. Levelwise search and borders of theories in knowledge discovery. Data Mining Knowledge Discovery 1997; 1(3):241–258.

20. Domingos P, Pazzani MJ. On the optimality of the simple bay esian classifier under zero-one loss. Machine Learning 1997; 29(2–3):103–130.

21. Jensen FV. Bayesian Networks and Decision Graphs. New York, Berlin, Heidelberg: Springer Verlag, 2001.

22. Clark P, Niblett T. The CN2 induction algorithm. Machine Learning 1989; 3:261–283.

23. Cohen WW. Fast effective rule induction. In: Proceedings of the 12th International Conference on Machine Learning. San Francisco, CA: Morgan Kaufmann, 1995; 115–123.

24. Cohen WW, Singer Y. A simple, fast, and effective rule learner. In: Proceedings of the the 16th National Conference on Artificial Intelligence (AAAI-99). Menlo Park, CA: AAAI Press, 1999.

25. Quinlan JR. C4.5: Programs for Machine Learning. San Mateo, CA: Morgan Kaufmann, 1993.

26. Breiman L, Friedman JH, Olshen RA, Stone CJ. Classification and Regression Trees. The Wadsworth Statistics/Probability Series. Belmont, CA: Wadsworth International Group, 1984.

27. Quinlan JR. Learning with continuous classes. In: Adams S, ed. Proceedings AI'92, Singapore: World Scientific, 1992: 343–348.

28. Hand D, Mannila H, Smyth P. Principles of Data Mining. Cambridge, MA: MIT Press, 2001.

29. Hastie T, Tibshirani R, Friedman J. The Elements of Statistical Learning. New York, Berlin, Heidelberg: Springer Verlag, 2001.

30. Jensen FV. An Introduction to Bayesian Networks. London: Taylor and Francis, 1996.

31. Ihaka R, Gentleman R. A language for data analysis and graphics. J Comput Graphical Stat 1996; 5:299–314.

32. Struyf A, Huber M, Rousseeuw PJ. Integrating robust clustering techniques in s-plus. Comput Stat Data Anal 1997; 26: 17–37.

GLOSSARY

Attribute: *See* **Feature.**

Classification: This represents the prediction of one of a finite number of discrete classes (e.g., carcinogens, non-carcinogens).

Clustering: This represents the task of finding groups of instances, such that the similarity within the groups is maximized and the similarity between the groups is minimized.

Complexity: *See* **Computational complexity.**

Computational complexity: The time complexity of an algorithm gives us an asymptotic upper bound on the runtime of the algorithm as a function of the size of the input problem. Thus, it specifies the worst-case behavior of algorithms. In ML and DM, we are often interested in the scalability of the algorithms in terms of the number of features and the number of instances in the dataset.

Data mining (DM): This represents the data analysis step in the process of knowledge discovery in databases. It consists of the application of statistics, ML, and database techniques to the dataset at hand:

Dependent variable: *See* **Target variable.**

Descriptive data mining: This represents the task to describe and characterize the data in some way, e.g., by finding frequently occurring patterns in the data.

Example: *See* **Instance.**

Feature: Features are used to describe instances. Strictly speaking, features are functions mapping the features an instance to its domain. The domain of a feature is the set of values it can take. For example, *amino group* might be a feature used to describe instances of compounds, the domain *of amino group* being {*present, absent*}. For each instance, the feature *amino group* gives us one of the values from its domain *(present* or *absent).* Features are used in the rules or equations of a model to predict the target variable (target class, dependent variable, output, response).

Hyperplane: This represents the decision boundary described by a linear combination of variables tested against a threshold. A hyperplane in a two-dimensional space is a straight line.

Independent variable: *See* **Feature.**

Instance: This represents the description of a single object in a dataset. Most ML and DM algorithms work on the level of instances, that is, it has to be clear what constitutes a single object in the application domain. For example, in a protein dataset, we have to decide whether the proteins themselves or the domains are counted as instances.

Joint probability distribution: This represents the probability distribution of all random variables under consideration viewed jointly.

Knowledge discovery in databases (KDD): This represents the non-trivial process of identifying valid, novel, potentially useful, and ultimately understandable patterns in data.

Linear classifier: This represents the classifier having a hypionplane as decision boundary, e.g., a linear perception or support vector machine.

Linear time complexity: Algorithms with linear time complexity have a runtime that scales linearly in the size of the input problem.

Log likelihood: This represents the: logarithm of the likelihood, which is the probability of the data given a particular model.

Machine learning (**ML**): This represents the study of computer algorithms that improve automatically through experience. One ML task of particular interest in DM is classification, that is, to classify new unseen instances on the basis of a known training instance.

Model: In a ML and DM context, a model is a mathematical object used to describe the data. A predictive model consists of rules or equations to predict the target variable from the given features. Models describe global regularities in the data, whereas patterns describe local regularities.

Pattern: A pattern is a local regularity in the data. In a DM context, we often assume a language of patterns (that is, a set of possible patterns), and look for patterns in that language of particular interest to the user (with respect to the dataset at hand).

Predictive data mining: This represents the task of making predictions for yet unseen data. Predictive DM mostly involves the search for classification or regression models.

Propositional (representation): The instances of a dataset in a propositional representation are feature vectors of fixed size. There is a fixed number of features, and for each feature, we know its value for a given instance. Datasets in propositional representations can be stored in a single relational of a relational database, where each tuple represents one instance.

Quadratic time complexity: The runtime of algorithms with quadratic time complexity is, in the worst case, quadratic in the size of the input problem.

Regression: This represents the prediction of a continuous, real-valued target variable.

Relational (representation): The instances of a dataset in a relational representation are collections of tuples from several relations.

Target variable: The target variable (target class, dependent variables, output or response) of a model is predicted by its equations or rules using the features (attributes, independent variables, inputs or predictors).

8

Neural Networks and Kernel Machines for Vector and Structured Data

PAOLO FRASCONI
Dipartimento di Sistemi e Informatica, Università degli Studi di Firenze, Firenze, Italy

1. INTRODUCTION

The linear models introduced earlier in the text offer an important starting point for the development of machine learning tools but are subject to important limitations that we need to overcome in order to cover a wider application spectrum.

Firstly, data in many interesting cases are not linearly separable; the need for a complex separation surface is not an artifact due to noise but rather a natural consequence of the representation. For the sake of illustration, let us

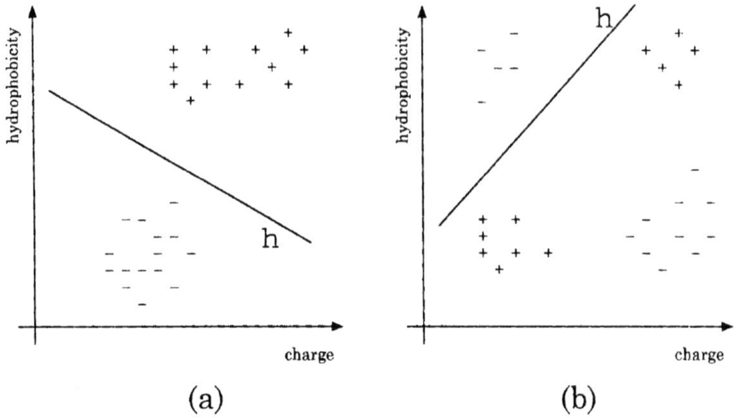

Figure 1 Artificial problems illustrating linear and nonlinear separability. Here the instance space $\mathcal{X}$ is $\mathbb{R}^2$ and the function *f* is realized by a hyperplane *h* that divides $\mathcal{X}$ into a positive and a negative semispace. If positive and negative points are arranged like they are in diagram (b) then no separating hyperplane exists.

construct an artificial problem involving nonlinear separation [our example is actually a rephrasing of the famous XOR problem (1)]. Suppose that we are given a problem involving the discrimination between active and nonactive chemical compounds and suppose that we are using as features that characterize each compound two physico-chemical properties expressed as real numbers (say, charge and hydrophobicity). In this way, each compound is represented by a two-dimensional real vector, as in Fig. 1. Now suppose that active compounds (points marked by +) have either low charge and high hydrophobicity, or low hydrophobicity and high charge, while nonactive compounds (−) have either high charge and low hydrophobicity, or high hydrophobicity and low charge, as in Fig. 1b. It is easy to realize that in this situation there is no possible linear separation between active and nonactive instances. By contrast, if active compounds had both high charge and high hydrophobicity (as in Fig. 1a), then linear separation would have been possible.

Secondly, data may not necessarily come in the form of real vectors or anyway in the form of attribute-value representations that can be easily converted into a fixed-size vectorial representation.[a] For example, a natural and appealing representation for a chemical compound is by means of an attributed graph with vertices associated with atoms and edges representing chemical bonds. Representing chemical structures in attribute-value form can be done, for example, using topological descriptors. However, while these "ad-hoc" representations may embody domain knowledge from experts, it is not obvious that they preserve all the information that would be useful to the learning process.

In this chapter, we focus on extending the basic models presented earlier to specifically address the two above issues. We will briefly review neural networks (in particular, the multilayered perceptron) and kernel methods for nonlinearly separable data. Then we will extend basic neural networks to obtain architectures that are suitable for several interesting classes of labeled graphs.

For several years, neural networks have been the most common (but also controversial) among an increasingly large set of tools available in machine learning. The basic models have their roots in the early cybernetis studies of the 1940s and 1950s (e.g., Refs. 2 and 3) but the interest in neural networks and related learning algorithms remained relatively low for several years. After Minsky and Papert (4) published their extensive critical analysis of Rosenblatt's perceptron, the attention of the artificial intelligence community mainly diverted toward symbolic approaches. In the mid 1980s, the popularity of neural networks boosted again, but this time as an interdisciplinary tool that captured the interest of cognitive scientists, computer scientists, physicists, and biologists. The backpropagation algorithm (5–8) is often mentioned as the main cause for this renaissance and the

---

[a] In attribute-value representations each instance consists of values assigned to a fixed repertoire of attributes or features. Since each value belongs to a specified set (discrete or continuous), conversion to fixed-size vectors is typically straightforward.

1990s witnessed a myriad of real-world applications of neural-networks, many of which quite successful. During the same decade, Cortes and Vapnik (9) developed previous ideas on statistical learning theory and introduced support vector machines and kernel methods, which still today represent a very active area of research in machine learning.

The material covered in this chapter is rather technical and it is assumed that the reader is knowledgeable about basic concepts of calculus, linear algebra, and probability theory. Textbooks such as Refs. 10–14 may be useful for readers who need further background in these areas.

2. SUPERVISED LEARNING

In supervised learning, we are interested in the association between some input instance $\boldsymbol{x} \in \mathcal{X}$ and some output random variable $Y \in \mathcal{Y}$. The input instance $\boldsymbol{x}$ is a representation of the object we are interested in making predictions about, while the output Y is the predicted value.

2.1. Representation of the Data

The set $\mathcal{X}$ is called the *instance space* and consists of all the possible realizations of the input portion of the data. In practice, if we are interested in making predictions about chemical compounds, then each instance $\boldsymbol{x}$: is a suitable *representation* of a particular compound. One possibility is to use a *vector-based* representation, i.e., $\boldsymbol{x} \in \mathbb{R}^n$. This means that, in our chemical example, each component of the input vector might be an empirical or a nonempirical descriptor for the molecule (see elsewhere in this book for details on how to represent chemicals by descriptors). However, since, in general, more expressive representations are possible at the most abstract level, we may assume that $\mathcal{X}$ is any set. Indeed, in Sec. 5, we will present architectures capable of exploiting directly graph-based representations of instances that could be interesting in chemical domains.

The type of $\mathcal{Y}$ depends on the nature of the prediction problem. For example, if we are interested in the prediction of the normal boiling points of halogenated aliphatics, then the output Y is a real-valued variable representing the boiling temperature. By contrast, if we are interested in the discrimination between potentially drug-like and nondrug-like candidates, then we will use $\mathcal{Y} = \{0,1\}$, i.e., the output Y is in this case a Bernoulli variable.[b] When $\mathcal{Y}$ is a continuous set, we talk about *regression* problems. When $\mathcal{Y}$ is a discrete set, we talk about *classification* problems. In particular, if $\mathcal{Y}=\{0,1\}$ we have binary classification, and more in general, if $\mathcal{Y} = \{1,2,\ldots, K\}$ we have multiclass classification.

Both regression and classification problems can be formulated in the framework of statistical learning that will be introduced in the next section.

2.2. Basic Ideas of Statistical Learning Theory

When introducing supervised learning systems we typically assume that some unknown (but fixed) probability measure p is defined on $\mathcal{X} \times \mathcal{Y}$ where $\times$ denotes the Cartesian product of sets.[c] The purpose of learning is, in a sense, to "identify" the unknown p. To this end, we are given a data set of i.i.d. examples drawn from p in the form of input output pairs: $D_m = \{(\boldsymbol{x}_i, y_i)\}_{i=1}^m$. The supervised learning problem consists of seeking a function $f(\boldsymbol{x})$ for predicting Y on new (unseen) cases. This function is sometimes referred to as the *hypothesis*.

In order to measure the quality of the learned function we introduce a nonnegative *loss function* L: $\mathcal{Y} \times \mathcal{Y} \rightarrow \mathbb{R}^+$, where $L(y, f(\boldsymbol{x}))$ is the cost of predicting $f(\boldsymbol{x})$ when the correct prediction is y. For example, in the case of regression, we may use the quadratic loss

$$L(y, f(\boldsymbol{x})) = (y - f(\boldsymbol{x}))^2 \tag{1}$$

[b] A Bernoulli variable is simply a random variable having two possible realizations, as in tossing a coin.

[c] Given two sets X and Y the Cartesian product $X \times Y$ is the set of all pairs (x,y) with $x \in X$ and $y \in Y$.

In the case of classification, we could use the 0–1 loss

$$L(y, f(\boldsymbol{x})) = \begin{cases} 0 & \text{if } y = f(\boldsymbol{x}) \\ 1 & \text{otherwise} \end{cases} \tag{2}$$

that can be easily generalized to quantify the difference between false positive and false negative errors.

The *empirical error* associated with f is the observed loss on the training data:

$$E_m(f) = \sum_{i=1}^{m} L(y_i, f(\boldsymbol{x}_i)) \tag{3}$$

For the 0–1 loss of Eq. (2), this is simply the number of misclassified training examples (i.e., the sum of false positives and false negatives).

The *generalization* error can be measured by the expected loss associated with f, where the expectation is taken over the joint probability p of inputs and outputs:

$$E(f) = E_p[L(Y, f(\boldsymbol{X}))] = \int L(y, f(\boldsymbol{x}))p(\boldsymbol{x}, y)\, d\boldsymbol{x}\, dy. \tag{4}$$

Thus, it immediately appears that learning has essentially and fundamentally to do with the ability of estimating probability distributions in the joint space of the inputs $\mathcal{X}$ and the outputs $\mathcal{Y}$.

To better illustrate the idea, suppose that $\mathcal{X} = \mathbb{R}$ and $\mathcal{Y} = \{0,1\}$ (a binary classification problem with inputs represented by a single real number). In this case, under the 0–1 loss, Eq. (4) reduces to

$$E(f) = \int_{x \in F_1} p(Y = 0, x) dx + \int_{x \in F_0} p(Y = 1, x)\, dx \tag{5}$$

where F_0 and F_1 are the regions that f classifies as negative and positive, respectively. Thus the first integral in Eq. (5) measures the error due to false positives (weighted by the probability $p(Y = 0, \boldsymbol{x})$ that $\boldsymbol{x}$ is actually a negative instance) and the second integral measures the error due to false negatives (weighted by the probability $p(Y = 1, \boldsymbol{x})$ that $\boldsymbol{x}$ is actually a positive instance). The error in Eq. (5) is also known as the

Bayes error of f. The *Bayes optimal classifier* f^* is defined as

$$f^*(x) = \begin{cases} 0 & \text{if } p(Y = 0|x) > p(Y = 1|x) \\ 1 & \text{otherwise} \end{cases} \tag{6}$$

and has the important property that it minimizes the error in Eq. (5). It can be seen as a sort of theoretical limit since no other classifier can do better. An example of Bayes optimal classifier is illustrated in Fig. 2. In this case, the decision function is simply

$$f(\boldsymbol{x}) = \begin{cases} 0 & \text{if } x < \omega \\ 1 & \text{otherwise} \end{cases} \tag{7}$$

where ω is a parameter. According to Eq. (6), the best choice ω^* satisfies $p(Y=0|\omega^*)=p(Y=1|\omega^*)$. Using Bayes theorem, we see that $p(Y|x)$ is proportional to $p(x|Y)p(Y)$ and assuming that $p(Y=0)=p(Y=1)=0.5$ the best choice also satisfies $p(\omega^*|Y=0)=p(\omega^*|Y=1)$. The densities $p(x|Y=0)$ and $p(x|Y=1)$ thus contain all the information needed to construct the best possible classifier. They are usually called *class conditional densities*. In Fig. 2 we have assumed that, they are normal distributions. The Bayes error [Eq. (5)] is measured

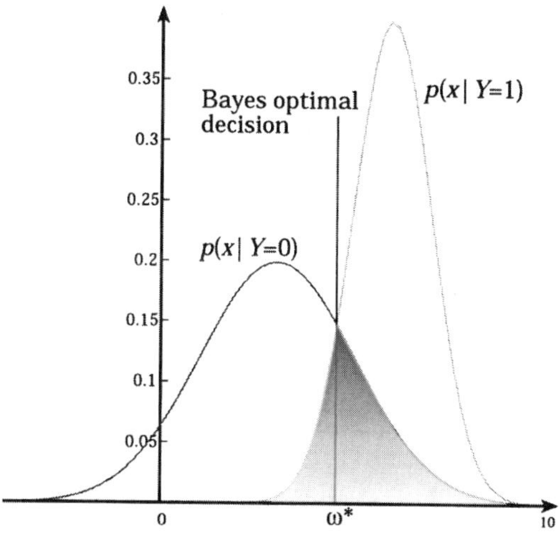

Figure 2 The Bayes optimal classifier.

in this case by the shaded area below the intersection of the two curves. Of course, in practice, we do not know $p(Y|x)$ and we cannot achieve the theoretical optimum error $E(f^*)$.

A learning algorithm then proceeds by searching its solution f in a suitable set of functions $\mathscr{F}$, called the *hypothesis space,* using the empirical error $E_m(f)$ as a guide. Unfortunately, the problem of minimizing the training set error $E_m(f)$ by choosing $f \in \mathscr{F}$ is not well posed, in the sense that the solution is not necessarily unique. For example, going back to our toy binary classification problem of Fig. 2, suppose we are given the following training set of five points $D_m = \{(1.1, 0), (1.9, 0), (2.3, 0), (8.2, 1)(8.9, 1)\}$ (two positive and three negative examples) and suppose we continue to use the decision function of Eq. (7). Clearly any ω in the interval (2.3, 8.2) will bring the training set error to 0, but the associated generalization error would depend on ω. Without additional information (i.e., besides the training data) or constraints, we have no way of making a "good" choice for ω (i.e., picking a value closer to the theoretical optimum ω^*).

Intuitively, complex models can easily fit the training data but could behave poorly on future data. As an extreme example, if $\mathscr{F}$ contains all possible functions on $\mathscr{X}$, then we can always bring $E_m(f)$ to 0 using a lookup table that just stores training examples, but this would be a form of *rote learning* yielding little or no chance of generalization to new instances. Indeed, one of the main issues in statistical learning theory consists of understanding under which conditions a small empirical error also implies a small generalization error. Intuitively, in order to achieve this important goal, the function f should be "stable," i.e., should give "similar" predictions for "similar" inputs.

To illustrate this idea, let us consider another simple example. Suppose we have a scalar regression problem from real numbers to real numbers. If we are given m training points, we can always find a perfect solution by fitting a polynomial of degree $m-1$. Again, this would resemble a form of rote learning. If data points are noisy, they can appear as if they had been generated according to a more complex mechanism than the real one. In such a situation, a polynomial of high degree is likely to be a wildly oscillating function that passes on the assigned points but varies too much between the examples and will yield large error on new data points. This phenomenon is commonly referred to as *overfitting*. Smoother solutions such as polynomial of small degree are more likely to yield smaller generalization error although training points may not be fitted perfectly (i.e., the empirical error may be greater than 0). Of course, if the degree of the polynomial is too small, the true solution cannot be adequately represented and *underfitting* may occur. In Sec. 3.4, we will discuss how the tradeoff between overfitting and model complexity can be controlled when using neural networks. Moreover, in general, *regularization* theory is a principled approach for finding smoother solutions (15,16) and is also one of the foundations of support vector machines that we will later discuss in Sec. 4.

Of course, we expect that better generalization will be achieved as more training examples are available. Technically, this fact can be characterized by studying the *uniform convergence* of a learner, i.e., understanding how the difference between empirical and generalization error goes to 0 as the number of training examples increases. Standard results [see, e.g., (Refs. 17–19)] indicate that uniform convergence of the empirical error to the generalization error essentially depends on the *capacity* of the class of functions $\mathscr{F}$ from which f is chosen. Intuitively, the capacity of a class of functions $\mathscr{F}$ is the number of different functions contained in $\mathscr{F}$. Going back to our scalar regression example where $\mathscr{F}$ is a set of polynomials, we intuitively expect that the capacity increases with the maximum allowed degree of a polynomial.

In the case of binary classification, $\mathscr{F}$ is a set of dichotomies (i.e., a set of functions that split their domain in two nonoverlapping sets) and it is possible to show that with probability at least $1-\delta$ it holds

$$E(F) \leq E_m(f) + \sqrt{\frac{8}{m} d\left(\ln\frac{2m}{d} + 1\right) - \ln\frac{\delta}{4}} \quad (8)$$

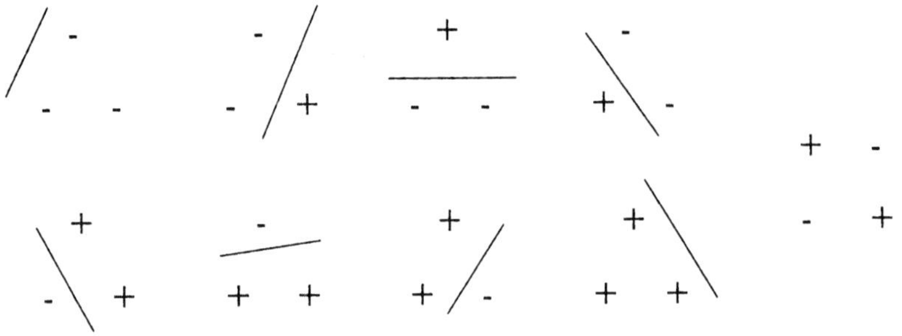

Figure 3 Illustration of the VC-dimension. Left: set of three points that can be classified arbitrarily using a two-dimensional hyperplane (i.e., a line). Right: no set of four points can be arbitrarily classified by a hyperplane.

In the above equation, the capacity of the learner is measured by d, an integer defined as the cardinality of the largest set of points that can be labeled arbitrarily by one of the dichotomies in $\mathscr{F}$; d is called the dimension of Vapnik and Cervonenkis (VC-dimension) of $\mathscr{F}$. For example, if $\mathscr{F}$ is the set of separating lines in the two-dimensional plane (as in Fig. 1), it is easy to see that the VC-dimension is 3. Note that for a given set of three points, there are $2^3 = 8$ possible dichotomies. It is immediate to see that in this case, there is a set of three points that can be arbitrarily separated by a line, regardless of the chosen dichotomy (see Fig. 3). However, given a set of 4 points, there are 16 possible dichotomies and two of them are nonlinearly separable.

Uniform convergence of $E_m(f)$ to $E(f)$ ensures consistency, that is, in the limit of infinite examples, the minimization of $E_m(f)$ will result in a generalization error as low as the error associated with the best possible function in $\mathscr{F}$. As Eq. (8) shows, the behavior of the learner is essentially controlled by the ratio between capacity and number of training examples and thus we cannot afford a complex model if we do not have enough data.

2.3. The Logistic Function for Classification

There are two general approaches for developing statistical learning algorithms. In the *generative* approach, we create a model for the joint distribution $p(\boldsymbol{x}, y)$ and then derive from it the conditional probability of the output given the input. In the *discriminant* approach, we model directly $p(y|\boldsymbol{x})$. Neural networks are typically conceived as discriminant models. In this section, we present one of the simplest conceivable discriminant models for binary classification, known as *logistic regression*. In binary classification, we will assume that the output Y has a Bernoulli distribution. In general, a variable has a Bernoulli distribution with parameter θ in $\theta \in [0, 1]$ if

$$p(Y = y) = \theta^y(1-\theta)^{1-y}, \quad y = 0, 1$$

It can be seen that the expected value of $E[Y] = \theta$. In the case of classification, we want to model the conditional probability $P(Y|\boldsymbol{x})$ and thus θ will depend on x. The form of this dependency is specified by the model. In the case of logistic regression, we will use

$$\theta = f(\boldsymbol{x}) = \sigma(\beta^{\mathrm{T}}\boldsymbol{x} + \beta_0) \tag{9}$$

where $\beta = (\beta_1, \ldots, \beta_{\mathrm{n}}) \in \mathbb{R}^{\mathrm{n}}$, $\beta_0 \in \mathbb{R}$ are adjustable parameters and

$$\sigma(a) = \frac{1}{1 + \exp(-a)} \tag{10}$$

is called the *logistic function*. Note that, since we are interpreting $f(x)$ as $p(Y = 1|\boldsymbol{x})$, the model actually gives us a continuous prediction that we can have, in practice, confidence about the predicted class. Suppose, again, that we are interested in the discrimination between active and nonactive compounds. Then, we could decide that the compound represented by $\boldsymbol{x}$ is active if $f(\boldsymbol{x}) > 0.5$. On the other hand, if the loss due to a false positive is higher than the loss due to a false negative, it might be more appropriate to assume that the compound is active only if $f(\boldsymbol{x})$ is closer to 1.

The model described by Eq. (9) actually corresponds to the simplest conceivable neural network that consists of a single output and n inputs, with $\beta_0, \beta_1, \ldots \beta_n$ playing the role of *connection weights* (see Fig. 4). Not surprisingly, Eq. (9) has also an interesting relation to linear discriminant

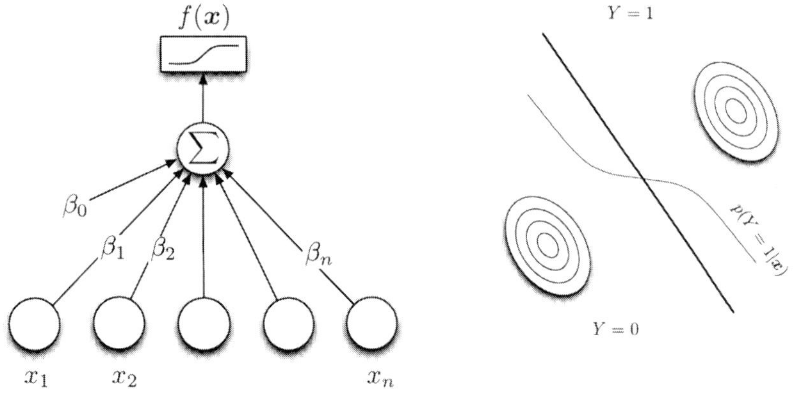

Figure 4 Left: The artificial neuron is also a logistic regression model. Right: The logistic function for binary classification.

analysis (LDA), a statistical model where predictions are obtained by using Bayes theorem to "reverse" the direction of modeling:

$$p(Y=1|\boldsymbol{x}) = \frac{p(\boldsymbol{x}|Y=1)p(Y=1)}{p(\boldsymbol{x}|Y=0)p(Y=0)+p(\boldsymbol{x}|Y=1)p(Y=1)} \tag{11}$$

The quantities $p(\boldsymbol{x}|Y=k)$ and $p(\boldsymbol{x}|Y=1)$ above are called *class conditional densities* and are typically determined by fitting some assigned distribution on the available data; $p(Y=k)$ are simply the class prior probabilities—see Ref. 20 for more details. Eq. (11) can be easily rewritten in terms of the logistic function as

$$p(Y=1|\boldsymbol{x}) = \frac{1}{1+\mathrm{e}^{-\log\frac{p(\boldsymbol{x}|Y=0)}{p(\boldsymbol{x}|Y=1)} - log\frac{p(Y=0)}{p(Y=1)}}} \tag{12}$$

where the argument to the logistic function is the log ratio between the negative and the positive class. A common choice for the class conditional densities is the normal distribution:

$$p(\boldsymbol{x}|Y=k) = \frac{1}{\sqrt{(2\pi)^n \Sigma_k}} \mathrm{e}^{-(1/2)(\boldsymbol{x}-\mu_k)^{\mathrm{T}}\Sigma_k^{-1}(\boldsymbol{x}-\mu_k)}, \quad k=0,1$$

In this case, if the covariance matrices are independent of the class ($\Sigma_0 = \Sigma_1$) then the log ratio is a linear combination of the inputs, hence the name *linear* discriminant analysis:

$$\begin{aligned} a(\boldsymbol{x}) &= \log\frac{p(Y=0|\boldsymbol{x})}{p(Y=1|\boldsymbol{x})} \\ &= \log\frac{p(Y=0)}{p(Y=1)} + \frac{1}{2}(\mu_1+\mu_0)^{\mathrm{T}}\Sigma^{-1}(\mu_1+\mu_0) \\ &+ x^{\mathrm{T}}(\mu_0+\mu_1) \end{aligned} \tag{13}$$

As shown in Fig. 4, the log ratio is thus the distance between x and the separating hyperplane. The conditional probability of the class given the input is finally obtained by applying the logistic function to the log ratio.

This analysis can be generalized to any class conditional distribution in the exponential family (which include most common distributions such as the Poisson, multinomial, gamma, etc.) provided that the *dispersion* parameters associated with the two classes are identical (for Gaussians, the dispersion parameter is the covariance matrix). As noted in Ref. 21, this in an interesting robustness property that may represent an advantageous trait of the discriminant approach with respect to the generative direction. In fact, we can remain undecided about the particular form of the class conditional densities, provided they belong to the exponential family and have a similar dispersion parameter. Additionally, even if we could spend some hypotheses about the form of the class conditional densities (e.g., assuming they are Gaussians), the resulting number of parameters would be greater than $n+1$ [which is the number of parameters in the simple model of Eq. (9)].

In the multiclass case ($K > 2$ classes), the model in Eq. (9) can be easily generalized as follows. First, let us introduce the linear combinations

$$a_k(\boldsymbol{x}) = \beta_k^{\mathrm{T}} + \beta_{k,0}\boldsymbol{x}, \quad k = 1,\ldots,K-1 \tag{14}$$

that play the role of the log ratio of class k over class K. Under the assumption that class conditional densities belong to the exponential family, the posterior probability of each class given x is proportional to $e^{a_k(x)}$ and thus

$$f_k(\boldsymbol{x}) = p(Y = k|\boldsymbol{x}) = \frac{e^{a_k(\boldsymbol{x})}}{1 + \sum_{i=1}^{K-1} a_i(\boldsymbol{x})} \tag{15}$$

which is based on a vectorial extension of the logistic function, often referred to as the *softmax* function in the neural networks community (22).

We note here that common linear regression models can also be seen as a single-unit neural network. In this case, Eq. (9) is simply replaced by

$$f(\boldsymbol{x}) = E[Y|\boldsymbol{x}] = \beta^{\mathrm{T}}\boldsymbol{x} + \beta_0 \tag{16}$$

3. THE MULTILAYERED PERCEPTRON

The simplest architecture that may realize a nonlinear input–output mapping is shown in Fig. 5. There are M units in the first layer, each computing a nonlinear function of the input vector, and K output units. As before, in the case of classifica-

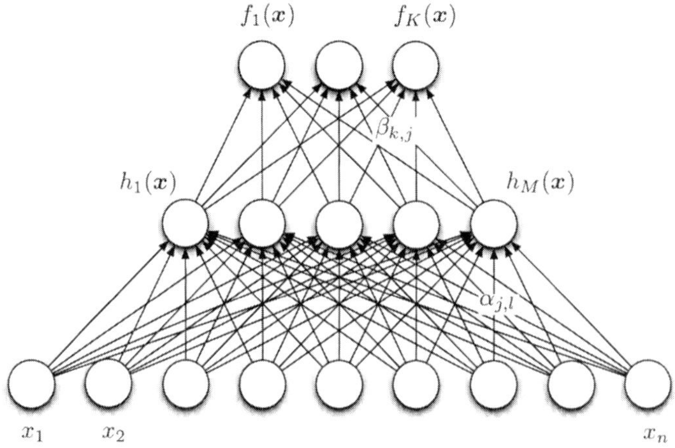

Figure 5 A multilayered perceptron.

tion, K is the number of categories, while in the case of regression, K is the dimension of the output vector. In most typical applications of regression, only a scalar function of the input will be required, so we assume $K=1$ for regression. Both the input vector x and the (desired) outputs $y_1,\ldots,y_K$ are assumed to be known during training. By contrast, the units in the middle layer are not observed and hence they are called *hidden units*. Note that the case of unobserved inputs (missing data) cannot be handled easily and this is often cited as a relative disadvantage of neural networks compared to generative models. The function computed by each hidden unit is typically chosen to be

$$h_j(\boldsymbol{x}) = \sigma(r_j) = \sigma(\alpha_{0,j} + \alpha_j^{\mathrm{T}}\boldsymbol{x}) \tag{17}$$

where $\sigma(\,)$ is the logistic function [see Eq. (10)] and $\alpha_{0,j} \in \mathbb{R}$, $\alpha_j \in \mathbb{R}^{\mathrm{n}}$, $j=1,\ldots, M$ are adjustable parameters or *connection weights*. A common alternative is to use the Gaussian-shaped function

$$h_j(\boldsymbol{x}) = \exp(-r_j) = \exp(-(\boldsymbol{x}-\mu_j)^{\mathrm{T}}\Sigma_j^{-1}(\boldsymbol{x}-\mu_j)) \tag{18}$$

with parameters $\Sigma_j \in \mathbb{R}^{n\times n}$ and $\mu_j \in \mathbb{R}^n$. The resulting architecture is called a *radial basis function* network.

Regardless of how an intermediate representation $\boldsymbol{h}=\boldsymbol{h}(\boldsymbol{x}) \in \mathbb{R}^M$ is computed, it will be useful to introduce the following quantities associated with each output unit:

$$a_k = \beta_{0,k} + \beta_k^{\mathrm{T}}\boldsymbol{h} \tag{19}$$

A suitable output function is finally applied to $a_1,\ldots, a_k$ to compute the prediction $f(\boldsymbol{x})$. We basically already know from Sec. 2.3 how to design the output function for binary and multiclass classification using the logistic and the softmax, respectively. For the MLP, we simply replace Eq. (9) by

$$f(\boldsymbol{x}) = E[Y|\boldsymbol{x}] = \sigma(\beta^{\mathrm{T}}\boldsymbol{h} + \beta_0) \tag{20}$$

and Eqs. (14) and (15) by

$$\begin{aligned} a_k(\boldsymbol{x}) &= \beta_{k,0} + \beta_k^{\mathrm{T}}\boldsymbol{h}, \quad k = 1,\ldots,K-1 \\ a_K(\boldsymbol{x}) &= 0 \end{aligned} \tag{21}$$

$$f_k(\boldsymbol{x}) = p(Y = k|\boldsymbol{x}) = \frac{e^{a_k(\boldsymbol{x})}}{\sum_{i=1}^{K} a_i(\boldsymbol{x})} \tag{22}$$

The case of regression is also a simple variant of Eq. (16):

$$f(\boldsymbol{x}) = E[Y|x] = \beta^{\mathrm{T}}\boldsymbol{h} + \beta_0 \tag{23}$$

3.1. TRAINING BY MAXIMUM LIKELIHOOD

We now discuss the most common approach for training neural networks. In the following, we denote by θ the whole set of parameters of a model [for example for the model of Eq. (9) $\theta = [\beta_0, \beta]$]. Let $D_m = \{(x_i, y_i)\}_{i=1}^m$ be the data set for training. Then consider the probability of D_m given the parameters θ of the model: $p(D_m|\theta)$. For a fixed training set, this quantity is only a function of the parameters and it is called the *likelihood function* of the data. Since we assume i.i.d. examples, we can factorize the probability of D_m as the product of the probabilities of single examples and taking the logarithm, we obtain

$$\begin{aligned} \ell(D_m|\theta) = \log p\,(D_m|\theta) &= \sum_{i=1}^{m} \log p(y_i, \boldsymbol{x}_i|\theta) \\ &= \sum_{i=1}^{m} \log p(y_i|\boldsymbol{x}_i, \theta) + c \end{aligned} \tag{24}$$

where c accounts for the likelihood of the input portion of the data (which is a constant with respect to the parameters).

According to the maximum likelihood principle, learning consists of maximizing $\ell(D_m|\theta)$ in Eq. (24) with respect to the parameters θ. Before we can do this, we need to fully specify the dependency of the likelihood on the parameters. We normally think of the neural network output, $f(\boldsymbol{x};\theta)$, as the expected value of the output variable Y, conditional to the input $\boldsymbol{x}$. But it remains to be decided what distribution to use for modeling such conditional probability. In the case of regression, the normal distribution is a natural choice. Assuming unitary variance and that $f(\boldsymbol{x};\theta)$ is computed as in Eq. (23), we have

$$p(y_i|\boldsymbol{x}_i,\theta) = \frac{1}{\sqrt{2\pi}} \mathrm{e}^{-1/2(y_i - f(\boldsymbol{x}_i;))^2} \tag{25}$$

which plugged into Eq. (24) yields

$$\ell(D_m|\theta) = -\frac{1}{2}\sum_{i=1}^{m} \left(y_i - f(\boldsymbol{x}_i;\theta)\right)^2 \tag{26}$$

It is interesting to note that in this way we have reobtained the well-known *least mean square* (LMS) criterion introduced by Widrow and Hoff (23) for the ADALINE, a simple learning machine that consists of a single adaptive linear element (i.e., a single artificial neuron). Note that training by maximum likelihood in this case also leads to the minimization of the empirical error [see Eq. (1) and (3)].

In the case of binary classification, the natural distribution of the output given the input is the Bernoulli:

$$p(Y = y_i|\boldsymbol{x}_i,\theta) = f(\boldsymbol{x}_i;\theta)^{y_i}\left(1 - f(\boldsymbol{x}_i;\theta)\right)^{1-y_i} \tag{27}$$

where in this case $f(\boldsymbol{x};\theta)$ is obtained by using a logistic function as in Eq. (20) to enforce a value in [0,1]. Plugging into Eq. (24), we obtain

$$\ell(D_m|\theta) = \sum_{i=1}^{m} y_i \log f(\boldsymbol{x}_i;\theta) + (1 - y_i)\log\left(1 - f(\boldsymbol{x}_i;\theta)\right) \tag{28}$$

which has the form of a *cross-entropy* function (a.k.a. Kullback Leibler divergence) (24) between the distribution obtained by reading the output of the network and the (degenerate) distribution that concentrates all the probability on the known targets y_i.

We can generalize the cross-entropy to the multiclass case by introducing the vector of indicator variables $z_i = [z_{i,1}, \ldots, z_{i,k}]$ with $z_{i,k} = 1$ if $y_i = k$ and $z_{i,k} = 0$ otherwise. In this case, we assume that the correct probabilistic interpretation is enforced using the softmax as in Eq. (22). The likelihood is based on the multinomial distribution

$$p(Y = y_i|\boldsymbol{x}_i, \theta) = \prod_{k=1}^{K} f_k(\boldsymbol{x}_i; \theta)^{z_{ik}} \tag{29}$$

that plugged into Eq. (24) yields

$$\ell(D_m|\theta) = \sum_{i=1}^{m} \sum_{k=1}^{K} z_{i,k} \log f(\boldsymbol{x}_i; \theta) \tag{30}$$

The optimization problem arising from this formulation can be solved using different numerical methods. One of the best approaches for the simple model based on a single neuron is the iteratively reweighted least squares (IRLS) algorithm, which can be applied to linear models and generalized linear models (GLIM) (25)—the logistic regression model of Eq. (9) is actually a GLIM. In the presence of hidden units, we need to resort to less efficient optimization algorithms, most of which are variants of steepest ascent. Note that while theoretically appealing, second-order methods (e.g., Newton) must be often discarded because of the prohibitive cost associated with the computation of the Hessian matrix (it is not uncommon to have of the order of 10^4 parameters in realistic applications of neural networks).

In the following, we show how to compute the gradients of the log-likelihood with respect to the parameters (which is anyway needed as a subroutine in most numerical optimization methods) and we restrict our discussion to the vanilla gradient ascent algorithm.

3.2. Backpropagation

Technically, *error backpropagation* refers to an efficient procedure for computing gradients of the likelihood function in an MLP. It was originally formulated in Ref. 5 and later rediscovered independently in Refs. 6–8. Subsequently, the term became very popular and today it is often used to mean, more in general, the whole learning algorithm for feed-forward neural networks. With some greater abuse, MLPs are sometimes referred to as "backpropagation networks."

We now illustrate how to compute gradients of the likelihood function, beginning with the case of networks with no hidden units. Initially, for simplicity, we restrict ourselves to the single output case ($K=1$) of regression or binary classification. Also, we shall shorten our notation writing ℓ for $\ell(D_m|\theta)$ and ℓ_i for the likelihood of the single example $(\boldsymbol{x}_i, y_i)$. Then we have

$$\frac{\partial \ell}{\partial \beta_j} = \sum_{i=1}^{m} \frac{\partial \ell_i}{\partial a_i} \frac{\partial a_i}{\partial \beta_j} \tag{31}$$

defining

$$\delta_i = \frac{\partial \ell_i}{\partial a_i} \tag{32}$$

where $a_i = \beta^{\mathrm{T}} \boldsymbol{x}_i + \beta_0$, we obtain immediately

$$\frac{\partial \ell}{\partial \beta_j} = \sum_{i=1}^{m} \delta_i x_{i,j} \tag{33}$$

and similarly

$$\frac{\partial \ell}{\partial \beta_0} = \sum_{i=1}^{m} \delta_i \tag{34}$$

To compute δ_i, we must proceed separately for regression and classification. In the case of regression, $\partial f(\boldsymbol{x}_i;\theta)/\partial a_i = 1$. Thus differentiating Eq. (26), we obtain

$$\delta_i = -\big(y_i - f(\boldsymbol{x}_i;\theta)\big) \tag{35}$$

In the case of binary classification, we have to differentiate Eq. (28). Note that for the logistic function, the simple differential equation holds: $\sigma'(a) = \sigma(a)(1-\sigma(a))$. Thus, we obtain

$$\begin{aligned}\delta_i &= y_i \frac{1}{\sigma(a_i)} \sigma(a_i)(1 - \sigma(a_i)) \\ &\quad - (1 - y_i) \frac{1}{1 - \sigma(a_i)} \sigma(a_i)(1 - \sigma(a_i)) \\ &= -(y_i - f(x_i; \theta))\end{aligned} \tag{36}$$

The case of multiclass classification is similar. Defining $a_{i,k} = \beta_k^{\mathrm{T}} \boldsymbol{x}_i + \beta_{0,k}$ and taking the derivatives of the softmax function, we see that

$$\delta_{i,k} = \frac{\partial \ell_i}{\partial a_{i,k}} = -(z_{i,k} - f_k(\boldsymbol{x}_i; \theta)) \tag{37}$$

The fact that in all these cases δ is the plain difference between targets and actual outputs is no coincidence and follows from the "right" match between the output function (linear, logistic, and softmax, in the cases we have seen) and the distribution of outputs given inputs (normal, Bernoulli, and multinomial, respectively) (26). Interestingly, if we had used a least squared error for classification instead of a cross-entropy, we would have obtained a different result:

$$\delta_i = -(y_i - f(\boldsymbol{x}_i; \theta)) f(\boldsymbol{x}_i; \theta)(1 - f(\boldsymbol{x}_i; \theta)) \tag{38}$$

The latter form may be less desirable. Suppose, for some weight assignment, we have $y_i = 1$ but $f(\boldsymbol{x}_i) \approx 0$, i.e., we are making a large error on the ith example because the logistic function is saturated in the wrong way. In this case, the derivatives of the quadratic cost would give $\delta_i \approx 0$ and the error would not be corrected.

We now describe how errors are backpropagated in multilayered networks of sigmoidal units. Note that the gradients for output weights are computed as in Eqs. (33) and (34). More precisely,

$$\frac{\partial \ell}{\partial \beta_{k,j}} = \sum_{i=1}^{m} \delta_{i,k} h_{i,j} \tag{39}$$

and

$$\frac{\partial \ell}{\partial \beta_{k,0}} = \sum_{i=1}^{m} \delta_{i,k} \tag{40}$$

Gradients with respect to the first layer weights $\alpha_{j,l}$ are obtained as follows:

$$\frac{\partial \ell}{\partial \alpha_{j,l}} = \sum_{i=1}^{m} \frac{\partial \ell}{\partial r_{i,j}} \frac{\partial r_{i,j}}{\partial \alpha_{j,l}} = \sum_{i=1}^{m} \delta_{i,j} x_{i,j} \tag{41}$$

having defined $\delta_{i,j} = \partial \ell / \partial r_{i,j}$. This quantity can be computed "recursively" from the delta's at the layer above as follows:

$$\delta_{i,j} = \sum_{k=1}^{K} \frac{\partial \ell}{\partial a_{i,k}}$$

Note the overall procedure for computing the gradients takes time linear in m and in the number of parameters (connection weights). The plain gradient ascent weight update rule is defined as follows:

$$\theta^{(t+1)} \leftarrow \theta^{(t)} + \epsilon \left. \frac{\partial \ell(D_m; \theta)}{\partial \theta} \right|_{\theta = \theta^{(t)}} \tag{43}$$

where t is the iteration index and $\theta^{(0)}$ is a small random vector.[d] A common variant to the above update rule is to compute the gradients of the likelihood associated with a single example, update the weights, then move to the next example, cycling through the data set. The latter approach, known as *online* or *stochastic* update [as opposite to the *batch* update described by Eq. (43)] may considerably speed up the overall optimization problem, especially for large data sets. When using this strategy, it is common to modify the update rule

[d] If we set $\theta^{(0)} = 0$ the optimization procedure would fail; because of network symmetries, the likelihood function has a saddle point in the origin of the parameters, space.

as follows:

$$\begin{aligned} \Delta^{(0)} &\leftarrow 0 \\ \Delta^{(t+1)} &\leftarrow \eta\Delta^{(t)} + \epsilon \left. \frac{\partial \ell_{1+(t \, mod \, m)(\theta)}}{\partial \theta} \right|_{\theta=\theta^{(t)}} \\ \theta^{(t+1)} &\leftarrow \theta^{(t)} + \Delta^{(t+1)} \end{aligned} \tag{44}$$

where the constant $\eta \in (0,1)$ is called the *momentum* and smoothes the steepest ascent process by "remembering" a discounted fraction of the previous updates.

Gradient-based optimization methods are local and therefore subject to fail if the likelihood function is not unimodal. Unfortunately, the absence of local optimal can be only guaranteed under very restrictive assumptions, such as having linearly separable examples in classification problem (27). Typically the likelihood function for neural networks has a complex shape and the gradient ascent procedure would converge towards one of several large plateaux rather than to a unique optimal solution. Although various techniques can help avoiding poor solutions (e.g., random restart, stochastic weight update, or global optimization, for example using evolutionary algorithms) the lack of a unique solution to the learning problem remains one of the weakest aspects of neural networks.

3.3. Feedforward Networks and Their Expressive Power

The single layer MLP introduced above may be generalized easily by adding more layers of hidden units. Actually a feedforward network is not limited to layered architectures: any arrangement of units as a directed acyclic graph (DAG) would be acceptable. However, it is not necessary to build too complex structures to achieve sufficient computational power. For example, it is easy to show that one hidden layer is sufficient to realize any Boolean function. Suppose $\boldsymbol{x}$ lay on a vertex of the unit hypercube. Then we immediately recognize that a hyperplane can realize any Boolean clause or any Boolean term.[e] Thus having any of the required terms (or clauses) in the hidden layer, we can design the output hyperplane in a way that it will realize the desired Boolean function in canonical form.[f] Other works have shown that a single hidden layer is sufficient to approximate continuous functions with arbitrary small error (measured by a finite norm), when the number of hidden units M is large enough (28,29). Thus neural networks are *universal approximators*, i.e., the associated hypothesis space $\mathscr{F}$ can be made arbitrarily large by letting the number of hidden units grow.

3.4 Controlling Overfitting and Generalization

The results mentioned in the previous section are good news since they tell us that underfitting will never occur if we use enough hidden units. However, if the capacity of the network is too large relative to the available number of examples, overfitting occurs: the error can be arbitrarily small on the training data but it gets large on unseen data points. Intuitively, when the hypothesis space $\mathscr{F}$ associated with the network is too large, the examples may not be sufficient to constrain the choice of our hypothesis $f \in \mathscr{F}$. While it is beyond the scope of this chapter to discuss this core issue in detail (the reader may consult for example Refs. 20 and 30), we will shortly explain how overfitting may be avoided using two common techniques: weight decay and early stopping.

Weight decay is best explained in terms of maximum a posteriori (MAP) parameter estimation. Under MAP, the objective function is the probability of the parameters given the data, that can be rewritten using Bayes theorem as

$$p(\theta|D_m) \propto p(D_m|\theta)p(\theta) \tag{45}$$

where the first term is the usual likelihood and the second

---

[e] A clause is the disjunction of literals and a term is the conjunction of literals; a literal is either a Boolean quantity or its negation. A data set of points labeled by a clause or by a term is always linearly separable.

[f] Any Boolean function can be equivalently written in one of the two canonical forms: conjunctive (CNF), which is a conjunction of clauses, or disjunctive (DNF), which is a disjunction of terms.

term is the so called *prior* on the parameters. Adding a prior can be seen as a form of *regularization* that biases toward simpler or smoother solutions to the learning problem (see, e.g., Ref. 31 for a more general discussion of the role of *inductive bias* in machine learning). Weight decay is obtained if we assume that $p(\theta)$ is a zero-mean Gaussian with covariance proportional to the identity matrix: $\Sigma = \lambda I$

$$p(\theta) \propto \mathrm{e}^{1/2\lambda\|\theta\|^2}$$

Taking the logarithms in Eq. (45), the new objective function is

$$J(D_m|\theta) = \ell(D_m|\theta) - \frac{1}{2}\lambda\|\theta\|^2 \qquad (46)$$

The update rule (43) becomes

$$\theta^{(t+1)} \leftarrow \theta^{(t)} + \epsilon \frac{\partial \ell(D_m;\theta)}{\partial \theta}\bigg|_{\theta=\theta^{(t)}} - \lambda\theta^{(t)} \qquad (47)$$

i.e., parameters are shrunk toward 0 (hence the name weight decay). The optimal value λ can be estimated by reserving part of the training data for validation.

Early stopping is an alternative technique that takes advantage of the peculiar way learning takes place in neural networks when using gradient descent. The procedure consists of computing the empirical loss on a validation set during the gradient ascent procedure. Then, rather than picking the parameters θ^* output by the optimization procedure after convergence (to a local optimum), we select the parameters $\theta^{(t)}$at the epoch t such that the empirical error on the validation set is minimized. An intuitive explanation of the method can be given as follows (assuming, for simplicity, we are interested in a binary classification task). When learning begins, weights are small random values and the sigmoidal units in the hidden layer(s) basically behave as linear functions (just remember that $\sigma(a) \approx a$ for small a). Therefore the effective number of parameters that are used is only $n+1$ and even a potentially very expressive network is equivalent to the single unit network of Eq. (9), only capable of linear separation. As learning proceeds, weight updates tend to increase the norm of θ and nonlinearities in the hidden layers begin to be exploited, forming a nonlinear separation surface. In other words, there is an almost continuous modification of the network during gradient ascent, from a very simple linear model, that will likely underfit the data, to an arbitrarily rich model that will likely overfit the. data. Early stopping is therefore an attempt to select the optimal model complexity. Not surprisingly, it can also be shown to be asymptotically equivalent to a form of regularization that favors small weights (32).

4. SUPPORT VECTOR MACHINES

Support vector machines (SVMs) (9) build on earlier foundational work by Vapnik on statistical learning theory (33). In particular, SVMs attempt to give a well-posed formulation to the learning problem through the principle of *structural risk minimization*, embodying a more general principle of parsimony that is also the foundation of Occam-Razor regularization theory, and the Bayesian approach to learning.

4.1. Maximal Margin Hyperplane

For simplicity, we shall focus on the binary classification problem and let $y_i \in \{-1,1\}$ denote the class (positive or negative) of the ith training example. A linear discriminant classifier is defined as

$$f(\boldsymbol{x}) = \beta^{\mathrm{T}}\boldsymbol{x} + \beta_0 \tag{48}$$

where $\beta \in \mathbb{R}^n$ and $\beta_0 \in \mathbb{R}$ are adjustable parameters just like in logistic regression. Let D_m be a linearly separable data set, i.e., $f(\mathrm{x}_i)y_i > 0$, $i = 1,\ldots,m$. We call the *margin* M of the classifier the distance between the separating hyperplane and the closest training example. The optimal separating hyperplane is defined as the one having maximum margin

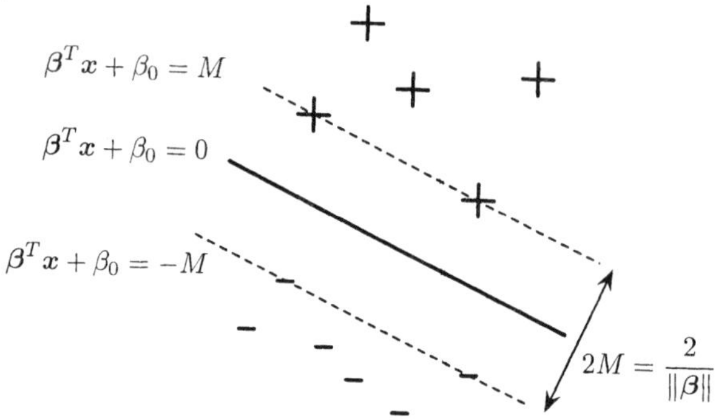

Figure 6 The maximum margin hyperplane.

(see Fig. 6). To compute it, we begin noting that the distance of any point $\boldsymbol{x}$ from the separating hyperplane is

$$\frac{1}{||\beta||}(\beta^{\mathrm{T}}\boldsymbol{x} + \beta_0)$$

This quantity should be maximized for the optimal hyperplane, leading to the following constrained optimization problem:

$$\begin{aligned} &\max_{\beta,\beta_0} \quad M \\ &\text{Subject to} \quad \frac{1}{||\beta||} y_i(\beta^{\mathrm{T}}\boldsymbol{x}_i + \beta_0) \geq M, \quad i = 1, \ldots, m \end{aligned} \tag{49}$$

where the constraints require that each training point should lie in the correct semispace and at a distance not less than M from the separating hyperplane. Since we can multiply β and β_0 by a scalar constant without changing the hyperplane, we arbitrarily set $||\beta|| = 1/M$ and rewrite the optimization problem (49) as

$$\begin{aligned} &\max_{\beta,\beta_0} \quad ||\beta|| \\ &\text{Subject to} \quad y_i(\beta^{\mathrm{T}}\boldsymbol{x}_i + \beta_0) \geq 1, \quad i = 1, \ldots, m \end{aligned} \tag{50}$$

The above problem can be transformed into its dual form by first introducing the vector of Lagrangian multipliers

$\alpha \in \mathbb{R}^n$, writing the Lagrangian function

$$\mathscr{L}(D_m) = \frac{1}{2}\|\beta\|^2 + \sum_{i=1}^{m} a_i[y_i(\beta^{\mathrm{T}}\boldsymbol{x} + \beta_0) - 1] \tag{51}$$

and finally setting to zero the derivatives of (51) with respect to β, β_0 obtaining

$$\begin{aligned} &\max_{\alpha} \quad -\frac{1}{2}\alpha^{\mathrm{T}} G \alpha + \sum_{i=1}^{m} \alpha_i \\ &\text{Subject to} \quad \alpha_i \geq 0, \quad i = 1, \ldots, m \end{aligned} \tag{52}$$

where G is an $m \times m$ matrix with entries

$$g_{ij} = y_i y_j \boldsymbol{x}_i^{\mathrm{T}} \boldsymbol{x}_j \tag{53}$$

This is a quadratic programming (QP) problem that can be solved, in principle, using standard optimization packages. For each training example i, the solution must satisfy the Karush–Kuhn–Tucker condition:

$$\alpha_i[y_i(\beta^{\mathrm{T}}\boldsymbol{x} + \beta_0) - 1] = 0 \tag{54}$$

and therefore either $a_i = 0$ or $y_i\ (\beta^{\mathrm{T}}\boldsymbol{x} + \beta_0) = 1$. In other words, if $\alpha_i > 0$ then point $\boldsymbol{x}_i$ has a distance M from the separating hyperplane. Such points are called *support vectors*. The decision function $f(\boldsymbol{x})$ can be computed via Eq. (48) or, equivalently, from the following dual form

$$f(\boldsymbol{x}) = \sum_{i=1}^{m} y_i \alpha_i \boldsymbol{x}^{\mathrm{T}} \boldsymbol{x}_i + \beta_0 \tag{55}$$

The maximum margin hyperplane has two important properties. First, it is unique for a given linearly separable D_m. Second, it can be shown that the VC-dimension of the class of "thick" hyperplanes having thickness $2M$ and support vectors $\|\boldsymbol{x}_i\| < R$ is bounded by $d < M^2/R^2 + 1$. When plugged into uniform convergence bounds like Eq. (8), this suggests that the complexity of the hypothesis space (and consequently the convergence of the empirical error to the generalization error) does not necessarily depend on the dimension n of the input space. In some application domains (e.g., gene expression data obtained from DNA microarrays, text categorization, etc.) where the number of training examples m is much smaller than the number of attributes n, this property is clearly of great appeal. The same property, however, may be very appealing also in other domains. Indeed, as we show in the next section, data requiring complex nonlinear separation surfaces or structured data must be first mapped into a high dimensional (or even infinite dimensional) space where linear separation takes place.

4.2. Kernels and Nonlinear Separation

The above method for finding a maximum margin hyperplane can be easily generalized for finding more complex decision surfaces $f(\boldsymbol{x}) = 0$ where f is a nonlinear function. Methods for constructing a nonlinear separation function by preprocessing of the input data have been around for several decades in the pattern recognition community—see, e.g., Ref. 34. For example, a decision function having the form

$$f(\boldsymbol{x}) = f(\boldsymbol{x}_1, \boldsymbol{x}_2, \ldots, \boldsymbol{x}_n) = \sum_{j=1}^{n} \sum_{i_j=0}^{p} \beta_{i_1, i_2, \ldots, i_n} \boldsymbol{x}_1^{i_1} \boldsymbol{x}_2^{i_2} \cdots \boldsymbol{x}_n^{i_n} \tag{56}$$

is based on a pth order *polynomial preprocessing* of the input vector and is equivalent to transforming $\boldsymbol{x}$ into a new vector $\phi(\boldsymbol{x}) \in \mathscr{F} = \mathbb{R}^{n'}$ with $n' = 2^p$. A linear separation surface in $\mathscr{F}$ (the so-called *feature space*) corresponds to a nonlinear separation surface in the original space $\mathscr{X}$. For large enough p, arbitrarily complex surfaces can be constructed.

While appealing because of its power, this technique would scale poorly from a computational point of view because it forces us to compute and store in memory vectors $\phi(\boldsymbol{x})$ with an exponential number of components. On the other hand, since we are only interested in computing $f(x)$, we may ask if there is a way of achieving the goal without explicitly performing the computations that are necessary to obtain $\phi(\boldsymbol{x})$. This may appear as impossible if we think of a direct version of f parameterized by β, but becomes easy if we consider a dual form like the one in Eq. (55). For example, consider again the main computations in dual form [see Eqs. (51), (52), and (55)]. We see that data points $\boldsymbol{x}_i$ are actually used only via dot products in the original vector space $\mathcal{X} = \mathbb{R}^n$. If we know how to evaluate the dot product between the images $\phi(\boldsymbol{x})$ and $\phi(\boldsymbol{y})$ of any two points $\boldsymbol{x}$ and $\boldsymbol{y}$ [without explicitly storing $\phi(\boldsymbol{x})$ and $\phi(\boldsymbol{y})$], we can carry out all the necessary computations. This approach is generally referred to as the *kernel trick*. The formalization of this approach requires mathematical notions related to Hilbert spaces that we only briefly sketch here.

A kernel is a function $K: \mathcal{X} \times \mathcal{X} \mapsto \mathbb{R}$ that embodies a notion of "similarity" between pairs of instances. Technically, a kernel corresponds to a inner product $\langle \cdot, \cdot \rangle$ in feature space:

$$K(\boldsymbol{x}, \boldsymbol{y}) = \langle \phi(\boldsymbol{x}), \phi(\boldsymbol{y}) \rangle = \sum_{i=0}^{\infty} \lambda_l \phi_l(\boldsymbol{x}) \phi_l(\boldsymbol{y}) \tag{57}$$

Intuitively, a "good" kernel should map dissimilar instances into nearly orthogonal vectors.

A common kernel function is the *polynomial kernel* defined as

$$K(\boldsymbol{x}, \boldsymbol{y}) = (\boldsymbol{x}^{\mathrm{T}} \boldsymbol{y})^d \tag{58}$$

for some integer d. It can be easily seen that the feature space representation $\phi(\boldsymbol{x})$ a vector consists in this case of all the d-degree products of the components of $\boldsymbol{x}$ [similar to Eq. (56)]. Another common choice is the Gaussian radial basis function kernel

$$K(\boldsymbol{x}, \boldsymbol{y}) = \mathrm{e}^{-\gamma \|\boldsymbol{x} - \boldsymbol{y}\|^2} \tag{59}$$

with $\gamma > 0$, whose associated feature space has infinite dimension. Interestingly, for appropriate choices of γ, the Gaussian kernel is capable of inducing arbitrarily complex separation surfaces in $\mathcal{X}$. This property is similar to the universal approximation, results discussed in Sec. 3.3 for neural networks. However, one should be careful that arbitrarily large values of γ completely destroy structure in the training data.

In fact, when γ is too large, $\phi(\boldsymbol{x}_i)$ and $\phi(\boldsymbol{x}_j)$ become orthogonal even for instances that may be similar. Clearly, such a poor notion of similarity would not allow us to go beyond rote learning.

Of course, the trivial function $K(\boldsymbol{x}, \boldsymbol{y}) = \boldsymbol{x}^{\mathrm{T}}\boldsymbol{y}$ is also a kernel whose feature space is simply $\mathcal{X}$ itself; this is sometimes called the linear kernel since it places a separating hyperplane in the original input space.

In general, Mercer's theorem states that $K(\cdot)$ is a valid kernel function (i.e., it does correspond to a inner product in some feature space) if and only if for any integer m and for any finite set of points $\boldsymbol{x}_1, \ldots, \boldsymbol{x}_m$, the Gram matrix G having entries $G_{\mathrm{ij}} = K(\boldsymbol{x}_i, \boldsymbol{x}_j)$ satisfies the following two conditions: symmetry, i.e., $G = G^{\mathrm{T}}$; and positive definiteness, i.e., $\alpha^{\mathrm{T}} G \alpha > 0$ for each $\alpha \in \mathbb{R}^m$. The class of valid kernels on $\mathcal{X} \times \mathcal{X}$ turns out to be closed under multiplication and linear combination using positive constant coefficients. This means that if $K_1(\boldsymbol{x}, \boldsymbol{y}), \ldots, K_t(\boldsymbol{x}, \boldsymbol{y})$ are kernels, then also $K_r(\boldsymbol{x}, y)K_s(\boldsymbol{x}, \boldsymbol{y})$ and $\Sigma_s c_s K_s(\boldsymbol{x}, \boldsymbol{y})$ with $c_{\boldsymbol{x}} > 0$ are all valid kernels.

In addition, it turns out that the space of functions having the form

$$f(\boldsymbol{x}) = \sum_{l=0}^{\infty} \beta_l \phi_l(\boldsymbol{x}) \tag{60}$$

is a reproducing kernel Hilbert space (RKHS) $\mathcal{H}$, i.e., a vector space endowed with a scalar product $\langle \cdot, \cdot \rangle_{\mathcal{H}}$ that satisfies

$$\langle f(\boldsymbol{x}), K(\boldsymbol{x}, \boldsymbol{y}) \rangle_{\mathcal{H}} = f(\boldsymbol{x}) \tag{61}$$

and the norm of f in this space is

$$\|f\|_K^2 = \sum_{i=0}^{\infty} \beta_l^2 \tag{62}$$

Note that in the simple case of the linear kernel, minimizing the norm of f in the RKHS subject to the constraints that $y_i\ f(x_i) \geq 1$ essentially corresponds to the optimization problem in Eq. (50) that yields the maximal margin hyperplane. In other words, seeking the maximal margin hyperplane for linearly separable data is simply the constrained minimization, of the norm of f in the Hilbert space induced by the dot product in instance space. More details and related results can be found in Refs. 18 and 19.

4.3. Support Vector Classifiers

If the training data set is not separable by the function f (for example because of noise or because of the presence of outliers), the analysis of Sec. 4.1 can be extended by allowing "soft" violations of the constraints associated with the training examples. To this purpose, we introduce m nonnegative *slack variables* ξ_i and relax the optimization problem in Eq. (50) as follows:

$$\begin{aligned} \min_{f \in \mathscr{H}, \xi} \quad & \|f\|_K^2 + c \sum_{i=1}^{m} \xi_i \\ \text{Subject to} \quad & y_i f(\boldsymbol{x}_i) \geq 1 - \xi_i \quad i = 1, \ldots, m \\ & \xi_i \geq 0 \quad i = 1, \ldots, m \end{aligned} \tag{63}$$

where the constant C controls the cost associated with the violations of the constraints (see Fig. 7).

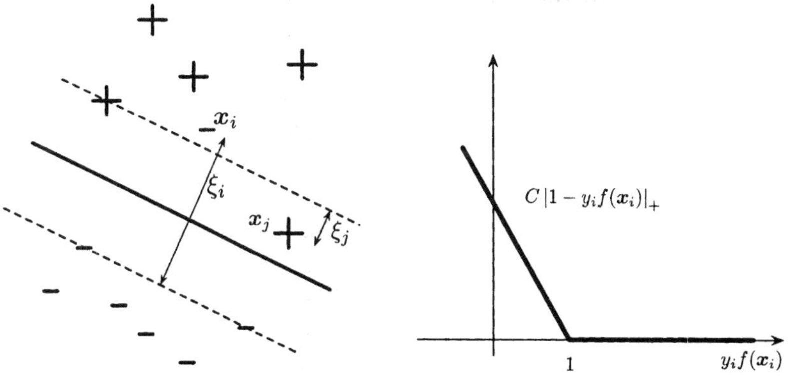

Figure 7 Left: Soft violations of the constraints. Right: The SVM loss function associated with violations.

The dual version of the problem is

$$\begin{aligned} &\max_{\alpha} \quad -\frac{1}{2}\alpha^{\mathrm{T}}G\alpha + \sum_{i=1}^{m}\alpha_i \\ &\text{Subject to} \quad 0 \le \alpha_i \le C/m, \quad i = 1, \ldots, m \end{aligned} \tag{64}$$

where in this case the entries of G are

$$g_{ij} = y_i y_j K(\boldsymbol{x}_i, \boldsymbol{x}_j) \tag{65}$$

The solution of the associated QP problem has the form

$$f(\boldsymbol{x}) = \sum_{i=1}^{m} y_i \alpha_i K(x, x_i) + \beta_0 \tag{66}$$

The classifier obtained in this way is commonly referred to as an SVM.

From a practical point of view, it may be useful to note that when using kernels the solution f cannot be written as a separating hyperplane because the feature space $\phi(\boldsymbol{x})$ where such a hyperplane would live is never explicitly represented and might have infinite dimension. For this reason, SVMs are *instance-based* classifiers that need to store all the support vectors $\boldsymbol{x}_i$, i.e., training points such that $\alpha_i \neq 0$. In practice, this may imply an increased computational cost for prediction, compared for example to neural network classifiers.

It can also be shown (35) that the solution of the problems in Eqs. (63) and (64) is a minimizer of the functional

$$H(f) = \sum_{i=1}^{m} \frac{C}{m} |1 - y_i f(\boldsymbol{x}_i)|_{+} + \frac{1}{2}\|f\|_K^2 \tag{67}$$

where $|a|_{+} = a$ if $a > 0$ and zero otherwise (see Fig. 7). This equation enlightens the important role played by C: it trades high classification accuracy on the training set (high C, potential overfitting) with high "smoothing" or regularization of f (low C, potential underfitting) that is achieved by minimizing its norm in the RKHS. Evgeniou et al. (35) show interesting links between SVMs, theory of regularization, and Bayesian approaches to learning. Intuitively, we can observe an important analogy between Eq. (67) for SVMs and Eq. (46) that is used to regularize neural networks. In the case of SVMs, we have a cost $\frac{C}{m}|1 - y_i f(\boldsymbol{x}_i)|_+$ associated with each example, while in the case of neural networks the cost is the negative log-likelihood for that example. In both cases, we want to keep "small" the class of functions that can be fitted to the data by minimizing the norm of f (in the case of SVMs) or the norm of the weight vector (in the case of neural nets). In both cases, the trade-off between fitting and regularization is controlled by a constant C or λ.

From a computational point of view, solving the QP problem using standard optimization packages would take time $O(m^3)$ (assuming the number of support vector grows linearly with the number of training examples). This time complexity is a practical drawback for SVMs. However, several approaches have been proposed in the literature that can reduce complexity substantially (36,37). Other researchers have proposed a divide-and-conquer technique that splits the training data into smaller subsets and subsequently recombines the solutions (38).

4.4. Support Vector Regression

The SVM regression can be formulated as the problem of minimizing the functional

$$H(f) = \sum_{i=1}^{m} \frac{C}{m} |y_i - f(\boldsymbol{x}_i)|_\epsilon + \frac{1}{2} \|f\|_K^2 \tag{68}$$

where $|a|_\epsilon = 0$ if $|a| < \epsilon$ and $|x| - \epsilon$ otherwise is the ϵ-insensitive loss. As depicted in Fig. 8 for the linear case, only the points that lie outside a "tube" of width 2ϵ around f contribute to the cost functional. This can be equivalently formulated as the problem

$$\begin{aligned} \min_{f \in \mathcal{H}, \xi, \xi^*} \quad & \frac{1}{2} \|f\|_K^2 + C \sum_{i=1}^{m} (\xi_i + \xi_i^*) \\ \text{Subject to} \quad & f(\boldsymbol{x}_i) - y_i \geq \epsilon + \xi_i, \quad i = 1, \ldots, m \\ & y_i - f(\boldsymbol{x}_i) \geq \epsilon + \xi_i^*, \quad i = 1, \ldots, m \\ & \xi_i \geq 0, \xi_i^* \geq 0, \quad i = 1, \ldots, m \end{aligned} \tag{69}$$

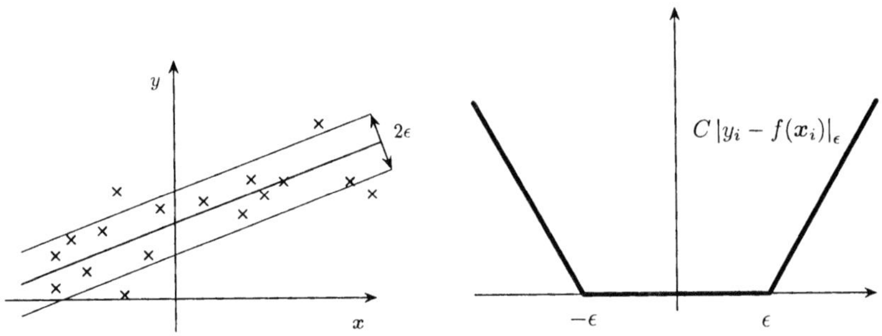

Figure 8 Left: The $\in$-tube of SVM regression. Right: The SVM loss function associated with violations.

whose QP dual formulation is

$$\begin{aligned} \max_{\alpha,\alpha^*} \quad & -\frac{1}{2}(\alpha-\alpha^*)^{\mathrm{T}}G(\alpha-\alpha^*) \\ & -\sum_{i=1}^{m}\epsilon(\alpha_i+\alpha_i^*)+y_i(\alpha_i-\alpha_i^*) \\ \text{Subject to} \quad & \sum_{i=1}^{m}(\alpha_i-\alpha_i^*) \\ & \alpha_i,\alpha_i^* \in [0,C], \quad i=1,\ldots,m \end{aligned} \tag{70}$$

where G is the $m \times m$ matrix with $g_{ij}=K(\boldsymbol{x}_i,\boldsymbol{x}_j)$. The solution of this problem has the form

$$f(\boldsymbol{x}) = \sum_{i=1}^{m}(\alpha_i-\alpha_i^*)K(x,x_i)+\beta_0 \tag{71}$$

5. LEARNING IN STRUCTURED DOMAINS

5.1. Introduction

Several algorithms for supervised learning (including the ones mentioned so far in this chapter) are conceived for data represented as real vectors or in the so-called attribute-value (or propositional) representation, where each instance is a tuple of attributes that may be either continuous or categorical. Mapping tuples to real vectors is straightforward and, in the case of neural networks, it is usually accomplished by transforming each categorical attribute into a subvector whose dimension equals the number of different values the attribute can take on, and representing the value using a sparse unary code.

However, in many applications, instances have a richer structure and converting them to vectors for subsequent learning is not necessarily straightforward. In the case of chemistry, for example, molecules are naturally described as undirected labeled graphs whose vertices are atoms and arcs represent chemical bonds. In general, mapping data structures to real vectors may result in some loss of information (i.e., it may not be possible to invert the transformation and to reconstruct the original structure from the vector). Different frameworks have been studied in machine learning in order to deal with rich structured representations:

- Inductive logic programming (ILP) (39) formulates the learning problem as the automatic construction of first-order logic programs from examples and background knowledge. The ILP has been shown to be very effective in several tasks related to computational chemistry including modeling structure–activity relationships (40), the prediction of mutagenicity (41), and the discovery of pharmacophores (42).
- Propositionalization or feature construction (43,44).
- Bayesian networks (45) and their learning algorithms (46) have been extended to handle relational data.
- Recurrent neural networks and hidden Markov models have been generalized from sequences to directed acyclic graphs (47–49). Examples of applications of this class of models include the classification of logical terms occurring in automated theorem proving (47), preference learning on forests of parse trees in natural language processing (50), classification of scanned image documents (51), classification of fingerprints (52), and prediction of protein contact maps (53).
- Kernel machines can operate on virtually any data type, provided a suitable kernel function can be

defined on the instance space. Early work in this area includes the unpublished paper Haussler (54) that introduce convolution kernels, and the paper by Jaakkola and Haussler (55) that introduces the Fisher kernel as a tool for turning generative models into discriminant classifiers.

The two latter frameworks are described in some detail in the remaining part of this chapter.

5.2. Recursive Connectionist Models for Data Structures

5.2.1. Structured Data

A directed graph is a pair $G=(V, E)$ where V is a finite set of *vertices* and E a binary relation on V, $E \subset V \times V$, called the *edge set*. A path is a sequence of vertices $v_0, v_1, \ldots, v_k$ such that $(v_i, v_{i+1}) \in E$. A path is a *cycle* if $v_0 = v_k$ and G is cyclic if it has at least one cycle. A vertex $u \in V$ is a *child* of v if $(v, u) \in E$. In this case, we also say that v is a *parent* of u. The number of children of a vertex v is called its out-degree and the number of parents its in-degree. The subgraph G' of G induced by a set of vertices $V' \subset V$ has edge set $E' = \{(u, v) \in E : u, v \in V'\}$. We say that v *dominates* a subgraph $G' = (V', E')$ of G if $v \in V'$ and all the vertices in V' can be reached following a path from v. A vertex q is a *supersource* for G if it dominates G. A DAG is ordered if for each vertex v a total order is defined in the set of edges $(v, u) \in E$. A graph is positional if there exists a total function $P : E \mapsto \mathbb{N}$ that assigns a position to each edge. The acronyms DOAG and DPAG are sometimes used when referring to ordered or positional DAGs. In order to explicitly account for the presence of missing children in DOAGs or DPAGs, we sometimes consider the extended graph $\hat{\mathrm{G}} = \hat{V}, \hat{\mathrm{E}}$ that is enriched with an extra nil node ν: $\hat{V} = V \cup \{\nu\}$, and whose edges are completed by connecting ν to fill-in missing children: $\hat{\mathrm{E}} = \mathrm{E} \cup \{(v, \nu) : (v, u) \notin E\}$.

Graphs can be labeled as follows. Consider an input space $\mathcal{X}$ and denote by $X : V \mapsto \mathcal{X}$ an *input labeling function* that associates to each vertex v its label $X(v)$. For simplicity, we may think that $\mathcal{X} = \mathbb{R}^n$ (alternatively $\mathcal{X}$ can be the Cartesian product of some continuous and categorical sets, as in the attribute-value representations).

In the following, we assume the instance space consists of labeled DOAGs or DPAGs with bounded out-degree. This set is denoted as $\mathcal{X}^{\#}$. Note that edges could be labeled as well, but we can safely ignore this issue in our discussion. For instance, let $L: E \mapsto \mathcal{X}'$ denote the edge labeling function (with edge label space $\mathcal{X}'$). Then we can always reduce a bounded out-degree graph with labeled edges to a graph having labels on the vertices only by "moving" the edge labels to the parent node. Note that a positional graph can also be seen as a graph with edge labels and a *finite* edge label space.

The above requirements are necessary for the model we are going to present next. Concerning the requirement of children ordering, we observe that in some domains this is a natural consequence of the data. For example, children in a syntactic tree are always ordered after the natural left-to-right reading of the sentence. In other cases (notably in the case of chemical compounds), the order must be obtained as the result of some conventional rules.

5.2.2. Structural Transductions

One straightforward extension of the supervised learning framework consists of mapping input graphs into output graphs. More precisely, let $\mathcal{X}$ and $\mathcal{Y}$ denote an input and an output space, respectively, and suppose a distribution p is defined on $\mathcal{X}^{\#} \times \mathcal{Y}^{\#}$. As in the vector-based case, we are given a data set of i.i.d. examples drawn from p in the form of input–output pairs: $D_m = \{(\boldsymbol{x}_i, \boldsymbol{y}_i)\}_{i=1}^{m}$ where $x_i \in \mathcal{X}^{\#}$ is the input graph and $\boldsymbol{y}_i \in \mathcal{Y}^{\#}$ is the output graph. The supervised learning problem consists again of seeking a function $f(\boldsymbol{x})$ for predicting y but since in this case x and y are graphs, the function f is called a *transduction*. Note that this formulation originates a formidably difficult learning problem since we put no restriction on the input–output mapping. In some applications, it makes sense to assume that the input and the output graph only differ in the labels, i.e., they have the same structure but different labeling functions. In this case, the output labeling function will be $f: \mathscr{X}^{\#} \times V \mapsto \mathscr{Y}$ and will associate a new label *f(x,v)* to each vertex v, in response to an input graph x.

We illustrate the concept of a structural transduction using the example shown in Fig. 9. Here the instance space consists of rooted ordered trees with labels in the finite set $\mathscr{X} = \mathscr{X}_n \cup \mathscr{X}_t$ where internal nodes have labels in $\mathscr{X}_n = \{a, b, \ldots, l\}$ and leaves have labels in $\mathscr{X}_t\{A, B, \ldots F\}$. Note that reading the leaves from left to right yields a sequence s with symbols in $\mathscr{X}_t$. Suppose that for each sequence s, there is a unique (but unknown) correct tree T spanning s, such as the one shown in Fig. 9a (to be a little more concrete, s could be a natural language sentence, being $\mathscr{X}_t$ a set of words, and T a parse tree for it). Our prediction task in this case consists of determining how well a candidate tree x for s approximates the true tree T. For this purpose, we can introduce a transduction that takes as input a tree x and produces an output tree $f(\boldsymbol{x})$ with labels in $\mathscr{Y} = \{0,1\}$ (output labels are drawn at the right of input labels in Fig. 9b) such that $f(\boldsymbol{x}, \mathrm{v}) = 1$ if and only if the subtree rooted at v is correct for the subsequence it spans (i.e., if it coincides with the subtree of T that spans the same subsequence).

The above example is also useful to understand a key property that enables us to realize transductions using a connectionist approach. A transduction $f(\boldsymbol{x})$ is said to be *causal* if

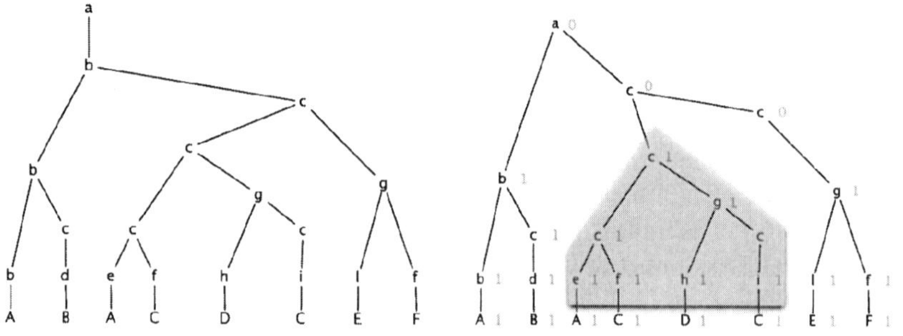

(a) (b)

Figure 9 Example of a causal transduction (see text for explanation).

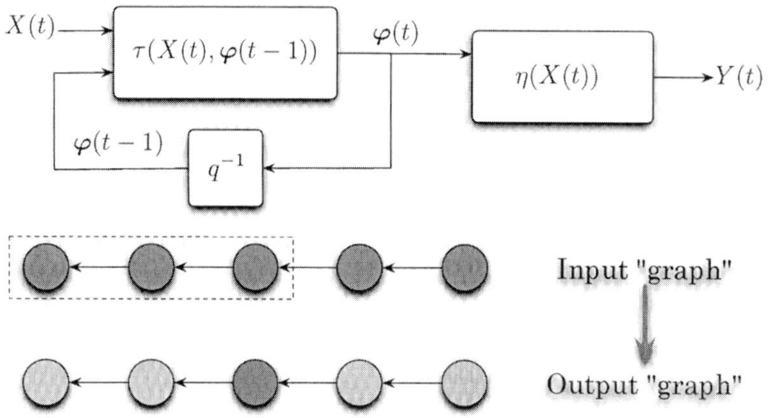

Figure 10 A causal discrete-time dynamical system described by its transition function and its output function.

for each input graph x and for each vertex v the output $f(\boldsymbol{x},\mathrm{v})$ only depends on the vertices of x that can be reached by following a directed path originating from v (see Fig. 9b). Note that causality defined in this way is actually a generalization of the concept of causality more commonly stated for temporal systems: a dynamical system is causal if the output at time t, $f(\boldsymbol{x},\mathrm{t})$, is independent of all the future inputs $\boldsymbol{x}(\tau)$ for $\tau > t$. In fact, we can imagine that discrete-time dynamical systems operate transductions on very special graphs that reduce to linear chains (see Fig. 10). In these systems, causality translates into the independence of the "current" output $f(\boldsymbol{x},\mathrm{t})$ on "future" inputs $\boldsymbol{x}(t+1), \boldsymbol{x}(t+2), \ldots$ When causality holds, past inputs $x(1), \ldots, x(t)$ can be summarized into a *state* vector $\varphi(t)$ that may have finite or infinite dimension. We shall assume $\varphi(t) \in \mathbb{R}^s$. States are computed recursively as follows:

$$\begin{aligned} \varphi(0) &= 0 \\ \varphi(t) &= \tau(\boldsymbol{x}(t), \varphi(t-1)) \end{aligned} \tag{72}$$

where τ is the transition function and $\varphi(0)$ is the *initial state* (note that when interpreting sequences as linear chains, time 0 essentially corresponds to the nil node ν). Outputs are obtained as a function of the current state:

$$f(x,t) = \eta(\varphi(t)) \tag{73}$$

Note that in this formulation the transition and the output function do not depend explicitly on t, a property known as *stationarity*. Causal transductions operating on directed acyclic graphs can be realized using a similar approach. In this case, the state φ(v) summarizes the information contained in the subgraph dominated by v following a directed path. Therefore $\varphi(v)$ depends on *x(v)* (i.e., a direct dependency on the information contained in the current node) and, denoting by $u_1, u_2, \ldots, u_k$ the children of v, on $\varphi(u_1), \varphi(u_2), \ldots, \varphi(u_k)$ (i.e., an indirect dependency on the contents of the subgraphs dominated by v's children). Thus, in the case of a DAG, Eqs. (72) and (73) are generalized by introducing a state labeling function $\varphi : \hat{v} \mapsto \mathbb{R}^s$ recursively computed as

$$\begin{aligned}\varphi(\nu) &= \varphi_0 = 0\\ \varphi(v) &= \tau(x(v), \varphi(u_1), \ldots, \varphi(u_k)) \quad \forall_v \in V,\ (v, u_i) \in \hat{E}\end{aligned} \tag{74}$$

The output labeling is then computed using an output function $\eta : \mathbb{R}^s \mapsto \mathcal{Y}$ as follows:

$$f(\boldsymbol{x}, v) = \eta(\varphi(v)) \tag{75}$$

$$\varphi(v) = \tau(\boldsymbol{x}(v), \varphi(u_1), \varphi(u_1); \theta)$$

These equations can be used as the basis to build a connectionist architecture for learning structural transductions from examples, as detailed below. We call this architecture *recursive* neural network (RNN).

The first step consists of introducing parameterized versions of the transition and output functions. Our solution consists of using a neural network with weights θ to realize $\tau(\boldsymbol{x}(v), \varphi(u_1), \ldots, \varphi(u_k);\ \theta)$ and a second network with weights θ to realize the output function $\eta(\varphi(v); \nu)$. The first network has n inputs associated with the input label $\boldsymbol{x}(v)$, $k \times s$ inputs

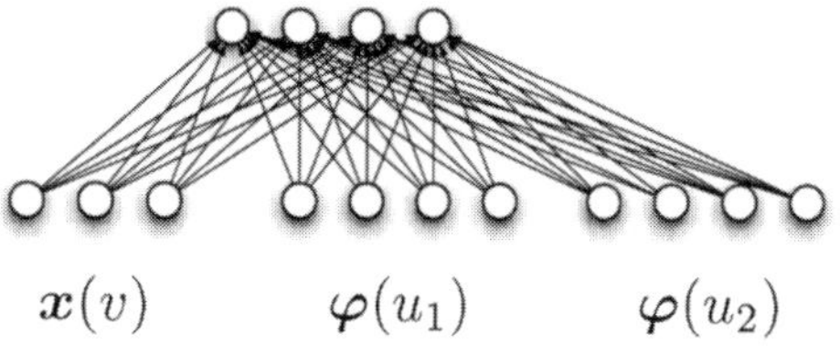

Figure 11 Neural network used to realize an adaptive transition function for structural transductions.

associated with the state labels of v's children, and s outputs representing the state $\varphi(v)$. Figure 11 shows an example. The output units of this network are typically nonlinear (e.g., using the hyperbolic tangent function). The second network has s inputs and K (the dimensionality of $\mathcal{Y}$) outputs. Both networks can use hidden layers to increase their expressive power.

The second step consists of creating a replica of the transition and the output network for each vertex of the input graph, and connecting the resulting units in a way that reflects the topology of the graph, as specified by Eqs. (74) and (75). This process is called *unfolding* and has been used earlier to explain recurrent neural networks for sequences (1). Figure 12 shows examples of unfolding in time and according to a DAG structure. Intuitively, recurrent networks for sequences have a simpler transition function since each "vertex" (i.e., discrete time step) in a sequence has a single "child" (i.e., the previous time step). The power of this simple trick is significant since the overall architecture we have obtained is a feedforward neural network with shared weights. States are propagated bottom up and then predicted outputs are computed. Desired outputs $\boldsymbol{y}(v)$ in this network are assigned to each vertex and comparing $\boldsymbol{y}(v)$ with the predicted $f(\boldsymbol{x},\mathrm{v})$, we obtain an error function to be minimized. Depending on the type of $\mathcal{Y}$ (e.g., continuous or categorical) we choose alternative error functions such as least squares [as in Eq. (26)] or cross-entropy [as in Eq. (28)]. Finally, we can derive a generalization of the backpropagation algorithm presented in Sec. 3.2 by simply noting that the derivatives of

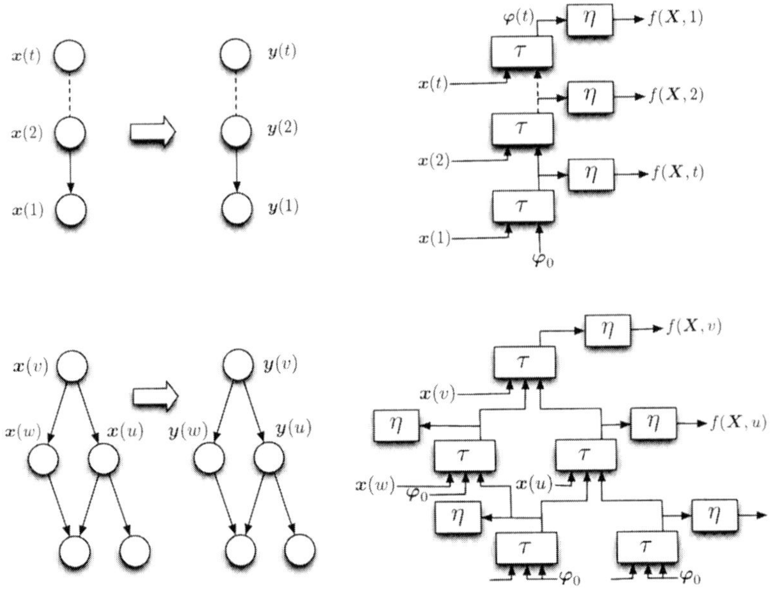

Figure 12 Unfolding recurrent networks in time (top) and unfolding recursive networks according to a DAG structure (bottom).

the error function with respect to local predictions must be propagated top-down in the unfolded network and that weight sharing is accounted to by summing up gradient contributions collected at each replica of the same local networks implementing τ and η.

As a special (but very important) case of the above framework, we can solve prediction tasks where a scalar quantity $f(\boldsymbol{x})$ (a real number in the case or regression or a category in the case of classification) is associated with the entire graph. The only requirement is that the DAG possess a supersource q so that after causal processing according to Eq. (74) the state $\varphi(q)$ depends on the entire graph. In this way, we obtain

$$f(\boldsymbol{x}) = \eta(\varphi(q)) \tag{76}$$

It may be interesting to note that this way of solving regression or classification tasks where the input data has a graphical structure offers important advantages over techniques based on descriptors. In practice, the adaptive function

τ is adjusted during learning to produce a suitable mapping from a graph x to the real vector $\varphi(q)$ at x's supersource. Each component of $\varphi(q)$ plays the role of a single descriptor but rather than being chosen using expert domain knowledge, these "adaptive descriptors" are determined by the particular learning task being considered and can potentially capture the relevant signals in the data even if their existence or role has not been fully detected or understood by the domain expert. It has been shown that this is actually the case in some practical applications related to computational chemistry. In particular, RNNs have been successfully applied to the QSPR of alkanes (prediction of the boiling point) and the QSAR of benzodiazepines (56,57). Alkanes are naturally represented as trees (their chemical structure has no cycles). Benzodiazepines do have cycles but can be represented as trees by associating the three main rings with the root of the tree and the substituents with subtrees of the root (see an example in Fig. 13). Bianucci et al. (56) show that the prediction accuracy of RNNs outperforms classic regression methods (58) thanks to the adaptive encoding of whole structures.

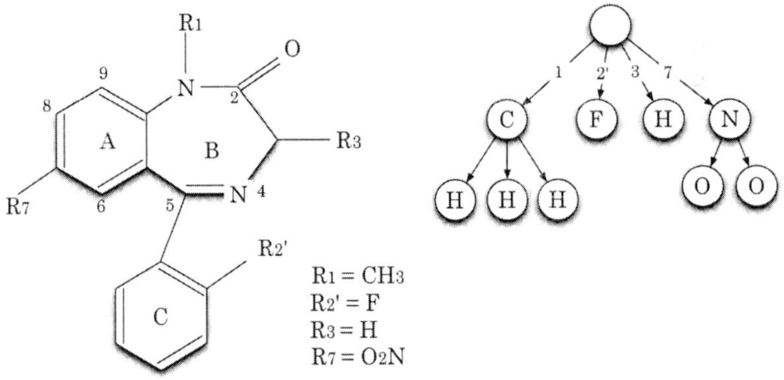

Figure 13 Tree representation of benzodiazepines. The basic template with three rings is shown at the left. The three rings are associated with the root while substituents are subtrees of the root. The example is a Flunitrazepam molecule and the four main substituents are in this case $R_1 = CH_3$, $R_{2'} = F$, $R_3 = H$, and $R_7 = O_2N$ (the remaining ones are all hydrogens).

5.3. Kernels for Discrete Structures

The theory on kernel we have reviewed in Sec. 4.2 never makes explicit assumptions on the nature of the instance space $\mathcal{X}$. We have seen examples of kernels [e.g., Eqs. (58) and (59)] that assume $\mathcal{X} = \mathbb{R}^n$. However, provided that we can find a suitable function $K: \mathcal{X} \times \mathcal{X} \mapsto \mathbb{R}$ that satisfies Mercer's conditions, we can solve classification and regression problems on virtually *any* input space $\mathcal{X}$. In particular, our instances could be discrete data structures such as sequences and labeled graphs. For lack of space, we limit here to a brief description of some important contributions in this area.

A general approach for defining the so-called *convolution* kernels on discrete structures was introduced by Haussler (59) and is based on a compositional approach for describing a structure as made of several substructures. Essentially a convolution kernel in these cases corresponds to a feature space representation $\phi($x where each of the component functions $\phi_i(\psi)$. For example, if x is a string (i.e., a sequence of symbols in a discrete set), then each component in the feature space counts the number of occurrences of each substring in x. Leslie et al. (60) have applied a similar kernel for sequences to the problem of protein classification. Collins and Duffy (61) have applied a particular convolution kernel in a natural language problem where instances are labeled parse trees. Lodhi et al. (62) have proposed string kernels for text classification.

A rather different approach for constructing kernels on strings is the Fisher kernel proposed by Jaakkola and Haussler (63). The method is based on the preliminary construction of a generative model $p(\boldsymbol{x};\theta)$ or $p(\boldsymbol{x},y;\theta)$ for the data. A feature space representation of $\boldsymbol{x}$ is then constructed by taking the gradient with respect to the parameters θ of the likelihood function evaluated on $\boldsymbol{x}$. The method can be easily applied to build kernels on strings using hidden Markov models as the generative model of the data.

6. CONCLUSION

Both neural networks and kernel machines can be applied in a wide range of classification and regression problems. Although they are not the only methods available in machine learning, they can be expected to be successful and useful when the problem at hand can be solved satisfactorily by properly choosing (or inferring) a measure of similarity between instances and using the trained model to make predictions on new data. If, on the other hand, there is a substantial interest in understanding and explaining the principles that regulate the behavior that is observed in the data, then one should resort to different approaches. As we have briefly mentioned in the paper, techniques based on belief networks and inductive logic programming can be successful for inferring unknown knowledge from data.

Comparing neural networks and kernel machines, we see that both have advantages and disadvantages. Kernel machines offer a more principled approach to learning and by properly choosing a kernel function they can handle a larger variety of instance types. In addition, they have been found to perform better than neural networks in many tasks, especially when few training examples are available or when the dimensionality of the input space is high. Finally, training neural networks is often made difficult by tricky problems related to the nonuniqueness of the solution to the underlying optimization problem. According to an informal but rather common folklore, training a neural net in the correct way may need a certain degree of witchery on some difficult problems. On the other hand, neural networks can learn much faster than support vector machines when one is faced with large data sets and their accuracy in these cases is typically comparable if not better. Finally, we are not aware of methods that can exploit support vector machines to solve more general problems like causal sequential (structural) transductions where predictions need to be associated with each discrete time step (vertex).

REFERENCES

1. Rumelhart DE, Hinton GE, Williams RJ. Learning internal representations by error propagation. In: Parallel Distributed Processing. Chap. 8. Cambridge, MA: MIT Press, 1986.
2. McCulloch WPWS. A logical calculus of ideas immanent in nervous activity. Bull Math Biophys 1943; 5:115–133.
3. Rosenblatt F. The perceptron: a probabilistic model for information storage and organization in the brain. Psychol Rev 1958; 65:386–408.
4. Minsky M, Papert S. Perceptrons: An Introduction to Computational Geometry. Cambridge, MA: MIT Press, 1969.
5. Werbos PJ. Beyond regression: New tools for prediction and analysis in the behavioural sciences. Ph.D. thesis, Harvard University, 1974.
6. LeCun Y. A learning scheme for asymmetric threshold networks. In: Proceedings of Cognitiva 85, Paris, France, 1985, 599–604.
7. Parker DB. Learning logic. Tech. Rep. TR-7, Center for Computational Research in Economics and Management Science, MIT, 1985.
8. Rumelhart DE, Hinton GE, Williams RJ. Learning representations by back-propagating errors. Nature 1986; 323:533–536.
9. Cortes C, Vapnik V. Support vector networks. Machine Learning 1995; 20:1–25.
10. Apostol TM. Calculus. Vol. I and II. John Wiley & Sons 1969.
11. Bhatia R. Matrix Analysis. New York: Springer Verlag, 1997.
12. Dudley RM. Real Analysis and Probability. Mathematics Series. Pacific Grove, CA: Wadsworth and Brooks/Cole, 1989.
13. Dautray R, Lions JL. Mathematical Analysis and Numerical Methods for Science and Technology. Vol. 2. Berlin: Springer-Verlag, 1988.
14. Feller W. An Introduction to Probability Theory and its Applications. Vol. 1. 3d ed. New York: John Wiley & Sons, 1968:1.

15. Poggio T, Girosi F. Regularization algorithms for learning that are equivalent to multilayer networks. Science 1990; 247:978–982.

16. Girosi F, Jones M, Poggio T. Regularization theory and neural networks architectures. Neural Comput 1995; 7:219–269.

17. Vapnik VN. Statistical Learning Theory. New York: Wiley, 1998.

18. Cristianini N, Shawe-Taylor J. An Introduction to Support Vector Machines. Cambridge University Press, 2000.

19. Schoelkopf B, Smola A. Learning with Kernels. Cambridge, MA: MIT Press, 2002.

20. Hastie T, Tibshirani R, Friedman J. Elements of Statistical Learning: Data Mining, Inference, and Prediction. New York: Springer Verlag, 2001.

21. Jordan MI. Why the logistic function? A tutorial discussion on probabilities and neural networks. Tech. Rep. 9503, Computational Cognitive Science MIT, 1995.

22. Bridle J. Probabilistic interpretation of feedforward classification network outputs, with relationships to statistical pattern recognition. In: Fogelman-Soulie F, Herault J, eds. Neurocomputing: Algorithms, Architectures, and Applications. New York: Springer-Verlag, 1989.

23. Widrow B, Hoff ME Jr. Adaptive switching circuits. In: Institute of Radio Engineers, Western Electronic Show and Convention, Convention Record. Vol. 4. 96–104. [Reprinted in (Anderson & Rosenfeld, 1988)].

24. Shore J, Johnson R. Axiomatic derivation of the principle of maximum entropy and the principle of minimum cross-entropy. IEEE Trans Inf TheoryIT-26, 1980:26—37.

25. McCullagh P, Nelder J. Generalized Linear Models. London: Chapman & Hall, 1989.

26. Rumelhart DE, Durbin R, Golden R, Chauvin Y. Back-propagation: the basic theory. In: Backpropagation: Theory, Architectures and Applications. Hillsdale, NJ: Lawrence Erlbaum Associates, 1995:1–34.

27. Gori M, Tesi A. On the problem of local minima in backpropagation. IEEE Trans Pattern Anal Machine Intell PAMI-14 (1),1992:76–86.

28. Cybenko G. Approximation by superpositions of a sigmoidal function. Math Control Signals Syst 1989; 2:303–314.

29. Hornik K, Stinchcombe M, White H. Multilayer feedforward networks are universal approximates. Neural Networks 1989; 2:359–366.

30. Anthony M, Bartlett PL. Neural Network Learning: Theoretical Foundations. Cambridge University Press, 1999.

31. Mitchell TM. Machine Learning. New York, US: McGraw Hill, 1996.

32. Sjöberg J, Ljung L. Overtraining, regularization, and searching for minimum in neural networks. In: Proceedings of IFAC Symposium on Adaptive Systems in Control and Signal Processing. Grenoble, France, 1992:669–674.

33. Vapnik VN. Estimation of Dependences Based on Empirical Data. Berlin: Springer-Verlag, 1982.

34. Greenberg H, Konheim A. Linear and nonlinear methods in pattern classification. IBM J Res Devel 1964; 3(8):299–307.

35. Evgeniou T, Pontil M, Poggio T. Regularization networks and support vector machines. Adv Comput Math 2000; 13:1–50.

36. Platt J. Fast training of support vector machines using sequential minimal optimization. In: Schölkopf B, Burges CJC, Smola AJ, eds. Advances in Kernel Methods—Support Vector Learning. Cambridge, MA: MIT Press, 1999:185–208.

37. Joachims T. Making large-scale SVM learning practical. In: Schölkopf B, Burges CJC, Smola AJ, eds. Advances in Kernel Methods—Support Vector Learning. Cambridge, MA: MIT Press, 1999:169–184.

38. Collobert R, Bengio S, Bengio Y. A parallel mixture of SVMs for very large scale problems. Neural Comput 2002; 14(5).

39. Muggleton S, Raedt LD. Inductive logic programming: theory and methods. J Logic Programming 1994; 19 & 20:629–680.

40. King R, Muggleton S, Lewis R, Sternberg M. Drug design by machine learning: The use of inductive logic programming to model the structure–activity relationships of trimethoprim analogues binding to dihydrofolate reductase. Proc Natl Acad Sci 1992; 89(23):11322–11326.

41. King R, Muggleton S, Srinivasan A, Sternberg M. Structure-activity relationships derived by machine learning: the use of atoms and their bond connectives to predict mutagenicity by inductive logic programming. Proc Natl Acad Sci 1996; 93:438–442.

42. Finn P, Muggleton S, Page D, Srinivasan A. Pharmacophore discovery using the Inductive Logic Programming system Progol Machine Learning 1998; 30:241–271.

43. Srinivasan A, King R. Feature construction with inductive logic programming: a study of quantitative predictions of chemical activity aided by structural attributes. In: Proceedings of the 6th International Workshop on Inductive Logic Programming, 1996.

44. Kramer S, Pfahringer B, Helma C. Stochastic propositionalization of non-determinate background knowledge. In: Proceedings of the 8th International Workshop on Inductive Logic Programming. Springer Verlag, 1998:80–94.

45. Pearl J. Probabilistic Reasoning in Intelligent Systems: Networks of Plausible Inference. Morgan Kaufmann, 1988.

46. Heckerman D. Bayesian networks for data mining. Data Mining and Knowledge Discovery 1997; 1(1):79–120.

47. Göller C, Kuchler A. Learning task-dependent distributed structure-representations by back-propagation through structure. IEEE International Conference on Neural Networks 1996:347–352.

48. Sperduti A, Starita A. Supervised neural networks for the classification of structures. IEEE Trans Neural Networks 1997; 8(3).

49. Frasconi P, Gori M, Sperduti A. A general framework for adaptive processing of data structures. IEEE Trans Neural Networks 1998; 9(5):768–786.

50. Costa F, Frasconi P, Lombardo V, Soda G. Learning incremental syntactic structures with recursive neural networks. In: Fourth International Conference on Knowledge-Based Intelligent Engineering Systems and Allied Technologies, 2000.

51. Diligenti M, Frasconi P, Gori M. Hidden tree Markov models for image document classification. IEEE Trans Pattern Anal Machine Intell 2003; 25(4):519–523.

52. Yao Y, Marcialis G, Pontil M, Frasconi P, Roli F. Combining flat and structured representations for fingerprint classification with recursive neural networks and support vector machines. Pattern Recognit 2003; 36(2):397–406.

53. Pollastri G, Baldi P, Vullo A, Frasconi P. Prediction of protein topologies using giohmms and grnns. In: Thrun S, Saul L, Schölkopf B, eds. Proceedings of Neural Information Processing Systems. MIT Press, 2003:16.

54. Haussler, D. Convolution Kernels on discrete structure. Tech. Rep. UCSC-CRL-99-10, University of California at Santa Cruz, Santa Cruz, CA, USA, 1990.

55. Jaakkola TS, Haussler D. Exploiting generative models in discriminative classifiers. In: Kearns MS, Solla SA, Cohn DA, eds. Advances in Neural Information Processing Systems Vol. 11. Cambridge, MA: The MIT Press 1999:11:487–493.

56. Bianucci A, Micheli A, Sperduti A, Starita A. Application of cascade correlation networks for structures to chemistry. Appl Intell 2000; 12(1–2):115–145.

57. Micheli A, Sperduti A, Starita A, Bianucci A. Analysis of the internal representations developed by neural networks for structures applied to quantitative structure–activity relationship studies of benzodiazepines. J Chem Inf Comput Sci 2001; 41:202–218.

58. Hadjipavlou-Litina D, Hansch C. Quantitative structure–activity relationships of the benzodiazepines. A review and reevaluation. Chem Rev 1994; 94(6):1483–1505.

59. Haussler D. Convolution kernels on discrete structures. Tech. Rep. UCSC-CLR-99-10, University of California at Santa Cruz, 1999.

60. Leslie C, Eskin E, Noble W. The spectrum kernel: a string kernel for SVM protein classification. In: Proceedings of Pacific Symposium on Bio-computing, 2002:564–575.

61. Collins M, Duffy N. Convolution kernels for natural language. In: Dietterich TG, Becker S, Ghahramani Z, eds. Proceedings of Neural Information Processing Systems. Vol. 14. MIT Press 2002.

62. Lodhi H, Shawe-Taylor J, Cristianini N, Watkins C. Text Classification using string kernels. In: Proceedings of Neural Information Processing Systems NIPS 13 2000:563–569.

63. Jaakkola T, Haussler D. Exploiting generative models in discriminative classifiers. In: Proceedings of Neural Information Processing Systems NIPS 10 1998.

64. Anderson JA, Rosenfeld E. Neurocomputing: Foundations of Research. Cambridge, MA: The MIT Press, 1988.

GLOSSARY

Backpropagation: A popular training algorithm for the multilayered perceptron. It allows us to compute analytically the gradients of a cost function with respect to the connection weights of the network.

Bayes optimal classifier: A theoretical optimal classifier that could be constructed if we knew exactly the joint probability distribution of inputs and outputs.

Capacity: Number of different predictive functions that can be realized by choosing in a given set of functions.

Classification: Form of supervised learning where predictions are discrete categories.

Connection weight: In an artificial neural network the strength of the connection between two units.

Cross-entropy: Also known as Kullback–Leibler (KL) divergence. A measure of dissimilarity between two probability distribution.

Empirical error: The prediction error (measured through a loss function) on the training set.

Generalization: Ability of a learning algorithm to make correct predictions on instances that are not part of the training set.

Hidden unit: Internal unit in an artificial neural network whose value is not specified in the data (hence hidden).

Hypothesis space: The set of predictive functions from which a supervised learning algorithm picks up its solution.

Instance space: The set comprising all possible input patterns.

Kernel function: A generalized dot product between two instances that allows us to measure their similarity. Using a kernel function in place of dot products is a simple way of achieving nonlinear separation.

Linearly separable: A set of vectors, each labeled as either positive or negative, is linearly separable if there exists a hyperplane that keeps positive vectors on one side and negative vectors on the other side.

Logistic regression: Simple binary classifier based on linear separation and the use of the logistic function to interpret the distance from the separating hyperplane as a conditional probability, to enable maximum likelihood estimation of the coefficient of the separating hyperplane.

Loss function: A measure of the discrepancy between predictions and actual (desired) outputs.

Margin: Distance from a separating hyperplane and the closest training example.

Multilayered perceptron (MLP) : A popular neural network architecture with hidden units arranged in a layered topology.

Overfitting: The behavior of a learning algorithm that has very small or no error on the training data but performs poorly on new instances.

Perceptron: Simple binary classifier based on linear separation and a learning algorithm based on iterative correction of classification errors. The perceptron algorithm fails to converge if the training points are not linearly separable.

Radial basis function: Type of internal unit in an artificial neural network that is based on a bell-shaped function like a Gaussian.

Regression: Form of supervised learning where predictions are continuous values.

Softmax: The normalized exponential function. It produces a probability distribution over n discrete values starting from an n-dimensional real vector.

Supervised learning: The problem of inducing from data the association between input patterns and outputs, where outputs represent discrete or continuous predictions.

Support vector machine (SVM): A popular machine learning tool based on two fundamental ideas: the use of a maximal margin separating hyperplane to avoid overfitting, and the use of a kernel function to measure the similarity between two instances.

Transduction: Generalized form of supervised learning where inputs and outputs are associated with the vertices of a graph. In other contexts transduction refers to a special version of the supervised learning problem in which the learner has access to the input portion of the data also for the test instances.

Underfitting: The behavior of a learning algorithm that has large error on both the training data and new instances, due to low capacity.

Unsupervised learning: The problem of inducing from data a suitable representation of the probability density function of patterns.

Vector-based representations: The representation of patterns for a learning problem in the form of real vectors.

9

Applications of Substructure-Based SAR in Toxicology

HERBERT S. ROSENKRANZ

Department of Biomedical Sciences, Florida Atlantic University, Boca Raton, Florida, U.S.A.

BHAVANI P. THAMPATTY

Department of Environmental and Occupational Health, Graduate School of Public Health, University of Pittsburgh, Pittsburgh, Pennsylvania, U.S.A.

1. INTRODUCTION

The increased acceptance of SAR techniques in the regulatory arena to predict health and ecological hazards (1–6) has resulted in the development and marketing of a number of SAR programs (7). The approaches are of optimal usefulness when they are employed as adjuncts to the appropriate

---

The authors have no commercial interest in any of the technologies described in this review.

human expertise. In addition to predicting specific toxicological endpoints, these methodologies, in the hands of an expert, can also be used to gain insight into the mechanistic basis of the action of toxicants and thereby allow a more refined health or ecological risk assessment (8,9).

This review deals with aspects of SAR methodologies that are based upon substructural analyses that are driven primarily by statistical constraints [e.g., MULTICASE (10–12)] as opposed to satisfying predetermined rules [e.g., DEREK, (13–15) ONCOLOGIC (16,17); "Structural Alerts" (18)]. It must, however, be made clear that human expertise is very much involved in most aspects of these knowledge-based substructural methods (8,9). Thus, the inclusion of experimental data into the "learning set" that forms the basis of any SAR model must adhere to previously agreed upon protocols and data handling procedures (Fig. 1). Moreover, prior to SAR modeling, the context in which the resulting model will be used has to be defined as it will affect the manner in which the biological/toxicological activities are encoded and the derived SAR model interpreted.

Thus, it is commonly recognized (7,19) that the induction of cancers in rodents is one of the most challenging phenomena to model by SAR techniques. Yet, bearing in mind the

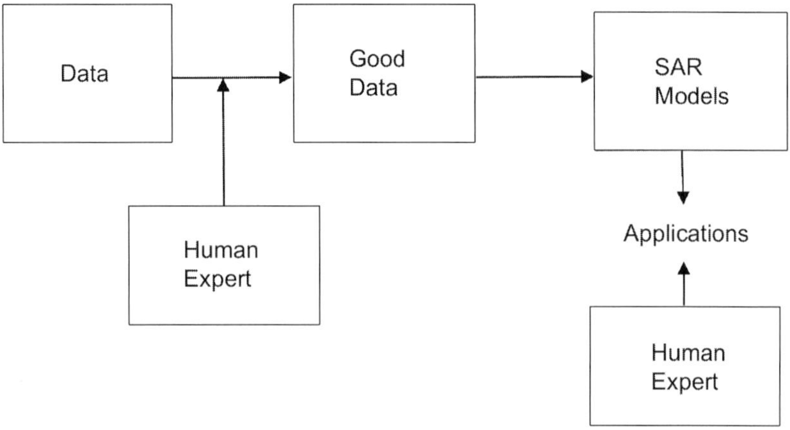

Figure 1 Outline of the SAR approach indicating the interactions with the human expert.

complexity of the phenomenon and the regulatory context in which SAR predictions were to be used, Matthews and Contrera (20) of the U.S. Food and Drug Administration—by encoding the spectrum of activities, i.e., carcinogenicity in male and/or female rats and/or mice and devising rules on how the predictions were to be used—were able to develop a highly predictive MULTICASE SAR model of rodent carcinogenicity. It needs to be stressed that the success in developing the model was primarily the result of the human insight brought by the investigators (20).

2. THE ROLE OF HUMAN EXPERTISE

Substructure-based SAR approaches can handle databases in which activities are expressed categorically, i.e., active, marginally active, inactive, or in a continuous scale. However, it is not always a matter of simply inserting data into the model. Thus, the database for the induction of unscheduled DNA synthesis is indeed categorical (21) and allows the derivation of a coherent SAR model (22). On the other hand, the *Salmonella* mutagenicity database generated under the aegis of the U.S. National Toxicology Program (23) requires insight into how to express activities with respect to SAR modeling. Essentially, in that data set, each chemical is reported with respect to its ability to induce mutations in five *Salmonella typhimurium* tester strains in the presence or in the absence of several postmitochondrial activation mixtures (S9) derived from rats, mice, and hamsters induced or uninduced with the polychlorinated biphenyl mixture Aroclor 1254 (24). Each of the tester strains has a different specificity with respect to its response to mutagens. Moreover, the exogenous S9 mixtures may contain different levels of cytochrome P450 activating and deactivating enzymes which may act on the test chemical and/or its metabolites. If the purpose for deriving a SAR model is to understand the basis of the mutagenicity of a class of chemicals, then the *Salmonella* strain that is the most responsive to that chemical class should be used [e.g., the mutagenicity of nitrated polycyclic aromatic hydrocarbons should be studied in *Salmonella typhimurium* TA98 in the absence of S9 (25–27)]. Similarly, if the aim is to understand the differences in mutagenicity in tester strains that respond to base substitution mutations vs. those that respond to frameshift mutations as a result of covalent binding to a DNA base, then one might model separately and then compare, for example, the responses of aromatic amines in *Salmonella* strains TA98 and TA100 in the presence of S9 (28). In such instances, for SAR modeling, the human expert would select the specific mutagenic potency (e.g., revertants/nmole/plate) reported for each chemical for the specific strain with or without S9. Moreover, based upon personal knowledge of the system and the specific class of chemicals, the expert would then have to select a cut-off value between mutagens and marginal mutagens, and between marginal mutagens and non-mutagens. The expert would then be able to derive an equation relating mutagenic potency to an SAR unit scale compatible with the SAR program being used (see below).

If, on the other hand, the purpose of deriving a SAR model is to identify potential "genotoxic" (i.e., mutagenic) carcinogens, which is the class of agents associated with risk to humans (29–33), then one might consider deriving a dozen or more separate SAR models (e.g., TA 100-S9, TA100 +S9, TA 98, TA 1537, etc.) and then devise an algorithm to combine the results of the different models into a single prediction [see Refs. (34) and (35)]. This, however, is a tedious and time-consuming process. Moreover, "genotoxic" carcinogenicity has not been associated with either a response in a specific tester strain or with the mutagenic potency in that strain. Rather, the association is a qualitative one between carcinogenicity and mutagenicity in any of the strains and carcinogenicity in rodents (36). Accordingly, consideration can then be given to the paradigm that a response in any one of the tester strains in the absence or the presence of a single S9 preparation will be sufficient to identify a carcinogenic hazard. Moreover, since different tester strains may respond differently qualitatively as well as quantitatively to individual chemicals, the indications of potencies that are used cannot be continuous. In fact, they must be categorical and the expert may designate specific criteria for defining a mutagen, e.g., twice the spontaneous frequency of mutations and a linear dose–response (37,38).

Depending upon an understanding of the mechanistic/biological basis of activity, there have been variations on the potency metrics. Thus, the Carcinogen Potency Data Base (CPDB) (39) reports results as TD_{50} values, i.e., the daily dose that in a lifetime study will permit 50% of the treated animals to remain tumor-free. The TD_{50} value is reported as mg/kg/day (39–41). However, given the widespread range in molecular weights of the chemicals in a data set (e.g., dimethylnitrosamine and benzo(a)pyrene, molecular weights 74 and 252 Da, respectively), for SAR studies that measure needs to be transformed into mmol/kg/day in order to yield a meaningful SAR model and the associated generation of "modulators" (see below) that affect the potency of the SAR projection.

The human expert has to make a further decision: the definition of a "marginal carcinogen" and a "non-carcinogen." Should only chemicals inducing no cancers even at the maximum tolerated dose (42–44) be considered non-carcinogens or should there be a cut-off dose, above which even if tumors are induced, they would not be considered biologically or toxicologically significant given the high dose needed? This would reflect Paracelsus' dictum "that it is the dose that makes the toxin" (45).

For the purpose of SAR modeling of CPDB, we chose cut-off values of 8 and 28 mmol/kg/day between carcinogens and marginal carcinogens, and between marginal carcinogens and non-carcinogens, respectively. Based upon the characteristics of the MULTICASE SAR methodology wherein SAR units ≤ 19 indicate non-carcinogenicity; 20–29 marginal carcinogenicity; and ≥ 30 carcinogenicity, this led to the relationship

$$\text{SAR activity} = (18.328 \log 1/TD_{50}) + 46.55 \qquad (1)$$

On the other hand, the rodent carcinogenicity database generated under the auspices of the NTP has been classified according to its spectrum of activities (29). The reason for that classification is derived from the realization that agents that are carcinogenic at multiple sites of both genders of rats and mice are generally found to be “genotoxic” (i.e., possess mutagenicity and/or structural alerts for DNA reactivity) (29,30). These characteristics are associated with a greater carcinogenic risk to humans than chemicals that are restricted to inducing cancers in a single tissue of a single gender of a single species (29,33).

That spectrum of carcinogenicity can be captured by having the scale of carcinogenic activities (i.e., SAR units) reflect it, i.e., 10 SAR units for non-carcinogens; 20 for “equivocal” carcinogens; 30 for chemicals carcinogenic at only a single site in a single sex of a single species; 40 for chemicals carcinogenic at a single site in both sexes of one species; 50 for chemicals carcinogenic to a single species but at two or more sites; and 60 for chemicals carcinogenic to mice and rats at one or more sites (46).

Because the spectrum of activities as well as the potencies reflect different aspects of the carcinogenic phenomenon, algorithms were developed to combine the results of the different SAR models of rodent carcinogenicity into a single prediction model (34,35). Although the approach used heretofore is a Bayesian one (47), there is no reason to suppose that other approaches (neural networks, genetic algorithm, rule learners) are not equally effective (e.g., see Refs. 48,49).

Obviously, this integrative approach is not restricted only to SAR models of rodent carcinogenicity. They could include projections obtained with other SAR models related to mechanisms of carcinogenicity, i.e., SAR projections of carcinogenicity combined with the prediction of the in vivo induction of micronuclei (50) and of inhibition of gap junctional intercellular communication (51). Finally, the same approach can be explored to combine SAR projections with the experimental results of surrogate tests for carcinogenicity (e.g., induction of chromosomal aberration and of mutations at the $tk^{+/-}$ locus of mouse lymphoma cells). Finally, combining the results from different SAR approaches, e.g., knowledge-based (e.g., MULTICASE) with rule-based [e.g., DEREK (13–15) or ONCOLOGIC] (16,17) is a promising avenue that is worthy of further investigation.

The point of the above examples is that human familiarity with an expertise in the biological phenomenon under investigation is essential for the maximal utilization of SAR techniques.

Another instance in which human expertise was essential for the development of a coherent SAR model involves allergic contact dermatitis (ACD) in humans. In that endeavor, initial human insight was needed at several crucial steps:

1. The recognition that in spite of common practice and assumption, human and guinea pig ACD data were not equivalent and could not be pooled to develop a coherent SAR model (52).
2. That the inclusion of "case reports" among experimentally determined human ACD data degraded the performance of the SAR model unless the number of independent "case reports" was greater than 7 (53).
3. That an ACD response calibration based upon the challenge dose, the extent of the response, and the proportion of responders among challenged humans had to be developed to provide a potency scale (54).

When these pre-SAR processing experimental data handling procedures were resolved, a coherent and highly predictive SAR model of human ACD was developed (54). But again, it required the participation and collaboration of experimental immunologists and SAR experts.

The same considerations entered in developing other models, e.g., human developmental toxicity which depended upon: (1) the acceptance of the results of an expert consensus panel, and (2) the rejection of results of borderline significance (55). Of course, it was also the reason for the success of the development of the aforementioned highly predictive SAR model of rodent carcinogenicity by Matthews and Contrera (20).

3. MODEL VALIDATION: CHARACTERIZATION AND INTERPRETATION

Irrespective of the SAR paradigm employed, knowledge and understanding of the performance of the resulting SAR model is crucial to its deployment. This is especially so as no SAR model is perfectly predictive. Yet, understanding a model's limitations is needed if it is to be used and interpreted. The most widely accepted measure of a model's performance is the concordance between experimentally determined results and SAR-derived predictions of chemicals external to the model. This parameter, in turn, is a function of a model's sensitivity (correctly predicted actives/total actives) and specificity (correctly predicted inactives/total inactives).

The most direct and preferable approach to determine these parameters is to randomly remove from the learning set a number of chemicals to be used as the "tester set." The remaining chemicals can be used to develop the SAR model. The resulting models' predictivity parameters and their statistical significance can then be determined by challenge with this external "tester set."

However, most frequently that approach cannot be taken with respect to SAR models describing toxicological phenomena. This derives from the fact that the performance of a SAR model depends upon its size (i.e., the number of chemicals in the database) (10,56–58). For most databases of toxicological phenomena, there is a paucity of experimental results for chemicals. Accordingly, the predictive performance of the model will be negatively affected by removal of chemicals to be used as the external "tester set." Because of this consideration, cross-validation and "leave-out one" approaches have been used (59). Thus, it has been demonstrated that the iterative random removal of chemicals (e.g., 5% of the total) and using the remaining ones (i.e., 95%) as the learning set and repeating the process (e.g., 20 times for a 5% removal), and determining the cumulative predictivity parameters are an acceptable approach (59).

In most substructure-based SAR approaches, the significant structural determinant (e.g., biophores and toxicophores) identified will be a substructure enriched among active chemicals. Accordingly, the presence of the toxicophore is associated with a probability of activity and a baseline potency (Table 1; Fig. 2).

While biophores/toxicophores are the significant as well as the principal determinants of biological and toxicological activity, toxicologists as well as health risk assessors are well aware that not all chemicals in a certain chemical class are toxicants even though the majority may be. Thus, only 83.3% of nitroarenes tested are *Salmonella* mutagens and only 74.4% of chloroarenes tested are reported to be rodent carcinogens (60). This situation is reflected in the fact that only 74% of the chemicals containing the toxicophore NH_2–c–cH= (Fig. 2) are rodent carcinogens. The question then arises whether SAR approaches can be used to explain this dichotomy as well as to provide a basis for the difference in projected potencies. In MULTICASE SAR, this discrimination is provided by modulators (10–12). Thus each biophore/toxicophore is associated with a probability of activity and a basal potency. For the illustration in Fig. 2, the presence of the toxicophore is associated with a 75% probability of carcinogenicity and a potency of 50.3 SAR units. Based upon Eq. (1), 50.3 SAR units correspond to a TD_{50} value of 0.62 mmol/kg/day. In MULTICASE, each biophore/toxicophore may be associated with a group of modulators (Table 2) which determine whether the potential for activity is realized and, if so, to what extent. Modulators are primarily structural elements that can either increase (Fig. 3), decrease (Fig. 4), or abolish (Fig. 5) the potential potency associated with a toxicophore. Additionally, the potential of a toxicophore can be negated by the presence in the molecule of deactivating moieties that are derived from inactive molecules in the data set. The latter are not associated with chemicals that are at the origin of the toxicophore (e.g., Fig. 6).

In addition to being substructural elements, modulators may also be physical chemical or quantum chemical in nature. Thus, the rat-specific carcinogenic toxicophore associated with the activity of the chloroaniline derivative shown in Fig. 7, which defines a non-genotoxic rat carcinogenic species,

Table 1 Some of the Major Toxicophores Associated with Rodent Carcinogenicity: Non-congeneric Data Base

	Number of				
Toxicophore 1–2–3–4–5–6–7–8–9–10	Fragments	Inactives	Marginals	Actives	Number
NH_2–c=cH–	65	15	3	47	1
NH–C=N–	9	1	0	8	2
[Cl–]⟨–4.0A–⟩ [Cl–]	21	2	0	19	3
CH_2–N–CH_2–	29	7	0	22	4
O–CH=	7	0	0	7	5
N–C=	5	0	0	5	6
O–C=	14	1	0	13	7
O^–CH_2–	6	0	0	6	8
Br–CH_2–	5	0	0	5	9
cH=cH–c=cH–cH=⟨3–Cl⟩	14	3	0	11	10
PO–O	11	1	0	10	11
CH_3–N–c=cH–⟨2–CH_3⟩	6	1	0	5	12
cH=c–cH=cH–c <=⟨2–NH⟩	6	1	1	4	13
Cl–CH_2–	26	4	1	21	14
c.″–CO–c.=	7	0	0	7	15
NO_2–C=CH–	14	0	0	14	16
cH=c–cH=cH–c=⟨2–CH_3⟩	5	0	0	5	17
CH_3–C=cH–cH=cH–	7	0	1	6	18

Toxicophore no. 1 is shown in Figs. 1–6, 18, and 19, no. 17 in Fig. 18.
"c" and "C" refer to aromatic and acyclic atoms, respectively; c. indicates a carbon atom shared by two rings; O^ indicates an epoxide; c″ indicates a carbon atom connected by a double bond to another atom. ⟨3–Cl⟩ indicates a chlorine atom substituted on the thrid non-hydrogen atom from the left. ←4.0 A→ indicates a 2-D 4 Angstrom distance descriptor.
In toxicophore no. 18, the second carbon from the left is shown as unsubstituted. This means that it can be substituted with any atom except hydrogen. On the other hand, for this toxicophore, the last carbon on the right is shown with an attached hydrogen. This means it cannot be substituted by any other atom but hydrogen. Finally, in toxicophore no. 10, the third non-hydrogen atom from the left is shown as unsubstituted. It can only be substituted by a chlorine atom.

```
The molecule contains the Toxicophore   (nr.occ.= 1):

  (A)        NH2 -c
                  \\
                   cH

   *** 48 out of the known  65 molecules ( 74%) containing such a Toxicophore
       are  Rodent  Carcinogens                     (conf.level=100%)


   *** QSAR Contribution :                         Constant is     50.28
                                                                  ------
   ** Total projected QSAR activity                                50.28

   *** The probability that this molecule is a Rodent Carcinogen is  75.0%  **

   ** The projected carcinogenic potency is   50.3     SAR units **
```

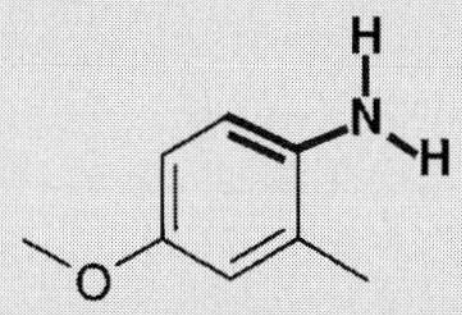

A

Figure 2 Prediction of the carcinogenicity in rodents of *m*-cresidine. The presence of toxicophore A is associated with a 75% probability of carcinogenicity and a basal potency of 50.3 SAR units which corresponds to a TD_{50} value of 0.62 mmol/kg/day [see Eq. (1)].

is modulated by -9^* (water solubility of the chemical). In effect, this can be interpreted (see legend to Fig. 7) that the greater the lipophilicity (i.e., the lower the water solubility) of a chemical containing that toxicophore, the greater its carcinogenic potency. Mechanistically, this may reflect that lipophilicity increases residence time in body tissues (e.g., storage in adipose tissues) and thus augments the effective dose.

An understanding of the nature of the toxicophores and associated modulators can provide insight regarding the mechanistic basis of the toxicity (see below). This knowledge can also be used to modify the chemical's structure in order to decrease or abolish the unwanted toxic effects inherent in

Table 2 Modulators Associated with the Toxicophore NH_2–c=cH–

1–2–3–4–5–6–7–8–9–10		OSAR	Number
2D [N–] ⟨–2.6A–⟩ [N=]		29.1	1
CO–NH_2		28.6	2
N=CH–C=		–18.9	3
NH–C=CH–		15.4	4
n=c–cH=	⟨2–NH_2⟩	19.0	5
c=cH–c=c–		–23.8	6
cH=c.–N=C–		–32.3	7
OH–CO–c=c <–		–20.1	8
Cl–c=cH–c <=		–23.2	9
cH=cH–c=c <–	⟨3–CH_3⟩	12.1	10
cH=c–cH=cH–c <=		17.7	11
CH_3–O–c=cH–c <=	⟨3–c=⟩	–0.7	12
CH_3–O–c=cH–c <=cH–		–0.7	13
NH_2–c=cH–cH=c–NH–		–20.1	14
NH_2–c=cH–cH=c–NH2		–25.5	15
NH_2–c=cH–cH=c–cH=		25.1	16
NH_2–c=cH–cH=cH–c=		–35.3	17
OH–CO–c=c <–	⟨3–cH=⟩	–5.0	18
OH–CO–c=c–cH=		–5.0	19
OH–CO–c=c <–cH=		–5.0	20
OH–CO–c=c <–cH=	⟨3–CH=⟩	–5.0	21

These modulators are associated with toxicophore no. 1 of Table 1 (i.e., non-congeneric database). Modulator no. 11 is shown in Fig. 3; no. 9 in Figs. 4 and 5; no. 6 in Fig. 5; no. 11 in Fig. 18.

For an explanation of the significance of the structural moieties, see the legend to Table 1.

a beneficial molecule without affecting the latter (also see below).

In addition to identifying toxicophores, MULTICASE also has the capability of identifying substructures that, although not statistically significant, may be indicative of biological or toxicological activity (Fig. 8). Such structures should be scrutinized by the human expert to determine whether they are relevant to a carcinogenic potential. Such an examination should include a search of databases to determine whether other chemicals containing that substructure are endowed with that or related potentials. An in-depth study of these “unique” structures is especially appropriate if it is

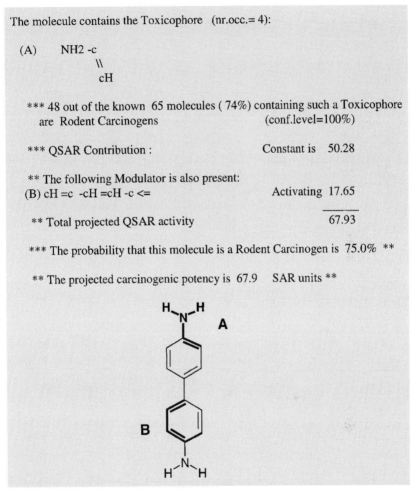

```
The molecule contains the Toxicophore   (nr.occ.= 4):

  (A)      NH2 -c
                  \\
                   cH

   *** 48 out of the known  65 molecules ( 74%) containing such a Toxicophore
       are  Rodent Carcinogens                              (conf.level=100%)

   *** QSAR Contribution :                              Constant is    50.28

   ** The following Modulator is also present:
  (B) cH =c  -cH =cH -c <=                              Activating  17.65
                                                                   -------
    ** Total projected QSAR activity                                 67.93

   *** The probability that this molecule is a Rodent Carcinogen is  75.0%  **

    ** The projected carcinogenic potency is  67.9     SAR units **
```

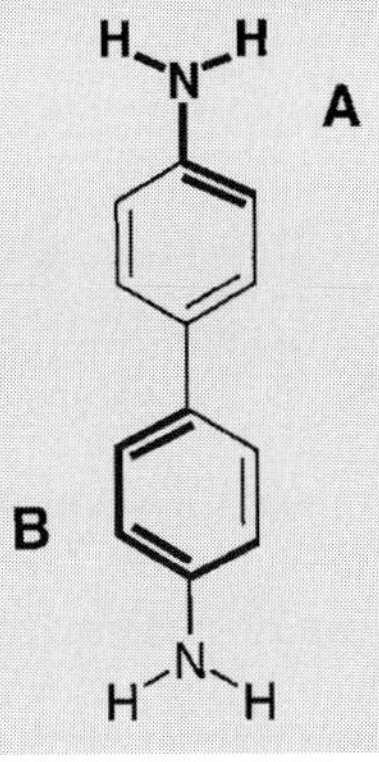

Figure 3 Prediction of the carcinogenicity in rodents of benzidine. The basal potency associated with toxicophore A (i.e., 50.3 units) is augmented by the presence of modulator B. The projected potency of 67.9 SAR units corresponds to a TD_{50} value of 0.07 mmol/kg/day.

derived from chemicals possessing great potency, e.g., tetrafluoro-*m*-phenylenediamine with a TD_{50} value of 0.50 mmol/kg/day (Fig. 8).

One of the characteristics that differentiates SAR methods used in drug discovery from those used in toxicology

```
The molecule contains the Toxicophore   (nr.occ.= 2):

    (A)     NH2 -c
                  \\
                   cH

  *** 48 out of the known  65 molecules ( 74%) containing such a Toxicophore
      are Rodent Carcinogens                          (conf.level=100%)


  *** QSAR Contribution :                          Constant is     50.28
   ** The following Modulator is also present:

    (B)   Cl -c  =cH -c <=                              Inactivating    -23.21
                                                                  ______
   ** Total projected QSAR activity                                27.07

  *** The probability that this molecule is a Rodent Carcinogen is  75.0%  **

   ** The projected carcinogenic potency is  27.1     SAR units **
```

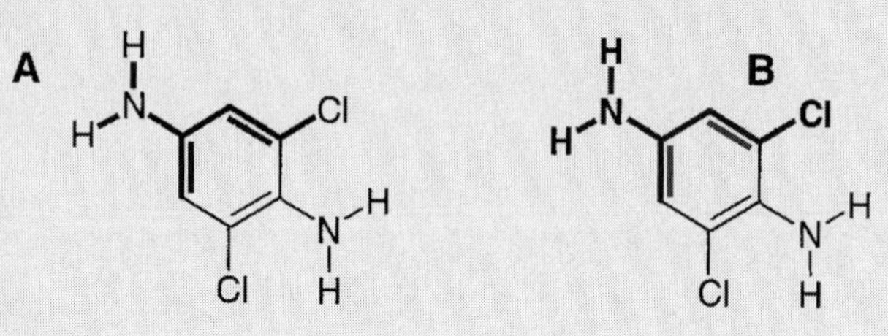

Figure 4 The projected marginal potency of 2,6-dichloro-*p*-phenylenediamine. The carcinogenic potency inherent in toxicophore A is greatly decreased by modulator B. A carcinogenic potency of 27.1 SAR units corresponds to a TD_{50} value of 11.5 mmol/kg/day. That potency is defined as "marginal" (see text).

derives from the fact that the former deal primarily with congeneric chemicals while the latter are concerned with noncongeneric ones. This is reflected by the fact that in medicinal chemistry one is most frequently dealing with a specific receptor or the active site of an enzyme (9). On the other hand, with respect to toxicological phenomena, the same endpoint can arise as a result of a multitude of pathways and can be caused

```
The molecule contains the Toxicophore   (nr.occ.= 2):

    (A)    NH2 -c
                 \\
                  cH

    *** 48 out of the known 65 molecules ( 74%) containing such a Toxicophore
        are Rodent Carcinogens                              (conf.level=100%)

    *** QSAR Contribution :                          Constant is      50.28
     ** The following Modulators are also present:
    (B) c  =cH -c  =c  -                             Inactivating   -23.77
    (C) Cl -c  =cH -c <=                             Inactivating   -23.21
                                                                   ------
     ** Total projected QSAR activity                                 3.31

    *** The probability that this molecule is a Rodent Carcinogen is  75.0%  **
```

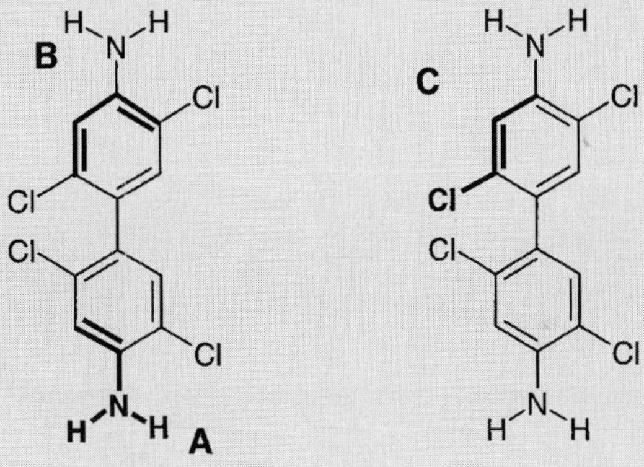

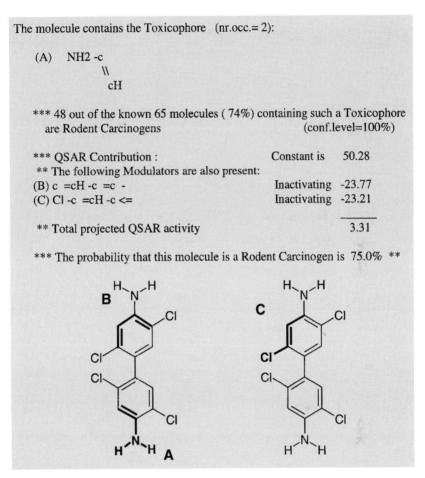

Figure 5 The prediction of the lack of carcinogenicity of 2, 2′, 5, 5′-tetrachlorobenzidine. Although the presence of toxicophore A endows the molecule with carcinogenic potential, the presence of the inactivating modulators B and C abolishes it.

by many different classes of chemicals (e.g., carcinogenesis, development toxicity). Given that SAR methods used in toxicology must be able to handle many different chemical classes within a single data basis, it is essential that the method must also be able to identify chemical structures that do not fall within the domain shared by chemicals that give rise to a common toxicophore. MULTICASE accomplishes this in two

```
The molecule contains the Biophore   (nr.occ.= 1):

              NH2 -c
     (A)            \\
                    cH

   *** 48 out of the known 65 molecules ( 74%) containing such a toxicophore
       are  Rodent carcinogens                    (conf.level=100%)

   *** QSAR Contribution :                            Constant is    50.28
                                                                     ------
    ** Total projected QSAR activity                                 50.28

    ** The molecule contains the DEACTIVATING Fragment:

              CO  -c         cH
                   \\       //
     (D)             c < -cH


   *** The probability that this molecule is a Rodent Carcinogen is 62.5%  **
```

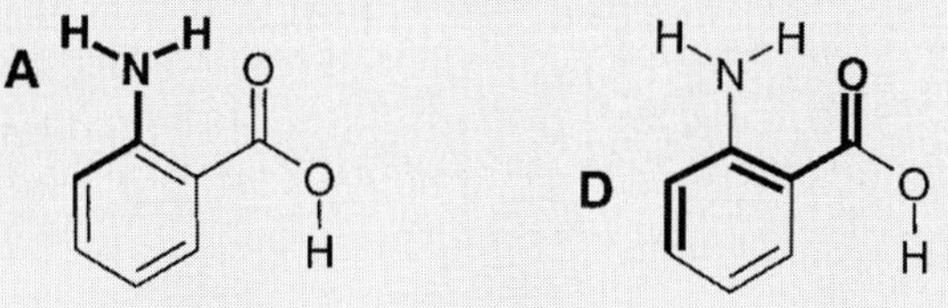

Figure 6 The projected lack of carcinogenicity of anthranilic acid. The carcinogenic potential associated with toxicophore A is negated by a deactivating moiety D derived from non-carcinogens external to the molecules associated with the toxicophore.

ways: (a) by identifying differences in the molecular environment, and (b) by recognizing ("unknown") structures that are not present in the learning set under investigation.

The presence of "unknown" moieties may be recognized in molecules that contain recognized toxicophores. In that situation, they have the potential of being modulators which

The molecule contains the toxicophore (nr.occ.= 1):

```
          cH"
             \
Cl -c          c <
     \\       //
      cH  -cH
```

*** 5 out of the known 5 molecules (100%) containing such a toxicophore are Rat Carcinogens (conf.level= 97%)

*** QSAR Contribution : Constant is 86.99
** The following Modulator is also present:

Water solubility = 2.66; WS contribution is <u>-23.88</u>

** Total projected QSAR activity 63.11

*** The probability that this molecule is a Rat Carcinogen is 85.7% **

** The projected carcinogenic potency is 63.1 SAR units **

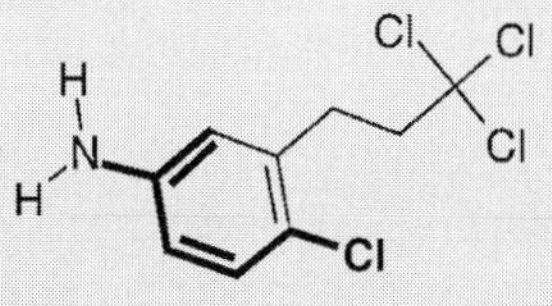

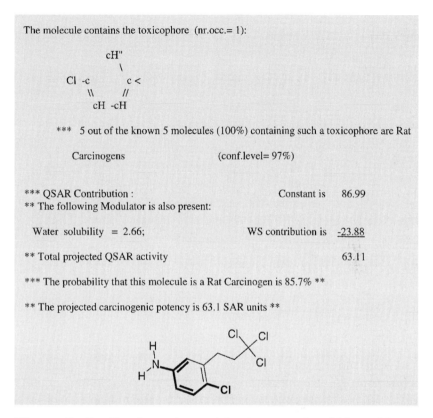

Figure 7 Predicted carcinogenicity in rats of 3-(l,l,l,-trichloro-) propyl-*p*-chloroaniline. The prediction is based on the toxicophore shown in bold. The potency is modulated by (-9^{*} [water solubility]). The potency of 63.1 units corresponds to a TD_{50} value of 0.12 mmol/kg/day. The analogous 3 propyl-*p*-chloraniline has a water solubility of 4.18 (i.e., it is less lipophilic) and this results in a contribution of −37.4 for a projected potency of 49.5 SAR units or a TD_{50} value of 0.54 mmol/kg/day, i.e., the decreased lipophilicity results in decreased carcinogenic potency.

either augment or decrease the potential toxicity. Hence, the presence of such a moiety might introduce an element of uncertainty in the prediction. However, overall, that type of uncertainty is taken into consideration when determining the predictive performance of the model, especially when a cross-validation approach is used.

The molecule also contains the fragment:
(A) F -c =
*** 1 out of the known 1 molecules (100%) containing such a fragment is a Rodent Carcinogen (conf.level= 50%)
*** This fragment is not statistically significant ***

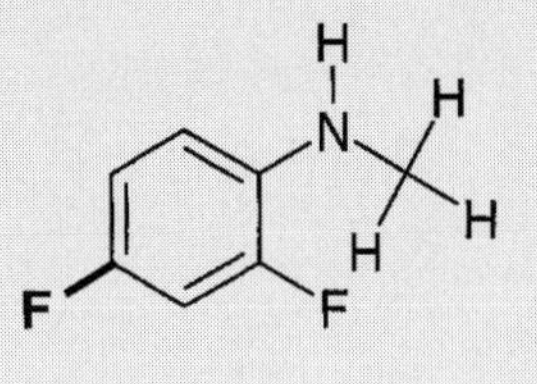

A

Figure 8 The identification of a moiety in 2,4-difluoro-*N*-methylaniline that is present once in the data set. However, the molecule containing it (tetrafluoro-*m*-phenylenediamine) is a carcinogen with a TD_{50} value of 0.50 mmol/kg/day. Accordingly, this *N*-methylaniline derivative must be examined further.

On the other hand, chemicals may be devoid of identifiable toxicophores and still possess an "unknown" moiety (Fig. 9). In that situation, the unknown could possibly be a toxicophore that might endow the molecule with toxicological potential. When faced with such a situation, it is advisable to conduct a search for molecules external to the data set that contains such a moiety and are also devoid of toxicophore to determine whether they have been tested in the same or a related assay system. Thus, for example, the chemical may not have been tested for mutagenicity in *Salmonella*, but it might have been tested for its ability to induce mutation in *E. coli* WP2 *uvr*A or error-prone DNA repair (37,38,61). Methods for determining the relatedness of such assays have been described (47,62). With respect to the molecule shown in Fig. 9, it has been reported that carcinogenic arylamine derivatives when substituted with sulfonates show decreased intestinal absorption and hence abolish carcinogenicity (63–66), thus decreasing the level of concern that the substance in Fig. 9 is a carcinogen.

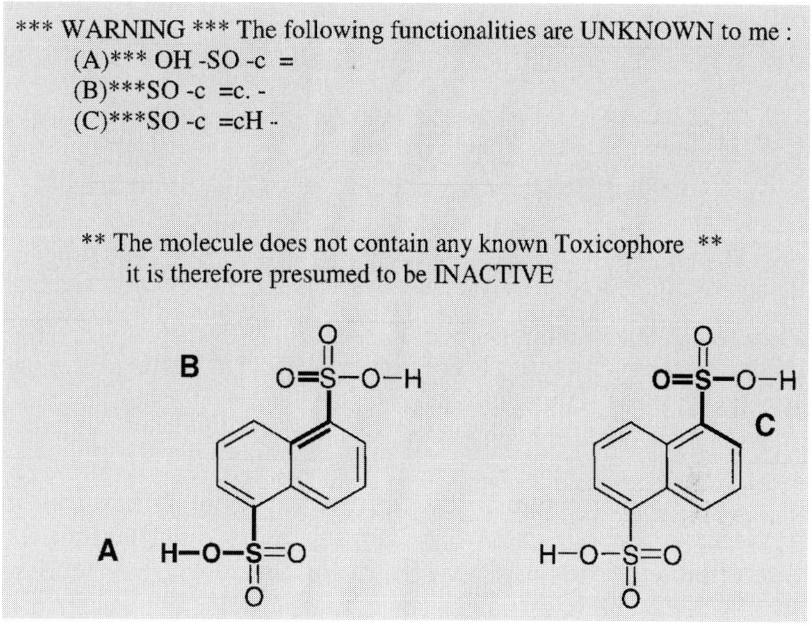
*** WARNING *** The following functionalities are UNKNOWN to me :
(A)*** OH -SO -c =
(B)***SO -c =c. -
(C)***SO -c =cH -
** The molecule does not contain any known Toxicophore **
it is therefore presumed to be INACTIVE
B
O=S–O–H
A
H–O–S=O
O=S–O–H
C
H–O–S=O

Figure 9 Prediction of the lack of carcinogenicity of 1,5 naphthalenedisulfonic acid. However, the prediction has an element of uncertainty because of the presence of the moieties "unknown" to the model. It is known, however, that in other instances the sulfonate moiety facilitates excretion and thereby inhibits carcinogenicity. (From Refs. 63 to 66.)

The identification of differences in the molecular environment is a more subtle exercise. It might derive from the presence of a toxicophore and a warning by the program that in the test substance it exists in a different milieu (Fig. 10). To ascertain the appropriateness of that determination requires the SAR system to be able to provide documentation, i.e., the nature of the chemicals that give rise to the toxicophore. SAR systems that cannot provide that information are at a disadvantage. Thus "human" examination of the difference in environments between the test chemical described in Fig. 10 and the chemicals that gave rise to that toxicophore indicates that indeed the environments are very different (Fig. 11) and the program's determination (Fig. 10) is warranted.

The molecule contains the toxicophore (wrong geometry):

CH2-O -CH2-CH2-

disqualified because of conformational differences

** The molecule is therefore presumed to be INACTIVE**

Figure 10 The prediction of the inability of 18-Crown ether-6 to induce sister chromatid exchanges in vitro. The structure of 18-crown ether-6 is shown in Fig. 11.

On the other hand, the determination of differences in environment reported in Fig. 12 may not be justified as the test chemical, 18-Crown-6 ether, can be biotransformed to an acyclic structure that bears similarities to the structures that gave rise to the toxicophore (Fig. 13). Thus, in this instance, the "human" expert overrules the SAR program's

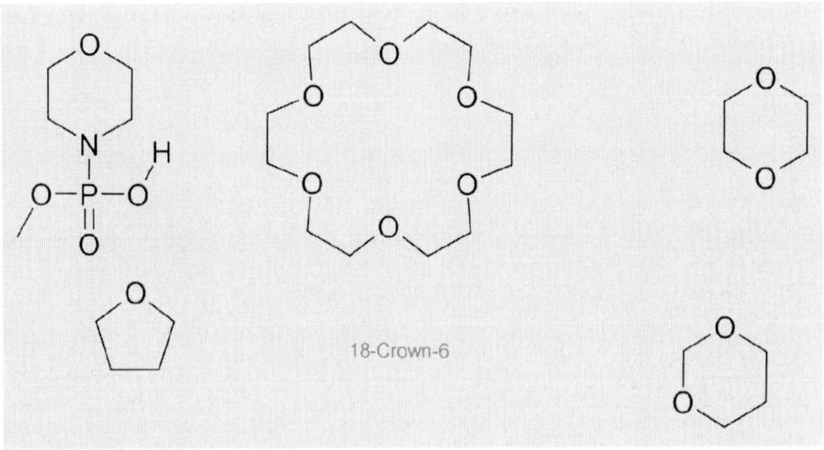

Figure 11 Structures which are the origin of the toxicophore associated with the induction of sister chromatid exchanges (see Fig. 10). The four structures are clearly different from 18-crown-6.

```
The molecule contains the toxicophore  (nr.occ.= 6):

        O  -CH2
             \
              CH2 -O

 ***  7 out of the known 7 molecules (100%) containing such Biophore
      are MLA mutagens with an average activity of  56

 ** This toxicophore exists in a significantly different environment
    than in the data base (i.e. 8.11);  It may not be relevant.

 *** QSAR Contribution :                    Constant is   59.00
                                                          ------
 ** Total projected QSAR activity                         59.00

 *** The probability that this molecule is a MLA mutagen is  88.9%  **

 ** The projected mutagenic potency is 59.0 SAR units **
```

Figure 12 The prediction of the potential of 18-crown ether-6 to induce mutations at the $tk^{+/-}$ locus of mouse lymphoma cells. The structure of 18-crown ether-6 as well as of the seven molecules that gave rise to the toxicophore are shown in Fig. 13. For an explanation of the structure of the toxicophore, see the legend in Table 1.

analysis and confirms the mutagenic potential of that chemical.

Finally, even when the SAR program does not recognize differences in environments, the "human" expert may do so. Thus curcumin is predicted to induce $\alpha_2\mu$-globulin associated nephropathy (67) by virtue of the presence of a toxicophore (Fig. 14), which is present in six molecules of the data set, all of which are inducers of $\alpha_2\mu$-globulin associated nephropathy. The SAR program does not detect a difference in environment (Fig. 14). Yet, a comparison of the molecules in the learning set with curcumin indicates that the molecular environments are quite dissimilar (Fig. 15). In the absence of experimental data regarding the induction of this nephropathy by curcumin or structurally related molecules, the

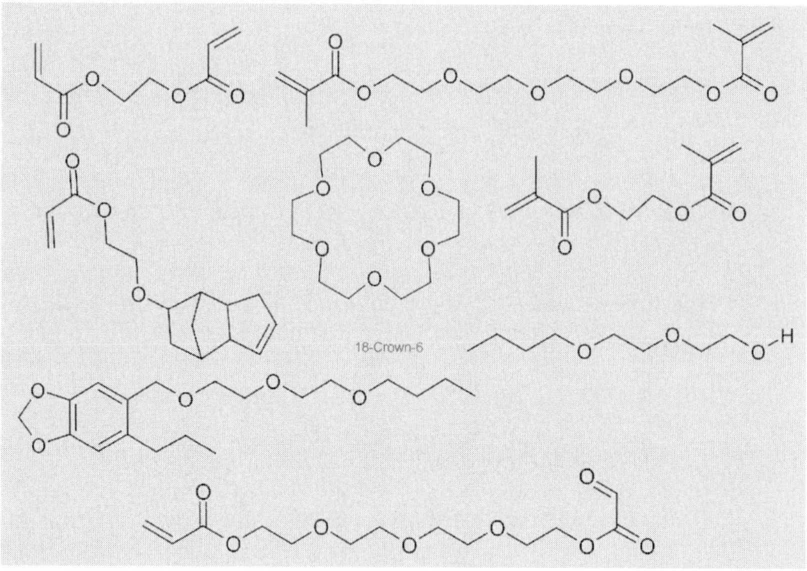

Figure 13 Structures of molecules that are at the origin of the toxicophore associated with the potential to induce mutations at the $tk^{+/-}$ locus of mouse lymphoma cells.

prediction (Fig. 14) is overruled. This illustrates the need to examine the basis of all SAR predictions.

As an additional example, we might examine the predicted carcinogenicity in mice of epitholone A (Fig. 16). Epitholone A, an inhibitor of tubulin polymerization, is a promising cancer chemotherapeutic adjunct that may have the potential to replace Taxol® in situations where tumors have become resistant to Taxol (68,69).

However, examination of the basis of the prediction of carcinogenicity (Fig. 16) indicates that the molecules in the learning set containing that toxicophore all contain other moieties (ammo, hydrazine, nitro) (Fig. 17). Each of these has been associated with carcinogenicity. Epitholone A does not contain any of them. Thus, in this instance, the toxicophore, albeit it is statistically significant, is in fact an artifact. Based upon these analyses, the "human expert" would agree with the SAR model-generated prediction which is accompanied by a warning regarding the "environment." Obviously, in the above

The molecule contains the Toxicophore (nr.occ.= 2):

CO -CH2

*** 6 out of the known 6 molecules (100%) containing such a Toxicophore
are inducers of $\alpha 2\mu$-Nephropathy (conf.level= 98%)

*** QSAR Contribution : Constant is 39.00

** Total projected QSAR activity 39.00

*** The probability that this molecule is an inducer of $\alpha 2\mu$-Nephropathy is 87.5% **

** The projected potency is 39.0 SAR units **

Figure 14 An example of a prediction subsequently overruled. The SAR model predicts that curcumin induces $\alpha_2\mu$-globulin associated nephropathy in male rats. However, a comparison of the structure of curcumin with the structures of the six chemicals at the origin of the toxicophore (see Fig. 15) indicates that they differ significantly. In this instance, the human expert overruled the model's prediction.

examples, the human expertise can only be maximally effective if the SAR method provides the necessary documentation.

As mentioned previously, the predictive performance of an SAR model is dependent upon the size and chemical diversity of the chemicals in the learning set (56–58). It follows that the number of predictions accompanied by "warnings" of the presence of "unknown" moieties will be a function of the size of the learning set (57,58). This relationship can be expressed as the informational content of an SAR model. It is defined as 100 Percent of Predictions Accompanied by "Warnings." In practice, this value is determined by challenging a SAR model with 10,000 chemicals representing the "universe of chemicals" and determining the number of predictions accompanied by such warnings (58). This also identifies the prevalence in the "universe" of moieties absent from the model and suggests that experimental data on such chemicals be identified and the data included in a future model.

Since SAR programs in use in toxicology may consist of prepackaged programs and include specific SAR models,

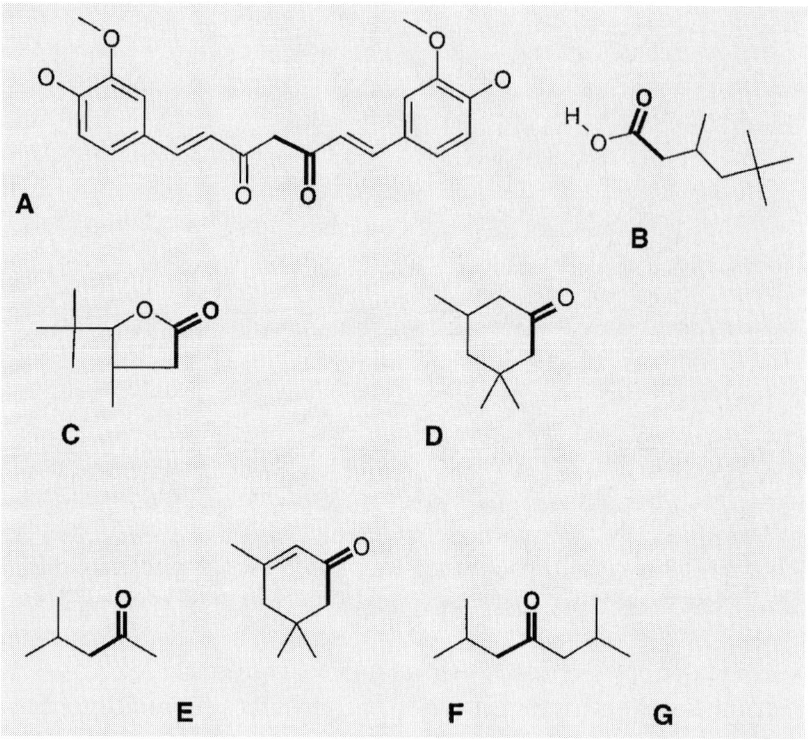

Figure 15 Comparison of curcumin with the structures of chemicals that contain the same toxicophore (see Fig. 14). The toxicophore is shown in bold. A: curcumin; B: 3,5,5-trimethylhexanoic acid (THMA); C: γ-lactone of TMHA; D: 3,5,5-trimethylcyclohexanone; E: methylisobutylketone; F: isophorone; G: isobutyl ketone. Chemicals B–G have been determined experimentally to induce $\alpha_2\mu$-globulin-mediated nephropathy.

there is a tendency among some users not to evaluate further either the SAR paradigm resident therein or the predictive performance of the resultant SAR model. This may negate the usefulness of the methodology, its applicability to a specific situation, and its regulatory acceptance (6). Thus, not only must the predictive performance of a model be known [i.e., concordance between experimental and predicted results; sensitivity and specificity (determined as previously described)], in order to make individual

The molecule contains the Toxicophore

```
S  -C
     \\
       N
      /
  C"
```

*** 5 out of the known 5 molecules (100%) containing such a Toxicophore

are Mouse carcinogens with an average activity of 62.

(conf.level=97%)

** This Toxicophore exists in a significantly different environment

than in the data base (*i.e.* 5.45); It may not be relevant

*** QSAR Contribution :	Constant is	64.00
** Total projected QSAR activity		64.00

*** The probability that this molecule is a Mouse carcinogen is 85.7%

** The projected Mouse carcinogenic potency is 64.0 SAR units **

Figure 16 Prediction of the carcinogenicity in mice of epitholone A. The structure of epitholone A (toxicophore shown in bold) is given in Fig. 17.

predictions, but also in applying the projections to hazard identification purposes or for the purpose of devising rational combinations of SAR models or of a SAR model coupled with certain experimental assays so as to make the exercise meaningful.

Moreover, in order to allow for maximal human input in the analyses, it is not sufficient to receive a message that the test molecule's structure or domain is not fully covered by the model. Even, if the program indicates that the test molecule falls with the domain, this may need verification. Accordingly, the human, expert must know the nature of the chemicals in the learning set, for example, in Figs. 10–17.

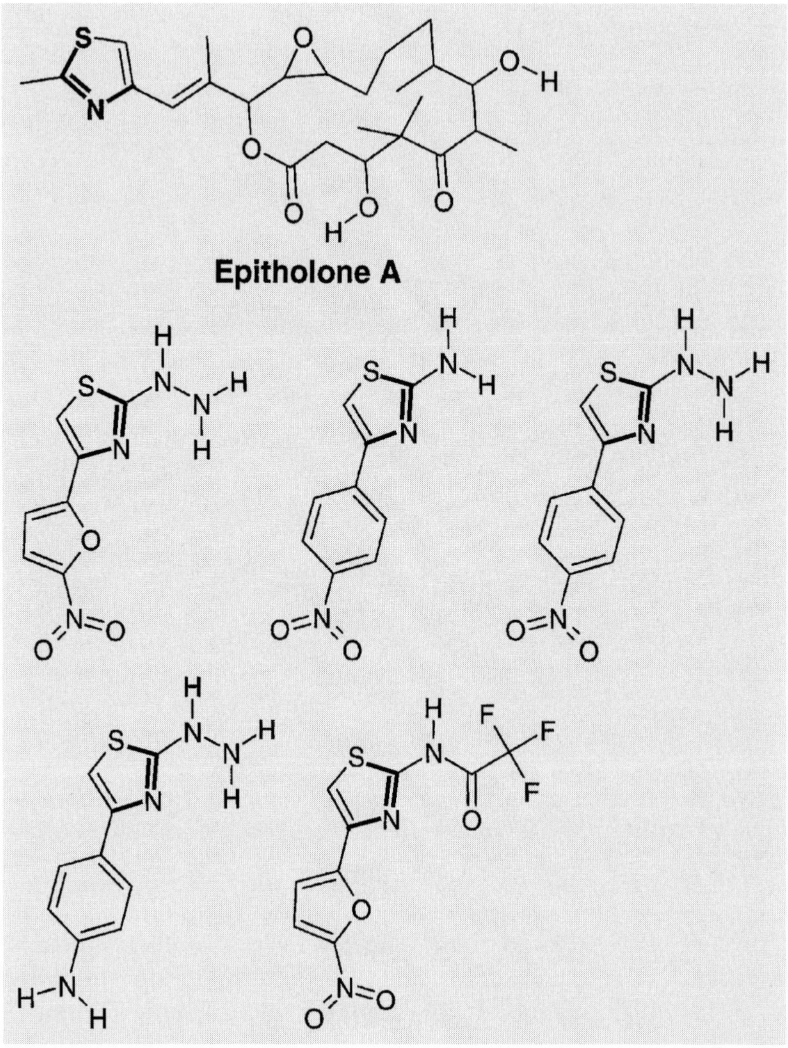

Figure 17 Structures of epitholone A and of chemicals which contain the toxicophore. The toxicophore (see Fig. 16) is shown in bold.

These considerations suggest that for optimal applicability, SAR methods may be most useful if they are used to evaluate one chemical at a time rather than by submitting batches of chemicals. This approach is reinforced when mechanisms of activity (see below) are also considered. The only time batchwise SAR analyses may be warranted is for priority setting but not for regulatory action (70,71).

In addition to being influenced by the number and nature of the chemicals in the learning set, the predictivity of an SAR model is also affected by the ratio of active to inactive molecules in the learning set. Generally, a ratio of unity is optimal (10,72,73). Transparency of the SAR paradigm and knowledge of the default assumptions may provide guidance on the optimal database. Yet, it must be realized that the experimental data used to obtain SAR models, in most instances, were not generated with SAR modeling in mind. Accordingly, most databases may not be fully optional for SAR model development. On the other hand, knowledge of the predictivity parameters of even less than perfect models makes their deployment for SAR analyses feasible. It is also of interest to note that some SAR methods may tolerate significant ambiguity in the experimental results used for model building and still be useful for a purpose such as high throughput screening (74).

4. CONGENERIC VS. NON-CONGENERIC DATA SETS

One of the strengths of the currently available substructure-based SAR approaches is their ability to handle non-congeneric databases (i.e., databases containing a mixture of classes). That is quite appropriate to the modeling of toxicological phenomena. Thus, a phenomenon such as carcinogenesis can be induced by many different chemicals (e.g., nitrosamine, and polycyclic aromatic hydrocarbons) and can proceed by a variety of different individual or sequential pathways. On the other hand, this diversity in causative agents as well as the multiplicity of mechanisms may "dilute" the learning set and result in SAR models of lower predictivity. So one might consider using these substructure-based approaches and applying them to congeneric data sets and possibly improve the predictive performance and refine the structural information to better elucidate mechanisms. This naturally requires a preliminary chemical class classification. Thus, should a chemical such as 4-amino-3-chloro-2-methylphenol be classified as an aminoarene, a chloroarene, a toluene, or a phenol? Such classifications might generate a large number of classes with individual classes containing too few chemicals to obtain meaningful SAR models.

One way of avoiding this problem is to use the toxicophores generated by the SAR model to identify classes that can be used as substrate for further analysis (see also Ref. 75). MULTICASE, for example, did not choose simply the aromatic amine moiety NH_2–C= but rather one with an ortho-unsubstituted NH_2–C=CH moiety (Table 1 and Fig. 2). Normally, that toxicophore is associated with a probability of activity and a basal potency, i.e., 75% and 50.3 SAR units (Fig. 2). Following the identification of the toxicophore, the program identifies modulators which augment, decrease, or abolish the activity associated with the toxicophore (Figs. 3–5 and Table 2).

An alternate approach would be to select the subset of chemicals containing the specific toxicophore and use it to initiate a fresh round of SAR model building. However, before investigating this approach, let us consider the possible advantage of the normative approach of using non-congeneric learning sets. Let us examine, for example, the aromatic amines illustrated earlier. Thus, some chemicals in addition to containing the toxicophore NH_2–C=CH– may contain a second toxicophore derived from non-arylamine-containing carcinogens (Fig. 18). Conceivably, it may be that second toxicophore is in fact the one responsible for the carcinogenic spectrum. Thus, most arylamine carcinogens induce cancers in multiple species and multiple tissues. This property, in addition to their genotoxicity, makes them suspect as human carcinogens (29). However, some arylamines have a much more restricted spectrum of carcinogenicity, i.e., a single tissue of a single gender of a single species (29). This makes them much less likely to be a potential risk to humans (29–31). This more restricted activity may be related to the second toxicophore (which, in fact, may be derived from such non-genotoxic single-species rodent carcinogens). That type of information will not be available when the learning set is

```
The molecule contains the Toxicophore   (nr.occ.= 2):

          NH2 -c
    (A)         \\
                cH

*** 48 out of the known  65 molecules ( 74%) containing such a Toxicophore
    are  Rodent Carcinogens                (conf.level=100%)

*** QSAR Contribution :                              Constant is    50.28
 ** The following Modulator is also present:
    cH =c  -cH =cH -c <=                             Activating  17.65
                                                                 -----
 ** Total projected QSAR activity                                67.93
    The molecule also contains the Toxicophore   :
 (B)    cH =c  -cH =cH -c <=               <2-CH3>
***  5 out of the known  5 molecules (100%) containing such a Toxicophore
       are  Rodent Carcinogens

*** QSAR Contribution :                              Constant is    75.93
 ** The following Modulators are also present:
    cH =c <-cH =cH -c  =cH -                         Inactivating   -2.47
    cH =cH -c  =cH -cH =c <-                         Inactivating   -4.95
    cH =cH -c <=cH -cH =c  -                         Inactivating   -2.47
    Ln Nr.Bi/Mol.Wt.   = -3.98 ;          Nr.Bioph/MW contrib.is   -16.93
                                                                   ------
 ** Total projected QSAR activity                                  49.11

*** The probability that this molecule is a Rodent Carcinogen is
    increased to 83% due to the presence of the extra Toxicophore
 ** The projected carcinogenic potency is  67.9    SAR units **
```

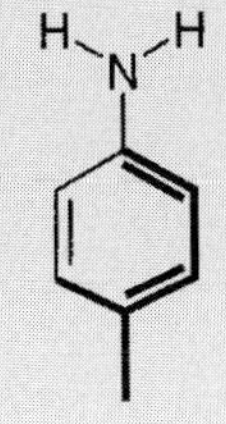

A

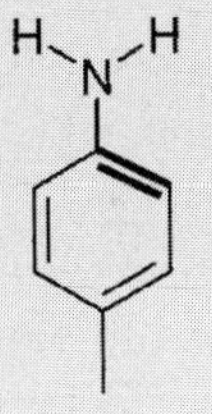

B

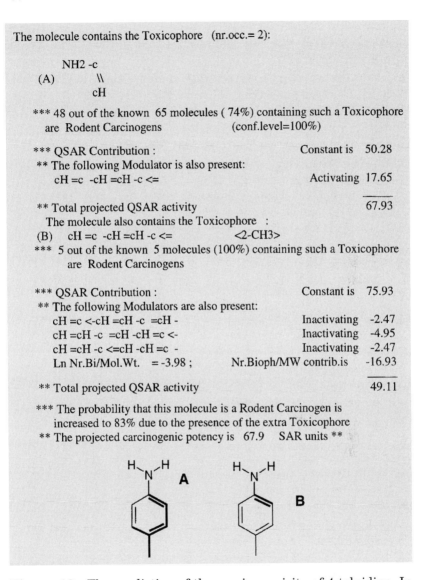

Figure 18 The prediction of the carcinogenicity of 4-toluidine. In addition to toxicophore A, this molecule contains toxicophore B which is derived from five non-arylamine carcinogens. Based upon toxicophore B, the potency is 49.1 SAR units or a TD_{50} value of 0.73 mmol/kg/day; i.e., the potency based upon the second toxicophore is lower. On the other hand, the probability of carcinogenicity has been increased due to the presence of the two toxicophores.

restricted to chemicals selected because they share a common toxicophore and whereby the second toxicophore is no longer apparent.

Similarly, the deactivating moiety that negates the potential activity associated with the NH_2–C–CH= toxicophore (Fig. 6) might not necessarily have been identified where a toxicophore-restricted learning subset was selected. On the other hand, when the non-congeneric data set was used, the resulting model predicted *p*-aminobenzoic acid (pABA) to be a carcinogen (Fig. 19). In all probability, this

The molecule contains the Toxicophore (nr.occ.= 2):

```
        NH2 -c
(A)           \\
               cH
```

*** 48 out of the known 65 molecules (74%) containing such Biophore are Rodent Carcinogens (conf.level=100%)

*** QSAR Contribution : Constant is 50.28

** Total projected QSAR activity 50.28

*** The probability that this molecule is a Rodent Carcinogen is 75.0% **

** The projected carcinogenic potency is 50.3 SAR units **

A
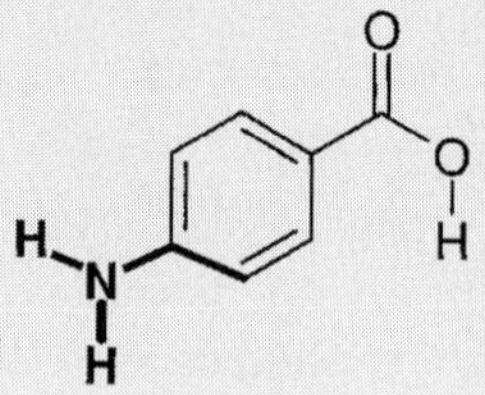

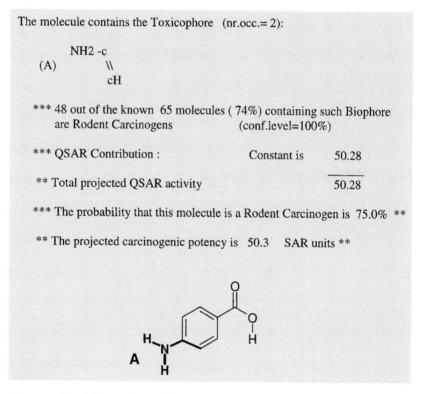

Figure 19 The projected "carcinogenicity" of *p*-aminobenzoic acid based on the non-congeneric SAR model. This physiological chemical is unlikely to be a carcinogen. A projection based upon the congeneric SAR model predicts this chemical to be non-carcinogenic (see text).

physiological vitamin component is unlikely to possess this attribute.

In order to determine whether the use of congeneric chemicals improves the performance of the resulting SAR model, we selected the 65 chemicals identified by MULTICASE (Table 1, toxicophore no. 1; the chemicals are listed in Table 6.4 of Ref. 8) that contain the NH_2–C=CH toxicophore (47 carcinogens, 3 marginal carcinogens, and 15 non-carcinogens) and used them as the learning set for a further MULTICASE model. It is to be noted that since all of the chemicals contain the NH_2–C=CH, it would not be expected that the aromatic amine moiety would be the only major toxicophore. This expectation was realized (Table 3). Thus, the prediction of the carcinogenicity of benzidine that is based upon the non-congeneric SAR model (Fig. 3) identifies the NH_2–c=CH toxicophore as responsible for a 75% probability of carcinogenicity; that model also used a modulator to increase the projected potency to 67.9 SAR units. On the other hand, the prediction based upon the non-congeneric model identified a different toxicophore (Fig. 20) that is associated with a much greater probability of carcinogenicity (i.e., 91% vs. 75% for the non-congeneric model). That is due to the fact that the toxicophore is derived from a population enriched with carcinogens. It is also interesting to note that the potency associated with this toxicophore (i.e., 67.1 SAR units) is close to that found with the non-congeneric model (67.9 SAR units, Fig. 3). The latter, however, depended upon the contribution of a modulator. Furthermore, the toxicophore derived from the congeneric model (Fig. 20) is in fact identical to the modulator associated with the prediction of benzidine based on the non-congeneric model (Fig. 3). This is not entirely unexpected given the MULTICASE paradigm. However, this does not apply to the other toxicophores associated with the congeneric model (i.e., compare Tables 2 and 3).

Interestingly, with this new SAR model pABA was predicted to be a non-carcinogen, i.e., none of the fragments derived from that molecule was a toxicophore. Moreover, there were no warnings of the presence of unrecognized moieties. With respect to anthranilate (Fig. 21), unlike the situa-

Table 3 Some of the Major Toxicophores Associated with Rodeat Carcinogenicity of Arylamines

	Number of				
Toxicophore 1–2–3–4–5–6–7–8–9–10	Fragments	Inactives	Marginals	Actives	Number
c <=cH–cH=c–	20	1	0	19	1
c.=cH–cH=c–	5	0	0	5	2
NH_2–c=c–cH=cH–	21	3	1	17	3
OH–c=c–cH=c–	2	0	0	2	4
CH_3–c=cH–c=	10	1	1	8	5

These toxicophores were obtained as a result of modeling the activity of 65 arylamines that contain the toxicophore NH_2–c=CH– (i.e., congeneric data base).
Toxicophore no. 1 is shown in Fig. 20 and no. 3 is shown in Fig. 21.
For an explanation of the significance of the structural moieties, see legend to Table 1.

The molecule contains the Toxicophore (nr.occ.= 4):

```
        c =cH
             \
(A)           cH
             //
        c
```

*** 19 out of the known 20 molecules containing such a Toxicophore are Rodent Carcinogens (conf.level=100%)

*** QSAR Contribution : Constant is 67.13

** Total projected QSAR activity 67.13

*** The probability that this molecule is a Rodent Carcinogen is 90.9% **

** The projected potency is 67.1 SAR units **

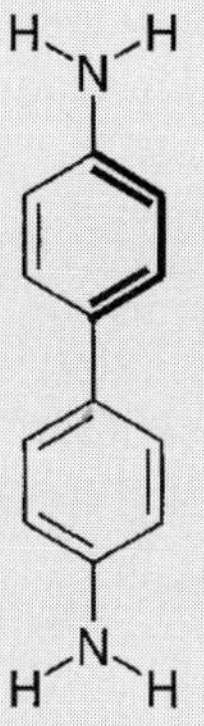

A

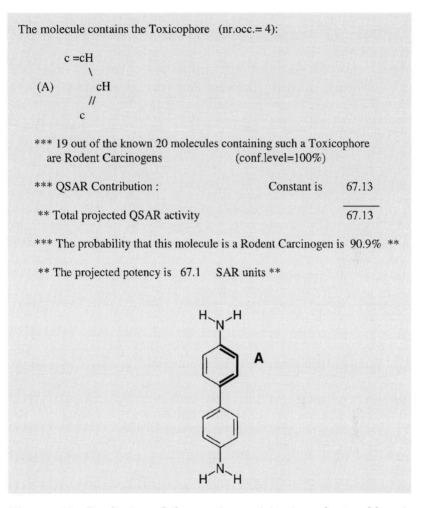

Figure 20 Prediction of the carcinogenicity in rodents of benzidine based upon an SAR model of congeneric arylamines. The toxicophore is shown in bold. A potency of 67.1 SAR units corresponds to a TD_{50} value of 0.08 mmol/kg/day. The probability of carcinogenicity (i.e., 91%) is greater than the prediction (75%) obtained with the non-congeneric model (Fig. 3).

tion with the non-congeneric model (Fig. 6), the program did not recognize deactivating moieties external to the data set. However, based upon the modulators associated with the new toxicophore, the molecule was predicted to be non-carcinogenic in its own right without resorting to the presence of the external deactivating moiety. This was a result of the fact that this toxicophore is associated with a number of inactivating biophores that cause a loss of potency (Fig. 21).

Overall, however, challenging the two models with a group of aromatic amines containing the NH_2–C=CH moiety that are external to the two models showed a concordance of 82.6% for the predictions obtained with the two models.

An in-depth SAR analysis of the carcinogenicity of arylamines optimally should include both types of SAR models, i.e., congeneric and non-congeneric. The latter may reveal alternate mechanisms of carcinogenesis, while the congeneric one, by identifying additional toxicophores (Table 3) that have greater statistical significance than the modulators associated with the parental NH_2–C=CH toxicophore, increases confidence in the prediction (Table 2). Such an approach allows a refinement in the understanding of the structural basis of activity. Moreover, by predicting pABA as inactive, it provides reassurance regarding the predictivity of the congeneric model. Moreover, it is possible to combine the outputs of the two models into a single prediction (62).

5. COMPLEXITY OF TOXICOLOGICAL PHENOMENA AND LIMITATIONS OF THE SAR APPROACH

A single toxicological phenomenon may often occur as a result of a series of independent and/or sequential events. In essence, this may have the net effect of having to model a series of separate phenomena using a single database. Thus, carcinogenicity may arise as a result of a somatic mutation induced by an electrophile; mitogenesis secondary to a toxic insult; tumor promotion by a variety of agents, some of which are receptor-mediated; and a variety of other mechanisms that are homeostatic or genetic in nature. When results obtained with agents that induce cancers by these various mechanisms are pooled into a single database, as is the practice, the question arises whether the complexity of the phenomenon may

```
The molecule contains the Toxicophore   (nr.occ.= 1):

              NH2 -c           cH
      (A)              \\     //
                       c -cH

  *** 17 out of the known  21 molecules ( 81%) containing such a Toxicophore
      are Rodent Carcinogens          (conf.level= 98%)
  *** QSAR Contribution :                                   Constant is   36.96
   ** The following Modulators are also present:
        CO -c  =c <-                                    Inactivating  -1.14
        cH =cH -c >=c <-                                Inactivating  -1.14
        CO -c  =c  -cH =                                Inactivating  -1.14
        CO c  =c  _NH2                                  Inactivating  -1.14
        CO -c  =c <-cH =                                Inactivating  -1.14
        CO -c  =cH -cH =                                Inactivating  -1.14
        cH =c >-c  =cH -                 <3-CO >        Inactivating  -1.14
        cH =cH -c  =c <-                 <3-CO >        Inactivating  -1.14
        NH2-c  =c  -cH =                 <3-CO >        Inactivating  -1.14
        CO -c  =c  -cH =                 <3-NH2>        Inactivating  -1.14
        cH =cH -c >=c <-cH =                            Inactivating  -1.14
        NH2-c  =c >-cH =cH -                            Inactivating  -1.14
        cH =cH -c  =c <-cH =             <3-CO >        Inactivating  -1.14
    (B) NH2-c  =c  -cH =cH -             <3-CO >        Inactivating -14.76
    Log partition coeff.=  0.91 ;                   LogP contribution is   4.32
                                                                        ------
   ** Total projected QSAR activity                                      11.75

  *** The probability that this molecule is a rodent carcinogen is   0.0%  **
   ** The projected potency is 11.7 SAR units **
```

A

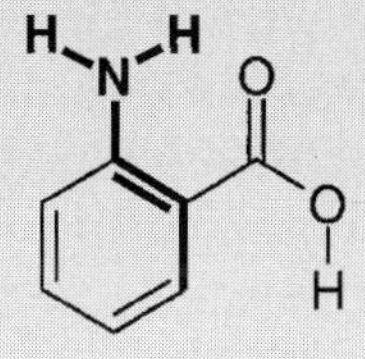

B

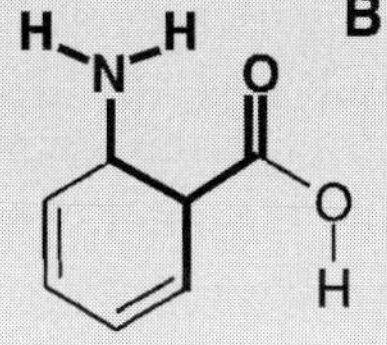

Figure 21 Prediction of the lack of carcinogenicity of anthranilic acid based upon a congeneric SAR model The 81% probability of activity and the 37 SAR units of basal potency are not realized due to the presence of inactivating modulators, one of which (B) is shown. These include an inactivating contribution due to the octanol:water partition coefficient. Thus, the probability is reduced to 0% and the potency to 11.7 SAR units (equivalent to 80 mmol/kg/day) which is considered non-carcinogenic.

overwhelm the methodology's ability to derive a coherent SAR model. It has been observed that a model's predictivity is influenced by the complexity of the phenomenon it seeks to model (10). Knowing that a model's performance is a function of its size, and realizing that we are in fact including multiple phenomena and/or pathways into a single SAR model, we could increase the number of experimental determinations in the expectation that we would increase the number of chemicals representing each contributing mechanism. However, a single cancer bioassay performed by currently accepted protocols may cost $4 million and requires 3 years to complete. Moreover, societal concerns regarding the welfare of animals would not permit such a use of animal resources and certainly not to improve the predictivity of SAR models. Thus, there is a need to explore other approaches to understand the limits of a SAR models. There are, in fact, other approaches to determine whether a toxicological phenomenon is at the limit of the informational content of an SAR method's resolution. One can mix, for example, databases describing rodent carcinogenicity and the induction of sensory irritation in mice, develop a single SAR model from the combined data set, and challenge it with external tester sets of carcinogens/non-carcinogens, and sensory irritants/non-irritants to determine the ability of the combined model to discriminate between these phenomena (76).

Using such an approach with respect to MULTICASE, it was demonstrated that there was sufficient reserve within the method and the currently available databases to model fairly complex phenomena (e.g., mutagenicity, allergic contact dermatitis). In fact, the system has the capacity to model phenomena twice (but not thrice) as complex as those currently modeled (76). Thus, it would be feasible, when investigating a toxicological phenomenon, to perform a similar exercise provided the SAR methodology allows the operator the option to input databases.

Of course, as mentioned earlier, there are other approaches to improve the predictive performance of SAR models, e.g., by a thorough calibration of the input data, such as was done by Matthews and Contrera (20), by combination of different SAR models describing different facets of a phenomenon (e.g., SAR models of rodent carcinogenicity, of unscheduled DNA synthesis and of the induction of chromosomal aberration), by combining SAR models that describe the same phenomena but use different approaches [e.g., ONCOLOGIC (16,17) and MULTICASE] or by combining the projection of SAR models with experimental results obtained with surrogate tests (e.g., a SAR model of carcinogenicity and the results of tests for the in vivo induction of micronuclei). There are a number of protocols for combining such results: rule makers (48,49), neural networks, genetic algorithms, and Bayesian approaches. We have obtained good results with the latter (34,35,47).

6. MECHANISTIC INSIGHT FROM SAR MODELS

Analysis of the nature of the toxicophores obtained from non-congeneric databases has shown that the prevalence of common toxicophores, i.e., identical toxicophores (e.g., CH_2–CH_2–Cl vs. CH_2–CH_2–Cl), or embedded toxicophores (e.g., CH_2–CH_2–CH_2–Cl vs. CH_2–CH_2–Cl) among SAR models describing seemingly "different" toxicological phenomena are indications of the extent of mechanistic commonalities between them (Table 4) (10).

Thus, there is extensive overlap between the toxicophores associated with the in vivo induction of sister chromated exchange (MoSCE) and carcinogenicity in rodents and mutations in *Salmonella* (*SalmM*) and no overlap with inhibition of gap junctional intercellular communication (iGJIC) (Table 4). This can be taken to indicate that the basis of the induction of MoSCE is a genotoxic event (related to *SalmM*) and that this, in turn, is related to carcinogenesis. On the other hand, there is no significant toxicophore overlap between MoSCE and iGJIC, the latter being an "epigenetic" (i.e., non-genotoxic) phenomenon *par excellence* (77) (Table 4). There is, however, also some overlap between MoSCE and cell toxicity. This suggests that MoSCE can also occur, albeit

Table 4 Commonality of Toxicophores Among SAR Models

Phenomena	Structural overlap (%)
MoSCE & RoCA	68
MoSCE & *SalmM*	63
MoSCE & iGJIC	4[a]
MoSCE & Celltox	21
Mnt & *SalmM*	53
Mnt & iTP	71
Mnt & CellTox	42
iTP & CellTox	71
iTP & *SalmM*	9[a]

Abbreviations: MoSCE, in vivo induction of sister chromatid exchanges; RoCA, carcinogenicity in rodents; iGJIC, inhibition of gap junctional intercellular communications; *SalmM*; mutagenicity in *Salmonella*; CellTox, cellular toxicity; Mnt, in vivo induction of micronuclei; iTP, inhibition of tubulin polymerization.
[a]Not significant, $p > 0.05$.

to a lesser extent, by a non-genotoxic mechanism involving cytotoxicity. Since, cell toxicity may lead to mitogenesis and subsequently to rodent carcinogenesis by a non-genotoxic mechanism (78–83), a positive MoSCE response does not exclusively indicate that a chemical poses a carcinogenic risk as a result of a genotoxic potential. This is, of course, relevant to a carcinogenic risk for humans. The latter is associated primarily with genotoxicity (30,31), i.e., the mutagenic activation of an oncogene or the inactivation of a suppressor gene (84,85).

In light of the above observations, a positive MoSCE response in the absence of a positive response in an assay designed to determine direct genotoxic potential [e.g., mutagenicity in *Salmonella*, induction of error-prone DNA repair (SOS Chromotest), unscheduled DNA synthesis] does not pose as great a human risk as when such an assay is positive. The latter result would indicate that the in vivo effect (i.e., MoSCE) is evidence of in vivo genotoxicity and therefore of a human carcinogenic risk.

Similarly, it was found (Table 4) that the in vivo induction of bone marrow micronuclei (Mnt) overlaps significantly with mutagenicity in *Salmonella* (*SalmM*) and with inhibition of tubulin polymerization (iTP). Both Mnt and iTP overlap with cell toxicity, but there was no significant overlap between iTP and the induction of mutagenicity in *Salmonella* (Table 4). These data indicate that Mnt can occur by two independent mechanisms, a genotoxic one (related to *SalmM*) and one related to microtubular integrity (detected by iTP) that can give rise to aneuploidy (86). iTP, in turn, may cause cell toxicity. The independence of the two mechanisms for inducing Mnt is confirmed by the absence of significant overlap between iTP and *SalmM* (Table 4).

With respect to human carcinogenic risk, this would suggest that the induction of Mnt in the absence of genotoxicity would indicate a lower level of concern especially if the agent is also endowed with beneficial potential. This is the situation with discodermolide (DISCO) (Fig. 22). This agent inhibits tubulin polymerization, which is also the target of Taxol, a proven cancer chemotherapeutic agent (68,69). Moreover, DISCO is more water soluble than Taxol, more effective than Taxol in iTP and is active against cells that have become resistant to Taxol (87).

DISCO is predicted to be neither a mutagen nor a genotoxicant (Table 5, nos. 2–5) nor to be a rodent carcinogen (Table 5, no. 1). It is, however, predicted to be iTP and Mnt positive (Table 5). In view of the earlier conclusion that Mnt could occur by a non-genotoxic mechanism related to iTP (in fact the substructure responsible for these phenomena is

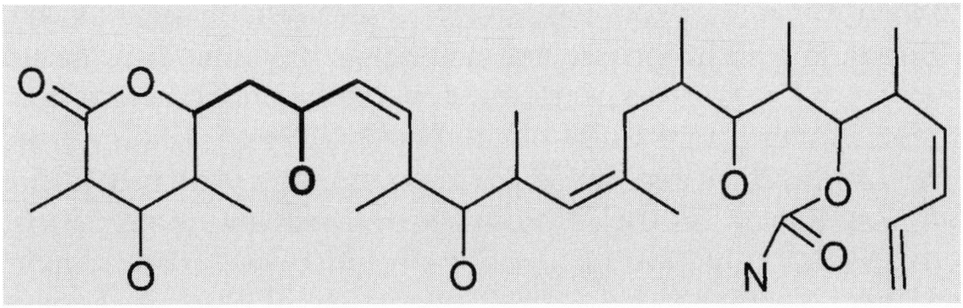

Figure 22 Region in discodermolide responsible for the inhibition of tubulin polymerization and for the in vivo induction of micronuclei.

Table 5 Projected Properties of Discodermolide

Analysis number	Phenomenon	Prediction
1	Rodent carcinogenicity	Neg
2	*Salm M*	Neg
3	Structural alert for DNA reactivity	Neg
4	UDS	Neg
5	SOS chromotest	Neg
6	iTP	Active
7	Mnt	Active

Abbreviations: *Salm M*, mutagenicity in *Salmonella*; UDS, unscheduled DNA synthesis; iTP, inhibition of tubulin polymerization; Mnt, *in vivo* induction of micronuclei; SOS chromotest, error-prone DNA repair; neg, inactive.

the same, Fig. 22), the fact that DISCO is predicted to be non-genotoxic (and non-carcinogenic) indicates that the positive Mnt does not suggest a carcinogenic risk for humans undergoing cancer chemotherapy. Obviously, the benefits far outweigh the risk, if any. The same conclusion (9,86) has been reached regarding Taxol which has been shown experimentally to induce Mnt (88).

7. APPLICATION OF SAR TO A DIETARY SUPPLEMENT

In the United States dietary supplementation to prevent diseases is widely practiced. However, due to current regulatory requirements, unlike therapeutics, such chemicals are not required to undergo premarking examination of their efficacy and/or safety (89,90). Accordingly, there have been reports of human toxicity as well as interference between dietary supplements and prescribed therapeutics and with medical procedures such as the administration of anesthetics. Moreover, demographic studies in the United States have identified the elderly as the prime consumers of these products (89,91). Due to the fact that the elderly population is considered at increased risk from the effects of toxicants, it behooves the public health and regulatory communities to devise approaches to develop protocols to establish the safety of these widely used agents.

In view of the fact that for many of these dietary supplements the efficacy has not been established, yet the intake of these agents will be for decades, the usual risk vs. benefit paradigm associated with recognized therapeutics is not appropriate. Accordingly, we must seek assurance of absolute safety. The application of SAR models to such a risk-averse situation might be an effective approach, especially as experimental data related to safety are unavailable.

For the present analysis, we selected curcumin (Fig. 23). This herbal agent has been proposed as a cancer chemopreventative agent. Curcumin is used extensively as a coloring and flavoring agent. It is the major yellow pigment extracted from turmeric and has been widely used for the treatment of various ailments because of the variety of the reported pharmacological and therapeutic properties (92,93). Curcumin and its related compounds inhibit free radical generation and act as free radical scavengers (94,95). As an antioxidant, it scavenges active oxygen species such as hydroxyl radical, superoxide anion, and singlet oxygen (96–98). It also interferes with lipid peroxidation by maintaining the activities of antioxidant enzymes at high levels (99). The ortho-substituted phenolic, methoxy group on the phenyl ring, and the 1,3-diketone system (Fig. 23) seem to be the important structural features that can contribute to these effects (100,102). Curcumin inhibits mutagenicity of certain chemical carcinogens (aflatoxin and benzo(a)pyrene) and it also inhibits their covalent DNA-binding in vivo and in vitro (103,104). Curcumin elevates the activities of phase II detoxification enzymes. This appears to be related to the presence of hydroxyl groups at the ortho position and the β-diketone functionality (105) (Fig. 23). Curcumin has shown potent chemopreventive activity in many animal tumor models during various stages of cancer formation. It has antiproliferative effects in various cultured cancer cells (103,106–113). It suppresses a number of key elements in the cellular signal transduction pathways pertinent to growth, differentiation, and malignant transformation. This is achieved by inhibiting protein kinases or

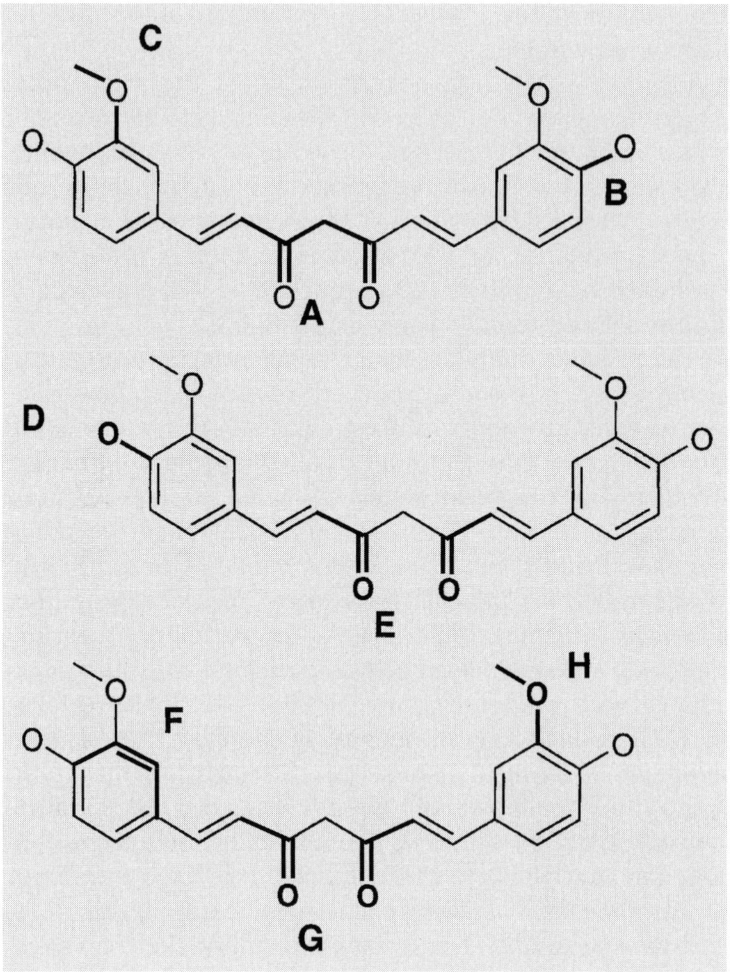

Figure 23 Structural moieties (shown in bold) associated with the presumed chemopreventative and toxicological potentials of curcumin. Substructure A: associated with inhibition of cell cycle, inhibition of cytochrome P450 (cyp), and antioxidant activity; B: inhibition of cyp and carcinogenicity; C: anticarcinogenicity; D and E: with developmental toxicity in humans; F and G: with carcinogenicity in rodents; H: with induction of mutations at the $tk^{+/-}$ locus and the in vivo induction of sister chromatid exchange and of micronuclei. It is obvious that there are significant overlaps between the putative beneficial and toxic attributes suggesting that these may reflect overlapping mechanisms.

ornithine decarboxylase, inducing phosphatase activity or inhibiting posttranslational modification of proteins (114–120). In addition, curcumin inhibits the activation of NFκB and the expression of several oncogenes (120,121). Curcumin also has been reported to cause cell cycle arrest, most often in G2/M phase and it induces apoptosis (120,122). The diketone moiety (Fig. 23) also appears essential for the latter inhibitory activity (123). Anti-inflammatory activity due to the inhibition of arachidonic acid metabolism is thought to be a key mechanism for the anticarcinogenic effect of curcumin. Curcumin is a potent inhibitor of reactive oxygen-generating enzymes such as lipoxygenase/cyclooxygenase, xanthine dehydrogenase, and nitric oxide synthase (124–127). Inhibition of COX-2 enzyme has been implicated in the chemopreventive activity of curcumin against colon carcinogenesis (128–130). Several curcumin analogues have been prepared and evaluated as potential androgen receptor antagonists against human prostate cancer cell lines. The dimethoxyphenyl and β-diketone moieties (Fig. 23) seem to be important factors related to the antiandrogenic activity (131). Additionally, an evaluation of synthetic curcumin analogues for in vitro toxicity against a panel of human tumor cell lines also implicated the β-diketone moiety (Fig. 23) (132).

Thus curcumin, a dietary supplement, is associated with a wide spectrum of biological effects, some of which are clearly also involved in toxicological response. These, in turn, are relevant to safety considerations especially in light of the recent β-carotene supplementation intervention trials to prevent lung cancers associated with cigarette smoking. These interventions had to be terminated prematurely due to an increased incidence of cancers among the treatment group (133–137).

In order to gain an appreciation of the possible health risks associated with the use of curcumin as a dietary supplement, we determined its toxicological potential using a number of validated SAR models (10). The predicted toxicological profile of curcumin (Table 6) indicates a number of toxicological potentials that raise concerns. Thus, there is a slight potential for inducing carcinogenicity in rodents by pre-

Table 6 Predicted Toxicological Profile of Curcumin

SAR model	Prediction probability (%)	SAR units
Structure alerts	Negative	
Salmonella mutagenicity	Negative	
SOS chromotest	Negative	
Carcinogenicity: rodents—NTP	83	54
Carcinogenicity: mice—NTP	75	18
Carcinogenicity: rats—NTP	Negative	
Carcinogenicity: rodents—CPDB	Negative	
Carcinogenicity: mice—CPDB	Negative	
Carcinogenicity: rats—CPDB	Negative	
Inhibition of gap junctional intercellular communication	Negative	
Binding to Ah receptor	Negative	
Mutations in mouse lymphoma [NTP]	91	39
Mutations in mouse lymphoma [GenTox]	92	59
Sister chromatic exchanges in vitro	81	35
Chromosomal aberrations in vitro	Negative	
Unscheduled DNA synthesis in vitro	Negative	
Cell transformation	Negative	
Sister chromatic exchanges in vivo	90	50
Induction of micronuclei in vivo	93	29
Yeast malsegregation	Negative	
Inhibition of tubulin polymerization	Negative	
Sensory irritation	95	102
Eye irritation	91	60
Respiratory hypersensitivity	Negative	
Allergic contact dermatitis	88	46
Rat lethality [LD_{50}]	Negative	
Mouse MTD	Negative	
Rat MTD	Negative	
Cellular toxicity [3T3]	78	22
Cellular toxicity [HeLa]	85	37
Nephrotoxicity: male rats ($\alpha_{2}\mu$)	88	39
Inhibition human cyt. P4502D	Negative	
Developmental toxicity: hamster	Negative	
Developmental toxicity: human	80	39
Aquatic toxicity (minnows)	94	99
Anticarcinogenicity: rodents	86	39
Biodegradability	89	39
Water solubility: 5.16	Log P (octanol:water): 3.65	
Hard/soft: 0.48		

sumably a non-genotoxic mechanism. That non-genotoxic potential is indicated by: (a) the chemicals contributing to the toxicophore are all non-mutagens; (b) curcumin is not predicted to induce frankly genotoxic effects, i.e., lack of mutagenicity in *Salmonella*, error-prone DNA repair, and unscheduled DNA synthesis; and (c) curcumin does not possess structural alerts for DNA reactivity (18). On the other hand, curcumin is predicted (Table 6) to induce mutations at the $tk^{+/-}$ locus of cultured mouse lymphoma cells. That effect can be the result of genomic or even cellular toxicity insults (138). Similarly, curcumin exhibits a potential to induce sister chromatid exchanges both in vitro as well as in vivo as well as micronuclei in vivo. These effects, also, may be due to non-genotoxic effects (139,140). In fact, the chemicals which contribute toxicophores associated with these potential effects of curcumin are primarily non-genotoxic. Curcumin has the potential to induce cellular toxicity and this has, in fact, been reported (141). Thus, the slight potential for carcinogenicity may be the result of cell toxicity resulting in mitogenesis (78–83). Curcumin is also predicted not to inhibit gap junctional intercellular communication, a mechanism associated with tumor promotion (77). Should curcumin be considered as a therapeutic agent to treat a serious medical condition, then, based upon a risk vs. benefit paradigm, the slight potential for carcinogenicity by a non-genotoxic mechanism might be acceptable as it is known that human cancer risks are associated primarily with "genotoxic" carcinogens and hormones (e.g., β-estradiol). However, as curcumin is under consideration as a cancer chemopreventative, of yet unproven efficacy, in humans, these toxicological potentials raise concerns. This is especially so, as dietary supplementation would be over decades. It is further recognized that the realization of these toxic potentials depends upon the dose, concomitant exposure to other agents, the response of homeostatic defense mechanisms including genetically determined activation, and detoxification enzymes and repair mechanisms. Still, these toxicological potentials, in the absence of proven chemopreventive efficacy, cast doubts about the advisability of using curcumin as a dietary supplement.

In addition to the potential for non-genotoxic carcinogenicity, curcumin also demonstrates a potential for developmental toxicity and for inducing sensory and eye irritation and allergic contact dermatitis (Table 6). Overall, these toxicological potentials suggest that prior to accepting curcumin as a dietary supplement, its toxicological effects need to be ascertained further and understood. Obviously, if subsequently curcumin were to be used as a supplement, it would be prudent to follow carefully the exposed population for prolonged periods.

In some instances, it has been demonstrated that SAR approaches can be used to enhance beneficial effects and reduce or eliminate unwanted (toxic) side effects (8,142,143). In the instance of curcumin, this might be difficult to accomplish. Thus, the structure–activity study of the relationship of the possible chemopreventative action of curcumin (see above) has indicated a major role of the diketone and methoxy functions in the antioxidant (100,102), inhibition of activation of procarcinogens (132), the augmentation of phase II detoxification enzymes (105), and the cell inhibition effects (123). These are also the structural moieties associated with curcumin's toxicological potentials (Fig. 23).

8. SAR IN THE GENERATION OF MECHANISTIC HYPOTHESES

The predictions of SAR can also be used to develop testable hypotheses related to explaining toxicological phenomena. Thus γ-butyrolactone (GBL) is presently a puzzle. GBL has been used as an anesthetic, a dietary supplement, and as an illicit recreational drug. GBL is relatively not toxic to animals yet it induces coma or even death in humans (144–147). Thus, GBL may be prototypical of other substances that have been found to be non-toxic in animals and yet are toxic and even lethal to humans (148,149).

The predicted toxicological profile of GBL matched the experimental results when these were available (150). The only significant finding with GBL was a potential for inhibiting human cytochrome P4502D6 (cyp2D6). Other toxicants known to inhibit that isozyme include 1,2,3,6-tetrahydro-1-methyl-4-phenylpyridine (MPTP). However, cyp2D6 is also inhibited by widely used therapeutics (e.g., nicardipine, budipine, haloperidol, alprenolol, and chlorpromazine) (151–153). Accordingly, it is unlikely that inhibition of cyp2D6 per se will result in serious toxicological sequences. This is borne out by the low systemic toxicity of GBL (150,154).

However, conceivably cyp2D6 might be implicated indirectly. Unlike the rodents used for toxicity testing, humans, especially those abusing illicit drugs (such as GBL), are likely to simultaneously abuse other drugs, alcohol, and cigarettes. These agents may "normally" be detoxified by cyp2D6. When, however, that pathway is blocked, as by GBL, or when the liver is compromised due to alcohol or drug abuse, neither the other drugs nor GBL will be detoxified and hence a toxicological effect, essentially reflecting a drug–drug interaction, may ensue.

These findings suggest the need to examine the proposed hypothesis experimentally, perhaps using human liver preparations, and illustrate the need to develop and validate additional SAR models of specific toxicities (e.g., neurotoxicity, hepatotoxicity, renal toxicity). They also illustrate the application of SAR methods in the generation of testable hypotheses.

9. MECHANISMS: DATA MINING APPROACH

We described earlier the derivation of mechanistic information from toxicophore commonalities. Obviously, this also requires an understanding of the nature of the biological phenomena under investigation. One of the constraints to that approach derives from the limitation on the number of chemicals in the various databases. Consequently, the number of toxicophores associated with a typical SAR model is frequently less than 15. Accordingly, some relationships are missed altogether while others do not achieve statistical significance. To address this shortcoming, the "chemical diversity approach" was devised (155). This procedure consists of using validated SAR models to predict the toxicological profiles of 10,000 chemicals representing the "universe of chemicals" (156). Essentially, the rationale is the same as for the approach using structural overlaps (see above and Table 4). Basically, the procedure consists of comparing the observed and expected joint prevalences of two potential activities based upon an assumption of independence. If the observed and expected joint prevalences for two phenomena are equal, then the two effects are independent of one another. If, on the other hand, the observed joint prevalence is greater than the expected one, then it can be concluded that the two phenomena are mechanistically related. Finally, if the observed joint prevalence is less than the expected one, then two effects can be assumed to be antagonistic or reflect competition for a substrate or a specific site (active site of an enzyme or receptor).

It is realized, of course, that no SAR model is perfectly predictive. However, on a chemical population basis (e.g., $N = 10{,}000$), as long as the model's sensitivity approximates its specificity, the prevalence of positive (or negative) predictions will reflect their true distribution.

Thus, iGJIC and Mnt are found to be unrelated (Table 7, no. 1), while iGJIC and the in vivo induction of sister chromatid exchanges (MoSCE) are slightly (but not significantly) antagonistic (Table 7, no. 3). On the other hand, MoSCE and Mnt (Table 7, no. 4) are synergistic, presumably indicating that these in vivo genotoxic events occur by one or more common mechanisms. Similarly, MoSCE and mutagenicity in *Salmonella* (*SalmM*) are related mechanistically, presumably reflecting their DNA-damaging genotoxic potentials (Table 7, no. 7). On the other hand, *SalmM* and iGJIC are independent of one another (Table 7, no. 8) which confirms the epigenetic (non-genotoxic) nature of the latter (77). With respect to carcinogenicity, both iGJIC (Table 7, no. 2) and MoSCE (Table 7, no. 5) show a strong mechanistic association. Finally, the joint association of iGJIC and MoSCE with rodent carcinogenicity (Table 7, no. 6), together with the independence of iGJIC and MoSCE (Table 7, no. 3), suggests

Table 7 Chemical Diversity Approach: Joint Prevalences

Analysis	Phenomena	Δ^*	100Δ/Expected*	p-Value
1	iGJIC &Mnt	−4	−0.2	0.5
2	iGJIC & RoCa	236	25.8	<0.00001
3	MoSCE & iGJIC	−66	−4.3	0.1
4	MoSCE & Mnt	1603	47.4	<0.00001
5	MoSCE & RoCa	494	26.0	<0.00001
6	iGJIC & MoSCE & RoCa	273	52.5	<0.00001
7	MoSCE & *SalmM*	473	26.5	<0.00001
8	iGJIC & *SalmM*	24	2.8	0.3

Abbreviations: *Salm M*, mutagenicity in *Salmonella*; UDS, unscheduled DNA synthesis; iTP, inhibition of tubulin polymerization; Mnt, in vivo induction of micronuclei; iGJIC, inhibition of gap junctional intercellular communication; MoSCE, in vivo induction of SCE; RoCA, carcinogenicity in rodents.
Δ is the difference between the observed and unexpected joint prevalences; 100Δ/expected is $100^*\Delta$/expected prevalence. The difference in joint prevalence can be expressed as Δ or 100Δ/expected.

that these effects may each affect the carcinogenic process independently of one another or sequentially, e.g., cancer initiation followed by tumor promotion as reflected by MoSCE and iGJIC, respectively.

10. A SAR-BASED DATA MINING APPROACH TO TOXICOLOGICAL DISCOVERY

The approaches of SAR to identify putative toxicants and to elucidate the basis of their activity have been useful tools in the hands of researchers with the appropriate expertise. Obviously, these applications are dependent upon the availability of validated SAR models (10). At the other extreme of the data spectrum is the daunting problem of the basis of human diseases for which an environmental or occupational etiology is suspected (e.g., autism, Parkinson's disease, progressive systemic sclerosis). For these conditions neither the offending agents nor the mechanisms of pathogenesis are known. In some instances, based upon epidemiological or laboratory studies, culprits have been proposed, e.g., MPTP and rotenone in Parkinsonism (157,158). In all of these situations, there are too few data to develop SAR models and moreover the differences in the chemical structures of the suspect chemicals are so dissimilar as to preclude the generation of mechanistic hypotheses. In order to investigate this vexing yet important public health problem, we developed a method to determine whether two or more agents that have been implicated in a specific pathology or health effect share toxicological properties. We designated this approach as the "virtual similarity index" (VSI) (159).

Virtual similarity index uses SAR-generated profiles (e.g., see Table 6) of the suspect chemicals and determines the extent of toxicities overlaps. It then compares these overlaps with the expected frequency of the same overlaps in a population of the "universe of chemicals" (155,156). There is no expectation that any of the SAR models used is directly associated with the pathology under investigation. However, since most disease-causing agents are not tissue- or cell-specific, it is likely that other tissues will be affected secondarily. Some of these may be represented among the SAR models. It is posited that the VSI overlaps will be sufficiently toxicognonomic to provide testable hypotheses.

The VSI procedure consists of generating the toxicological profiles of the chemicals of interest using 27–30 validated SAR models (Table 8). The specific overlaps of putative toxicities are identified and enumerated and the occurrence of the same overlaps among 10,000 chemicals representative of the "universe of chemicals" is ascertained. To illustrate the VSI paradigm, we chose pyrethrin I, veratridine, and DDT, three structurally very different chemicals that are central nervous system toxicants, presumably as a result of their ability to inhibit voltage-gated sodium channel closure. Their virtual toxicological profiles indicate that pyrethrin I and DDT share eight properties (Table 8). Among 10,000 chemicals, 0.74% are expected to share these properties (Table 9) (the lower the expected prevalence, the greater the uniqueness and significance of the shared properties). Similarly, veratrine and

Table 8 Virtual Toxicological Profiles

SAR model	Pyrethrin I	DDT
Carcinogenicity in rodents	Inact	Inact
Mutagenicity: *Salmonella*	Inact	Inact
Mutagenicity: $tk^{+/-}$ locus	Inact	Active
Sensory irritation in mice	Inact	Active
Allergic contact dermatitis	Inact	Inact
Respiratory hypersensitivity	Inact	Active
Eye irritation	Inact	Active
MTD: mice	Inact	Active
Cell toxicity: Balb3T3	Active	Active
Toxicity: minnow	Active	Active
Lethality: rat LD_{50}	Active	Active
Developmental toxicity: hamsters	Active	Active
Developmental toxicity: humans	Inact	Inact
SCE in vitro	Active	Active
SCE in vivo	Active	Active
Chromosomal aberrations	Inact	Inact
Micronuclei in vivo	Active	Active
UDS	Inact	Inact
SOS chromotest	Inact	Inact
$\alpha_2\mu$-globulin associated nephropathy	Active	Active
Ah receptor binding	Inact	Active
Cell toxicity: hela	Inact	Inact
Biodegradation	Active	Inact
IGJIC	Inact	Active
Skin permeability	Inact	Active
Inhibition of cytochrome P4502D6	Inact	Inact

Inact and active indicate predictions of inactivity and activity, respectively.

DDT share nine properties, the expected occurrence of which among 10,000 chemicals is 0.005% (Table 9). On the other hand, the putative properties shared by these sodium channel blocks and the unrelated streptomycin, an ototoxic antibiotic, are much less unique (Table 9, nos. 3–5). Finally, the related kanamycin and streptomycin share properties that have an expected prevalence of <0.01% (Table 9, no. 6), i.e., a highly significant commonality.

We applied the same approach to examine the properties of dietary supplements. Both curcumin (see above) and

Table 9 Virtual Similarity Indices

Analysis	Combination	Expected prevalence (%)
1	DDT and pyrethrin I	0.74
2	Veratridine and DDT	0.009
3	Pyrethrin I and streptomycin	10.4
4	Veratridine and streptomycin	6.0
5	DDT and streptomycin	10.4
6	Kanamycin and streptomycin	<0.01
7	Curcumin and genistein	0.6
8	Genistein and rofecoxib	5.6
9	Curcumin and rofecoxib	5.4
10	Rofecoxib and penicillin	22.9

Expected prevalence for a population of 10,000 chemicals respresentative of the "universe of chemicals." The lower the expected prevalence the more unique the similarity index.

genistein have been proposed as cancer chemopreventative agents (160). Albeit, they both possess a broad spectrum of potential biological and toxicological properties (see above and Refs. 141,160) that suggest that they be examined further prior to usage as supplements. The structures of the two agents differ greatly (Fig. 24). Analysis by VSI (Table 9, no. 7) indicates a significantly unique series of properties that suggests that they share common properties. Because genistein and curcumin have been reported to exhibit COX-2 inhibiting potentials (124,125,128,161), we compared them to Rofecoxib (Vioxx®, a COX-2 inhibitor used therapeutically (Table 9, nos. 8 and 9). Those prevalences were significantly less unique, which suggests that overall the shared toxicities of genistein and curcumin are not primarily due to their COX-2 inhibiting properties. However, one of the components may well include their COX-2 inhibiting properties as the overlap between Rofecoxib and either genistein or curcumin (Table 9, nos. 8 and 9) is more unique than between Rofecoxib and Penicillin G (Table 9, no. 10).

These examples suggest that the VSI approach might well provide a tool for hypothesis generation that can be tested experimentally.

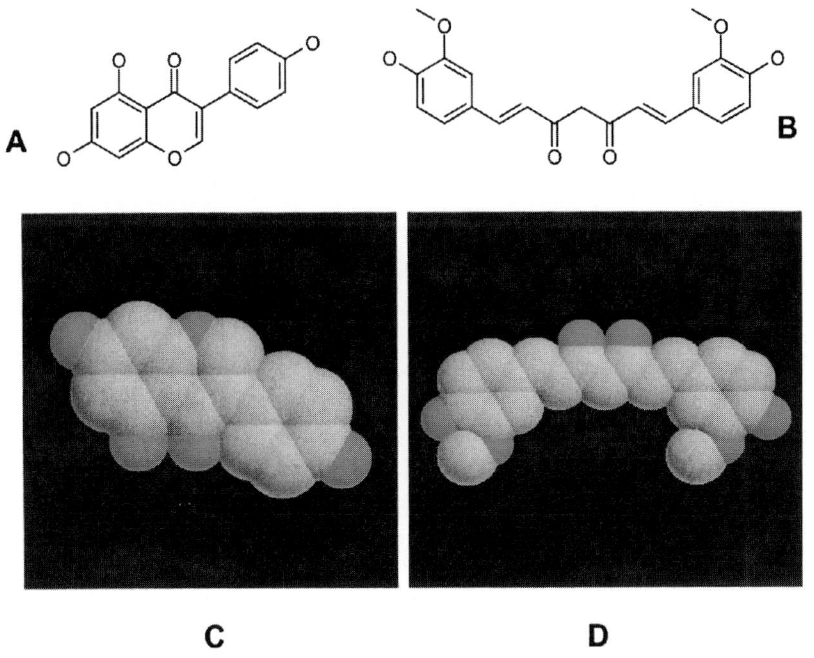

Figure 24 Structures of genistein (A and C) and of curcumin (B and D).

11. CONCLUSION

It was the aim of this review to demonstrate that the current state of substructure-based SAR allows the methodologies to be used not only to predict toxicological phenomena but also to gain an understanding of the mechanistic basis of these manifestations. The technology can also be used to generate testable hypotheses. Throughout this presentation, we sought to stress the importance of the interaction of the biologist/toxicologist/health risk assessor with the computational toxicologist.

It is the authors' opinion that an understanding of the performance characteristics of SAR models, even though they may be less than perfect, enables their deployment. It is also our opinion that at this time, the rate-limiting step of SAR modeling is not the software but rather the paucity of reliable experimental data. This bottleneck is likely to remain for the foreseeable future. Since different SAR paradigms each have their advantages and thus can complement one another, it would seem that methods to combine the results of different SAR approaches should be developed further and undergo rigorous validation. Once such "batteries" of SAR approaches have been developed, it is suggested that they be used with similar learning sets and thus refine the predictions obtained with them.

Finally, the application of SAR methodologies is a dynamic process that needs to take cognizance of newer emerging understanding of biological phenomena as well as new techniques in the biomedical sciences and bioinformatics. Thus, preliminary evaluation of the results of DNA microarrays from cells and tissues subjected to toxic challenge indicates that those data are an ideal substrate for SAR analyses, albeit the coupling of the huge information output of microassays into the SAR-generating engine will require the development and/or adaption of data management and mining protocols.

ACKNOWLEDGMENTS

The authors are grateful to Ms. Wanda Dominger for shepherding and editing this manuscript. Without her supervision and attention to details, this review would not have been completed.

The support of the Vira Heinz Endowment and the National Institute of Environmental Health Sciences Training Grant No. 5T32ES07318-02 is gratefully acknowledged.

REFERENCES

1. NRC. Science and Judgment in Risk Assessment. Washington, DC: National Research Council, National Academy Press, 1994.
2. McKinney ID, Richard A, Waller C, Newman MC, Gerberick F. The practice of structure–activity relationships (SAR) in toxicology. Toxicol Sci 2000; 56:8–17.

3. CEC. White Paper: Strategy for a Future Chemicals Policy, Commission of the European Communities, 2001. http://europa.eu.int/comm/environment/chemicals/whitepaper.htm.
4. Cronin MTD, Jaworska JS, Walker JD, Comber MHI, Watts CD, Worth AP. (2003). Use of QSARs in international decision-making frameworks to predict health effects of chemical substances. Environ Health Pers. In press.
5. ATSDR. Chemical-Specific Health Consultation Toxicological Information on Substances Identified by the State of New Jersey. Atlanta, GA: Dept. of Health and Human Services, Agency for Toxic Substances and Disease Registry, Div. of Toxicology, 2001:83.
6. Walker JD, Carlsen L, Jaworska J. Improving opportunities for regulatory acceptance of QSARS: the importance of model domain, uncertainty, validity and predictability. Quant Struct Activ Rel 2003; 22:1–5.
7. Richard AM, Benigni R. AI and SAR approaches for predicting chemical carcinogenicity: survey and status report. SAR QSAR Environ Res 2002; 13:1–19.
8. Rosenkranz HS. SAR in the assessment of carcinogenesis: the MULTICASE approach. In: Benigni R, ed. Quantitative Structure–Activity Relationship (QSAR) Models of Mutagens and Carcinogens. Boca Raton, FL: CRC Press, LLC, 2003:175–206.
9. Rosenkranz HS. Structural concepts in the prediction of the toxicity of therapeutical agents. Burger's Medicinal Chemistry and Drug Discovery. New York, NY: John Wiley & Sons, 2003:827–847.
10. Rosenkranz HS, Cunningham AR, Zhang YP, Claycamp HG, Macina OT, Sussman NB, Grant SG, Klopman G. Development, characterization and application of predictive-toxicology models. SAR QSAR Environ Res 1999; 10:277–298.
11. Klopman G, Rosenkranz HS. Prediction of carcinogenicity/mutagenicity using MULTICASE. Mutation Res 1994; 305:33–46.

12. Klopman G, Rosenkranz HS. Toxicity estimation by chemical substructure analysis: the Tox II program. Toxicol Lett 1995; 79:145–155.
13. Sanderson DM, Earnshaw CG. Computer prediction of possible toxic action from chemical structure; the DEREK system. Human Exp Toxicol 1991; 10:261–273.
14. Ridings JE, Barratt MD, Cary R, Earnshaw CG, Eggington CE, Ellis MK, Judson PN, Langowskin JJ, Marchant CA, Payne MP, Watson WP, Yih TD. Computer prediction of possible toxic action from chemical structure: an update of the DEREK system. Toxicology 1996; 106:267–279.
15. Greene N. Computer software for risk assessment. J Chem Inf Comput Sci 1996; 37:148–150.
16. Woo YT, Lai DY, Argus MF, Arcos JC. Development of structure–activity relationship rules for predicting carcinogenic potential of chemicals. Toxicol Lett 1995; 79:219–228.
17. Woo YT, Lai DY. Mechanisms of action of chemical carcinogens and their role in structure–activity relationships (SAR) analysis and risk assessment. Benigni R, ed. Quantitative Structure–Activity Relationship (QSAR) Models of Mutagens and Carcinogens. Boca Raton, FL: CRC Press, 2003:41–80.
18. Ashby J. Fundamental structural alerts to potential carcinogenicity or non-carcinogenicity. Environ Mutag 1985; 7: 919–921.
19. Benigni R. SARs and QSARs of mutagens and carcinogens: understanding action mechanisms and improving risk assessment. Benigni R, ed. Quantitative Structure–Activity Relationship (QSAR) Models of Mutagens and Carcinogens. Boca Raton, FL: CRC Press, 2003:259–282.
20. Matthews EJ, Contrera JF. A new highly specific method for predicting the carcinogenic potential of pharmaceuticals in rodents using enhanced MCASE QSAR-ES software. Regulatory Toxicol Pharmacol 1998; 28:242–264.
21. Williams GM, Mori H, McQueen CA. Structure–activity relationships in the rat hepotocyte DNA-repair test for 300 chemicals. Mutation Res 1989; 272:111–124.

22. Zhang YP, Praagh Av, Klopman G, Rosenkranz HS. Structural basis of the induction of unscheduled DNA synthesis in rat hepatocytes. Mutagenesis 1994; 9:141–149.

23. Zeiger E. Genotoxicity database. Gold LS, Zeiger E, eds. Handbook of Carcinogenic Potency and Genotoxicity Databases. Boca Raton, FL: CRC Press, 1997:687–729.

24. Maron DM, Ames BN. Revised methods for the *Salmonella* mutagenicity test. Mutation Res 1983; 113:173–215.

25. Mermelstein R, Kiriazides DK, Butler M, McCoy EC, Rosenkranz HS. The extraordinary mutagenicity of nitropyrenes in bacteria. Mutation Res 1981; 89:187–196.

26. Klopman G, Rosenkranz HS. Structural requirements for the mutagenicity of environmental nitroarenes. Mutation Res 1984; 126:227–238.

27. Rosenkranz HS, Mermelstein R. Mutagenicity and genotoxicity of nitroarenes: all nitro-containing chemicals were not created equal. Mutation Res 1983; 114:217–267.

28. Klopman G, Frierson MR, Rosenkranz HS. Computer analysis of toxicological databases: mutagenicity of aromatic amines in *Salmonella* tester strains. Environ Mutagen 1985; 7:625–644.

29. Ashby J, Tennant RW. Definitive relationships among chemical structure, caicinogenicity and mutagenicity for 301 chemicals tested by the U.S. NTP. Mutation Res 1991; 257:229–306.

30. Ashby J, Morrod RS. Detection of human carcinogens. Nature 1991; 352:185–186.

31. Ennever FK, Noonan TJ, Rosenkranz HS. The predictivity of animal bioassays and short-term genotoxicity tests for carcinogenicity and non-carcinogenicity to humans. Mutagenesis 1987; 2:73–78.

32. Bartsch H, Malaveille C. Prevalence of genotoxic chemicals among animal and human carcinogens evaluated in the IARC Monograph Series. Cell Biol Toxic 1989; 5:115–127.

33. Shelby MD. The genetic toxicity of human carcinogens and its implications. Mutation Res 1988; 204:3–15.

34. Macina OT, Zhang YP, Rosenkranz HS. Improved predictivity of carcinogens: the use of a battery of SAR models. Kitchen KT, ed. Testing, Predicting and Interpreting Carcinogenicity. New York, NY: Marcel Dekker, 1998:227–250.

35. Zhang YP, Sussman N, Macina OT, Rosenkranz HS, Klopman G. Prediction of the carcinogenicity of a second group of chemicals undergoing carcinogenicity testing. Environ Health Perspec 1996; 104:1045–1050.

36. Zeiger E. Carcinogenicity of mutagens: predictive capability of the *Salmonella* mutagenesis assay for rodent carcinogenicity. Cancer Res 1987; 47:1287–1296.

37. Dunkel VC, Zeiger E, Brusick D, McCoy EC, McGregor D, Mortelmans K, Rosenkranz HS, Simmon VF. Reproducibility of microbial mutagenicity assays: I. Tests with *Salmonella typhimurium* and *Escherichia coli* using standardized protocol. Environ Mutagen 1984; 6(suppl 2):1–254.

38. Dunkel VC, Zeiger E, Brusick D, McCoy E, McGregor D, Mortelmans K, Rosenkranz HS, Simmon VF. Reproducibility of microbial mutagenicity assays: II. Testing of carcinogens and noncarcinogens in *Salmonella typhimurium* and *Escherichia coli*. Environ Mutagen 1985; 7(suppl 5):1–248.

39. Gold LS, Slone TH, Ames BN. Overview and update of analyses of the carcinogenic potency database. Gold LS, Zeiger E, eds. Handbook of Carcinogenic Potency and Genotoxicity Databases. Boca Raton, FL: CRC Press, 1997:661–685.

40. Gold LS, Sawyer CB, Magaw R, Backman GM, deVeciana M, Levinson R, Hooper NK, Havender WR, Bernstein L, Peto R, Pike MC, Ames BN. A carcinogenic potency database of the standardized results of animal bioassays. Environ Health Perspect 1984; 58:9–319.

41. Peto R, Pike MC, Bernstein L, Gold LS, Ames BN. The TD50: a proposed general convention for the numerical description of the carcinogenic potency of chemicals in chronic-exposure animal experiments. Environ Health Perspect 1984; 58:1–8.

42. Huff JE, Haseman JK, Rail DP. Scientific concepts, value, and significance of chemical carcinogenesis studies. Ann Rev Toxicol Pharmacol 1991; 31:621–652.

43. Swenberg JA. Bioassay design and MTD setting" old methods and new approaches. Regulatory Toxicol Pharmacol 1995; 21:44–51.

44. Bucher JR, Portier CJ, Goodman JI, Faustman EM, Lucier GW. Workshop overview. National Toxicology Program Studies: principles of dose selection and applications to mechanistic based risk assessment. Fundamental Appl Toxicol 1996; 31:1–8.

45. Gallo MA. History and scope of toxicology. Casarett and Doull's Toxicology: The Basic Science of Poisons. New York, NY: McGraw-Hill, 1996:3–11.

46. Rosenkranz HS, Klopman G. Structural basis of carcinogenicity in rodents of genotoxicants and non-genotoxicants. Mutation Res 1990; 228:105–124.

47. Chankong V, Haimes YY, Rosenkranz HS, Pet-Edwards J. The carcinogenicity prediction and battery selection (CPBS) method: A Bayesian approach. Mutation Res 1985; 153:135–166.

48. Lee Y, Rosenkranz HS, Buchanan BG, Mattison DR, Klopman G. Learning rules to predict rodent carcinogenicity of non-genotoxic chemicals. Mutation Res 1995; 328:127–149.

49. Lee Y, Buchanan BG, Rosenkranz HS. Carcinogenicity predictions for a group of 30 chemicals undergoing rodent cancer bioassays based on rules derived from subchronic organ toxicities. Environ Health Perspect 1996; 104:1059–1063.

50. Yang W-L, Klopman G, Rosenkranz HS. Structural basis of the in vivo induction of micronuclei. Mutation Res 1992; 272:111–124.

51. Rosenkranz M, Rosenkranz HS, Klopman G. Intercellular communication, tumor promotion and non-genotoxic carcinogenesis: relationships based upon structural considerations. Mutation Res 1997; 381:171–188.

52. Graham C. Development and comparisons of structure–activity relationship models of chemical respiratory and dermal hypersensitivity. PhD dissertation, Pittsburgh, PA, University of Pittsburgh, 1998.

53. Gealy R, Graham C, Sussman N, Macina OT, Rosenkranz HS, Karol MH. Evaluating clinical case report data for SAR modeling of allergic contact dermatitis. Human Exp Toxicol 1996; 15:489–493.
54. Graham C, Gealy R, Macina OT, Karol MH, Rosenkranz HS. QSAR for allergic contact dermatitis. Quant Struct Activ Rel 1996; 15:224–229.
55. Ghanooni M, Mattison YP, Zhang YP, Macina OT, Rosenkranz HS, Klopman G. Structural determinants associated with risk of human developmental toxicity. Am J Obstetrics Gynecol 1997; 76:799–806.
56. Takihi N, Zhang YP, Klopman G, Rosenkranz HS. An approach for evaluating and increasing the informational content of mutagenicity and clastogenicity databases. Mutagenesis 1993; 8:257–264.
57. Takihi N, Zhang YP, Klopman G, Rosenkranz HS. Development of a method to assess the informational content of structure–activity databases. Quality Assurance: Good Practice Regulation Law 1993; 2:255–264.
58. Liu M, Sussman N, Klopman G, Rosenkranz HS. Estimation of the optimal data base size for structure–activity analyses: the *Salmonella* mutagenicity data base. Mutation Res 1996; 358:63–72.
59. Zhang YP, Sussman N, Klopman G, Rosenkranz HS. Development of methods to ascertain the predictivity and consistency of SAR models: application to the US National Toxicology Program rodent carcinogenicity bioassays. Quant Struct Activ Rel 1997; 16:290–295.
60. Rosenkranz HS. Organische Nitroverbindungen. Mersch-Sundermann V, ed. Umweltmedizin. Stuttgart, Germany: Georg Thieme Verlag, 1999:200–206.
61. Rosenkranz HS, Poirier LA. An evaluation of the mutagenicity and DNA modifying activity of microbial systems of carcinogens and non-carcinogens. J Natl Cancer Inst 1979; 62:873–892.
62. Pet-Edwards J, Rosenkranz HS, Chankong V, Haimes YY. Cluster analysis in predicting the carcinogenicity of

chemicals using short-term assays. Mutation Res 1985; 153:167–185.

63. Combes RD, Haveland-Smith RB. A review of the genotoxicity of food, drug and cosmetic colours and other azo, triphenylmethane and xanthene dyes. Mutation Res 1982; 98:101–248.

64. Hueper WD, Conway WD. Chemical Carcinogenesis and Cancers. Thomas Charles: Springfield, IL, 1964:294–318.

65. Ariens EJ. Domestication of chemistry by design of safer chemicals: structure–activity relationships. Drug Metabol Rev 1984; 15:425–504.

66. Ashby J, Styles JA, Paton D. In vitro evaluation of some derivatives of the carcinogen butter yellow: implications for environmental screening. Br J Cancer 1978; 38:34–46.

67. Swenberg JA, Short B, Borghoff S, Strasser I, Charbonneau M. The comparative pathobiology of $\alpha_2\mu$-globulin nephropathy. Toxicol Appl Pharmacol 1989; 97:35–46.

68. Wilson L, Jordan MA. Microtubule dynamics: taking aim at a moving target. Chemical Biol 1995; 2:569–573.

69. Hamel E. Antimitotic natural products and their interactions with tubulin. Med Res Rev 1996; 16:207–231.

70. Johnson R, Macina OT, Graham C, Rosenkranz HS, Cass GR, Karol, MH. Prioritizing testing of organic compounds detected as gas phase air pollutants: structure–activity study for human contact allergens. Environ Health Perspect 1997; 105:986–992.

71. Cunningham AR, Rosenkranz HS. Estimating the extent of health hazard posed by high production volume chemicals. Environ Health Perspect 2001; 109:953–956.

72. Rosenkranz HS, Cunningham AR. SAR modeling of genotoxic phenomena: the effect of supplementation with physiological chemicals. Mutation Res 2001; 476:133–137.

73. Rosenkranz HS, Cunningham AR. SAR modeling of unbalanced data sets. SAR QSAR Environ Res 2001; 12:267–274.

74. Thampatty BP, Rosenkranz HS. SAR modeling: effect of experimental ambiguity. Combinat Chem High Throughput Screening 2003; 6:129–132.

75. Rosenkranz HS, Cunningham AR. Chemical categories for health hazard identification: a feasibility study. Regulatory Toxicol Pharmacol 2001; 33:313–318.

76. Rosenkranz HS. SAR modeling of complex phenomena: probing procedural limitations. ATLA 2003; 31:393–399.

77. Trosko JE, Chang CC, Upham B, Wilson M. Epigenetic toxicology as toxicant-induced changes in intracellular signaling leading to altered gap junctional intercellular communication. Toxicol Lett 1998; 102–103:71–78.

78. Ames BN, Gold LS. Too many rodent carcinogens: mitogenesis increases mutagenesis. Science 1990; 249:970–971.

79. Ames BN, Profet M, Gold LS. Dietary pesticides (99.99% all natural). Proc Natl Acad Sci 1990; 87:7777–7781.

80. Cohen SM, Ellwein LB. Cell proliferation in carcinogenesis. Science 1990; 249:1007–1011.

81. Cohen SM, Ellwein LB. Genetic errors, cell proliferation and carcinogenesis. Cancer Res 1991; 51:6493–6505.

82. Preston-Martin S, Pike MC, Ross RK, Jones PA, Henderson BE. Increased cell division as a cause of human cancer. Cancer Res 1990; 50:7415–7421.

83. Butterworth BE. Consideration of both genotoxic and nongenotoxic mechanisms in predicting carcinogenic potential. Mutation Res 1990; 239:117–132.

84. Reynolds SH, Stowers SJ, Patterson RM, Maronpot RR, Aaronson SA, Anderson MW. Activated oncogenes in B6C3F1 mouse liver tumors: implications for risk assessment. Science 1987; 237:1309–1316.

85. Harris CC. p53: at the crossroad of molecular carcinogenesis and risk assessment. Science 1993; 262:1980–1981.

86. ter Haar E, Day BW, Rosenkranz HS. Direct tubulin polymerization perturbation contributes significantly to the induction of micronuclei in vivo. Mutation Res 1996; 350:331–337.

87. ter Haar E, Kowalski RJ, Hamel E, Lin CM, Longley RE, Gunasekera SP, Rosenkranz HS, Day BW. Discodermolide, acytotoxic marine agent that stabilizes microtubules more potently than taxol. Biochemistry 1996; 35:243–250.

88. Tinwell H, Ashby J. Genetic toxicity and potential carcinogenicity of taxol. Carcinogenesis 1994; 15:1499–1501.

89. GAO. Health Products for Seniors: “Anti-Aging”: Products pose potential for physical and economic harm. GAO-01-1129 2001; 7–11.

90. Woteki C. The extraordinary case of dietary supplements. In: The Richard and Hinda Rosenthal Lectures in “Exploring Complementary and Alternative Medicine” Institute of Medicines. National Academies Press, 2003:25–32.

91. Eskin SB. Dietary supplements and older consumers. Data Digest 66. Washington, DC: AARP Public Policy Institute, 2001:1–8.

92. Ammon HP, Wahl MA. Pharmacology of *Curcuma longa* L. Planta Medica 1991; 57:1–7.

93. Araujo CC, Leon LL. Biological activities of *Curcuma longa* L. Memorias do Instituto Oswaldo Cruz 2001; 96: 723–728.

94. Reddy AC, Lokesh BR. Studies on the inhibitory effects of curcumin and eugenol on the formation of reactive oxygen species and the oxidation of ferrous iron. Mol Cellul Biochem 1994; 137:1–8.

95. Reddy AC, Lokesh BR. Effect of dietary turmeric (*Curcuma longa*) on iron-induced lipid peroxidation in the rat liver. Food Chem Toxicol 1994; 32:279–283.

96. Ruby AJ, Kuttan G, Babu KD, Rajasekharan KN, Kuttan R. Anti-tumour and antioxidant activity of natural curcuminoids. Cancer Lett 1995; 94:79–83.

97. Zhao BL, Li XJ, He RG, Cheng SJ, Xin WJ. Scavenging effect of extracts of green tea and natural antioxidants on active oxygen radicals. Cell Biophys 1989; 14:175–185.

98. Subramanian M, Sreejayan Rao MN, Devasagayam TP, Singh BB. Diminution of singlet oxygen-induced DNA damage by curcumin and related antioxidants. Mutation Res 1994; 311:249–255.

99. Rajakumar DV, Rao MN. Antioxidant properties of dehydrozingerone and curcumin in rat brain homogenates. Mol Cellul Biochemis 1994; 140:73–79.

100. Sugiyama Y, Kawakishi S, Osawa T. Involvement of the beta-diketone moiety in the antioxidative mechanism of tetrahydrocurcumin. Biochem Pharmacol 1996; 52:519–525.

101. Ahsan H, Parveen N, Khan NU, Hadi SM. Pro-oxidant, anti-oxidant and cleavage activities on DNA of curcumin and its drivatives demethoxycurcumin and bisdemethoxycurcumin. Chemico-Biol Interactions 1999; 121:161–175.

102. Rajakumar DV, Rao MN. Antioxidant properties of dehydrozingerone and curcumin in rat brain homogenates. Mol Cellul Biochem 1994; 140:73–79.

103. Huang MT, Wang ZY, Georgiaidis CA, Laskin JD, Conney AH. Inhibitory effects of curcumin on tumor initiation by benzo[a]pyrene and 7,12-dimethylbenz[a]anthracene. Carcinogenesis 1992; 13:2183–2186.

104. Nagabhushan M, Amonkar AJ, Bbide SV. In vitro antimutagenicity of curcumin against environmental mutagens. Food Chem Toxicol 1987; 25:545–547.

105. Dinkova-Kostova AT, Talalay P. Relation of structure of curcumin analogs to their potencies as inducers of Phase 2 detoxification enzymes. Carcinogenesis 1999; 20:911–914.

106. Rao CV, Rivenson A, Simi B, Reddy BS. Chemoprevention of colon cancer by dietary curcumin. Ann NY Acad Sci 1995; 768:201–204.

107. Huang MT, Lou YR, Ma W, Newmark HL, Reuhl KR, Conney AH. Inhibitory effects of dietary curcumin on forestomach, duodenal, and colon carcinogenesis in mice. Cancer Res 1994; 768:201–204.

108. Boone CW, Steele VE, Kelloff GJ. Screening for chemopreventive (anticarcinogenic) compounds in rodents. Mutation Res 1992; 267:251–255.

109. Mehta K, Pantazis P, McQueen T, Aggarwal, BB. Antiproliferative effect of curcumin (diferuloylmethane) against human breast tumor cell lines. Anti-Cancer Drugs 1997; 8.

110. Conney AH, Lysz T, Ferraro T, Abidi TF, Manchand PS, Laskin JD, Huang MT. Inhibitory effect of curcumin and some related dietary compounds on tumor promotion and arachidonic acid metabolism in mouse skin. Adv Enzyme Regulation 1991; 31:385–396.

111. Shao ZM, Shen ZZ, Liu CH, Sartippour MR, Go VL, Heber D, Nguyen M. Curcimin exerts multiple suppressive effects on human breast carcinoma cells. Int J Cancer 2002; 98:234–240.

112. Dorai T, Cao YC, Buttyan R, Katz AE. Therapeutic potential of curcumm in human prostate cancer. III. Curcumin inhibits proliferation, induces apoptosis, and inhibits angiogenesis of LNCaP prostate cancer cells in vivo. Prostate 2001; 47: 293–303.

113. Kawamori T, Lubet R, Steele VE, Kelloff GJ, Kaskey RB, Rao CV, Reddy BS. Chemopreventive effect of curcumin, a naturally occurring anti-inflammatory agent, during the promotion/progression stages of colon cancer. Cancer Res 1999; 59:597–601.

114. Liu JY, Lin SJ, Lin JK. Inhibitory effects of curcumin on portein kinase C activity induced by 12-0-tetradecanoylphorbol-13-acetate in NIH 3T3 cells. Carcinogenesis 1993; 14:857–861.

115. Chen YR Tan TH. Inhibition of the c-Jun N-terminal kinase (JNK) signaling pathway by curcumin. Oncogene 1998; 17:173–178.

116. Chen HW, Huang HC. Effect of curcumin on cell cycle progression and apoptosis in vascular smooth muscle cells. Br J Pharmacol 1998; 124:1029–1040.

117. Han SS, Chung ST, Robertson DA, Ranjan D, Bondada S. Curcumin causes the growth arrest and apoptosis of B cell lymphoma by downregulation of egr-1, c-myc, bcl-XL, NF-kappa B, and p53. Clin Immunol 1999; 93:152–162.

118. Manson MM, Holloway KA, Howells LM, Hudson EA, Plummer SM, Squires MS, Prigent SA. Modulation of signal-transduction pathways by chemopreventive agents. Biochem Soc Transac 2000; 28:7–12.

119. Choudhuri T, Pal S, Agwarwal ML, Das T, Sa G. Curcumin induces apoptosis in human breast cancer cells through p53-dependent Bax induction. FEBS Lett 2002; 512:334–340.

120. Manson MM, Gescher A, Hudson EA, Plummer SM, Squires MS, Prigent SA. Blocking and suppressing mechanisms of chemoprevention by dietary constituents. Toxicol Lett 2000; 112–113:499–505.

121. Li JK, Lin-Shia SY. Mechanisms of cancer chemoprevention by curcumin. Proc Natl Sci Council Republic of China—Part B Life Sci 2001; 25:59–66.

122. Jiang MC, Yang-Yen HF, Yen JJ, Lin JK. Curcumin induces apoptosis in immortalized NIH 3T3 and malignant cancer cell lines. Nutrition Cancer 1996; 26:111–120.

123. Simon A, Allais DP, Duroux JL, Basly JP, Durand-Fontanier S, Delage C. Inhibitory effect of curcuminoids on MCF-7 cell proliferation and structure–activity relationships. Cancer Lett 1998; 129:111–116.

124. Plummer SM, Hill KA, Festing MF, Steward WP, Gescher AJ, Sharma RA. Clinical development of leukocyte cyclooxygenase 2 activity as a systemic biomarker for cancer chemopreventive agents. Cancer Epidemiol Biomarkers Prevention 2001; 10:1295–1299.

125. Cuendet M, Pezzuto JM. The role of cyclooxygenase and lipoxygenase in cancer chemoprevention. Drug Metabol Drug Interactions 2000; 17:109–157.

126. Chan, MM, Huang HI, Fenton MR, Fong D. Invivo inhibition of nitric oxide synthase gene expression by curcumin, a cancer preventive natural product with anti-inflammatory properties. Biochem Pharmacol 1998; 55:1955–1962.

127. Pan MH, Lin-Shiau SY, Lin JK. Comparative studies on the suppression of nitric oxide synthase by curcumin and its hydrogenated metabolites through down-regulation of Ikap-

paB kinase and NFkappaB activation in macrophages. Biochem Pharmacol 2000; 60:1665–1676.

128. Reddy BS, Rao CV. Novel approaches for colon cancer prevention by cyclooxygenase-2 inhibitors. J Environ Pathol Toxicol Oncol 2002; 21:155–164.

129. Rao CV, Kawamori T, Hamid R, Reddy BS. Chemoprevention of colonic aberrant crypt foci by an inducible nitric oxide synthase-selective inhibitor. Carcinogenesis. 1999; 20:641–644.

130. Surh YJ. Anti-tumor promoting potential of selected spice ingredients with antioxidative and anti-inflammatory activities: a short review. Food Chem Toxicol 2002; 40:1091–1097.

131. Ohtsu H, Xiao Z, Ishida J, Nagai M, Wang HK, Itokawa H, Su CY, Shih C, Chiang T, Chang E, Lee Y, Tsai MY, Chang C, Lee KH. Antitumor agents. Part 217: curcumin analogues as novel androgen receptor antagonists with the potential as anti-prostate cancer agents. J Med Chem 2002; 45:5037–5042.

132. Ishida J, Ohtsu H, Tachibana Y, Nakanishi Y, Bastow KF, Nagai M, Wang HK, Itokawa H, Lee KH. Antitumor agents. Part 214: synthesis and evaluation of curcumin analogues as cytotoxic agents. Bioorg Med Chem 2002; 10:3481–3487.

133. Albanes D, Heinonen OP, Taylor PR, Virtamo J, Edwards BK, Rautalahati M, Hartman AM, Palmgren J, Freedman LS, Haapaskoki J, Barrett MJ, Pietinen P, Malila N, Tala E, Liippo K, Salomaa ER, Tangrea JA, Teppo L, Askin FB, Taskinen E, Erosan Y, Greenwald P, Huttunen JK. Alpha-tocopherol and beta-carotene supplements and lung cancer incidence in the alpha-tocopherol, beta-carotene cancer prevention study; effects of base-line characteristics and study compliance. J Natl Cancer Inst 1996; 88:1560–1570.

134. ATBC (1994) Alpha-tocopherol beta-carotene cancer prevention study group.The effect of vitamin E and beta-carotene on the incidence of lung cancer and other cancers in male smokers. N Engl J Med 330:1029-1035.

135. Omenn GS, Goodman GE, Thornquist MD, Barnhart S, Balmes J, Cherniak M, Cullen M, Glass A, Keogh J, Liu D, Meyskens FJ, Perloff M, VAlanis B, Williams JJ.

Chemoprevention of lung cancer: the beta-carotene and retinol efficacy trial (CARET) in high-risk smokers and asbestos-exposed workers. IARC Sci Publ 1996; 136:67–85.

136. Omenn GS, Goodman GE, Thornquist MD, Balmes J, Cullen MR, Glass A, Keogh JP, Meyskens FL, Valanis B, Williams JHJ, Barnhart S, Cherniack MG, Brodkin CA, Hammar S. Risk factors of lung cancer and for intervention effects in CARET, the beta-carotene and retinol efficacy trial. J Natl Cancer Inst 1996; 88:1550–1559.

137. Vainio H. Chemoprevention of cancer: lessons to be learned from beta-carotene trials. Toxicol Lett 2000; 112–113: 513–517.

138. Grant SG, Zhang YP, Klopman G, Rosenkranz HS. Modeling the mouse lymphoma forward mutational assay: the Gene-Tox Program database. Mutation Res 2000; 465:201–229.

139. Rosenkranz HS, Zhang YP, Klopman G. Evidence that cell toxicity may contribute to the genotoxic response. Regulatory Toxicol Pharmacol 1994; 19:176–182.

140. Rosenkranz HS, Cunningham AR. The High Production Volume Chemical Challenge Program: the relevance of the in vivo micronucleus assay. Regulatory Toxicol Pharmacol 2000; 31:182–189.

141. Surh YJ. Molecular mechanisms of chemopreventive effects of selected dietary and medicinal phenolic substances. Mutation Res 1999; 428:305–327.

142. Klopman G, Fercu D, Li JY, Rosenkranz HS, Jacobs MR. Antimycobacterial quinolones: a comparative analysis of structure–activity and structure–cytotoxicity relationships. Res Microbiol 1996; 147:86–96.

143. Rosenkranz HS, Klopman G, Macina OT. Evaluation of therapeutic benefits and toxicological risks using structure–activity relational expert systems. The Benefit/Risk Ratio, A Handbook for the Rational Use of Potentially Hazardous Drugs. Boca Raton, FL: CRC Press, 1998:29–55.

144. Higgins TF, Borron SW. Coma and respiratory arrest after exposure to butyrolactone. J Emerg Med 1996; 14:435–437.

145. Bradsher K. Daughter's death prompts fight on 'date rape' drug. New York Times. New York, NY, Oct 16, 1999.

146. Bernasconi R, Mathivet P, Bischoff S, Marescaux C. Gamma-hydroxybutyric acid: an endogenous neuromodulator with abuse potential. Trends Pharmacol Sci 1999; 20:135–141.

147. Stout D. New drug can induce coma or death, F.D.A. warns. New York Times. New York, NY, Jan 22, 1999.

148. McKenzie R, Fried MW, Sallie R, Conjeevaram H, DiBisceglie AM, Park Y, Savarese B, Kleiner D, Tsokos M, Luciano C, Pruett T, Stotka JL, Straus SE, Hoofnagle JH. Hepatic failure and lactic acidosis dueto fialuridine (FIAU), an investigational nucleoside analogue for chronic hepatitis B. N Eng J Med 1995; 333:1099–1105.

149. Olson H, Betton G, Robinson D, Thomas K, Monro A, Kolaja G, Lilly P, Sanders J, Sipes G, Bracken W, Borato M, Van Duen K, Smith P, Berger B, Heller A. Concordance of the toxicity of pharmaceuticals in humans and in animals. Regulatory Toxicol Pharmacol 2000; 32:56–67.

150. IARC. IARC Monographs on the Evaluation of Carcinogenic Risks to Humans. Vol. 71. Re-Evaluation of Some Organic Chemicals, Hydrazine and Hydrogen Peroxide (Part Two). Lyon, France: International Agency for Resarch on Cancer, 1999:367–382.

151. Strobl GR, von Kruedener S, Stockigt J, Guengerich FP, Wolff T. Development of a pharmacophore for inhibition of human liver cytochrome P-450 2D6: molecular modeling and inhibition studies. J Med Chem 1993; 36:1136–1145.

152. de Groot M, Ackland MJ, Home VA, Alex AA, Jones BC. Novel approach to predicting P450-mediated drug metabolism: development of a combined protein and pharmacophore model for CYP2D6. J Med Chem 1999; 42:1515–1524.

153. de Groot M, Ackland MJ, Home VA, Alex AA, Jones BC. A novel approach to predicting P450 mediated drug metabolism. CYP2D6 catalyzed N-dealkyalation reactions and qualitative metabolite predictions using a combined protein and pharmacophore model for CYP2D6. J Med Chem 1999; 42:4062–4070.

154. Merck. The Merck Index: An Encyclopedia of Drugs-, Chemicals-, and Biologicals. 12th ed. Whitehouse Station, 1996:262.

155. Pollack N, Cunningham AR, Klopman G, Rosenkranz HS. Chemical diversity approach for evaluating mechanistic relatedness among toxicological phenomena. SAR QSAR Environ Res 1999; 10:533–543.

156. NAS. Toxicity Testing. Strategies to Determine Needs and Priorities ed. Washington, DC: National Academy of Sciences, National Academy Press, 1984.

157. Langston JW, Ballard P, Tetrud JW, Irwin I. Chronic Parkinsonism in humans due to a product of meperidine-analog synthesis. Science 1983; 219:979–980.

158. Betarbet R, Sherer TB, MacKenzie G, Garcia-Osuna M, Panov AV, Greenamyre JT. Chronic systemic pesticide exposure reproduces features of Parkinson's disease. Nature Neurosci 2000; 3:1301–1306.

159. Rosenkranz HS. A data mining approach for the elucidation of the action of putative etiological agents: application to the non-genotoxic carcinogenicity of genistein. Mutation Res 2003; 526:85–92.

160. Steele VE, Pereira MA, Sigman CC, Kelloff GJ. Cancer chemoprevention agent development strategies for genistein. J Nutrition 1995; 125:713S–716S.

161. Mutoh M, Takahashi M, Fukuda K, Komatsu H, Enya T, Matsushima-Hibiya Y, Mutoh H, Sugimura T, Wakabayashi K. (2000). Suppression by flavonoids of cyclooxygenase-2 promoter-dependent transcriptional activity in colon cancer cells: structure–activity relationship. Jpn J Cancer Res 2000; 91:686–691.

INTRODUCTORY LITERATURE

1. Klaassen C. Casarett & Doull's Toxicology, The Basic Science of Poisons. 6th ed. McGraw Hill, 2001.

2. Marquardt H, Schafer S, McClellan R, Welsch F, eds. Toxicology. Academic Press, 1999.

GLOSSARY

Allergic contact dermatitis (ACD): This is also known as "delayed hypersensitivity" and is caused by chemicals. The phenomenon consists of two phases: (a) an initial sensitization due to contact with an allergen, and (b) the elicitation of an adverse inflammatory dermal response upon subsequent exposure to the allergen.

Aneuploidy: This involves the gain or loss of one or more chromosomes. This process is the result of exposure to a chemical. Aneuploidy may result from disturbance of tubulin or microtubules.

Apoptosis: Also known as "programmed cell death," this is a cellular mechanism that is involved in cell repair or in the deletion of injured cells.

Base substitution mutation: This refers to a mechanism of gene mutations whereby one base pair in DNA is replaced by another one (e.g., guanine-cytosine to adenine-thymine).

Biophore: This refers to a SAR-derived substructure that is significantly associated with a specific biological activity.

Carcinogen: This refers to a chemical that induces cancers in mammals, including humans.

Case report: This is a description of defined symptoms associated with exposure to a chemical. These reports are usually limited to a single or several individuals. A causal relationship between exposure and effect can rarely be shown based upon case reports.

Chromosomal aberrations: These are changes in chromosome structure (breakages, rearrangements) detectable with the light microscope. Chemicals that induce chromosomal aberrations are potential carcinogens.

Congeneric chemicals: This refers to a group of chemicals that belong to a common chemical class, e.g., chloroarenes, phenols, aromatic amines.

COX-2 (cyclooxygenase-2): This is an enzyme that is involved in the inflammatory process.

Cyp2P6 (cytochrome P4502D6): This is an enzyme that is involved in the biotransformation of a number of drugs. Inhibition of this enzyme may lead to enhanced responses to certain drugs.

Developmental toxicity: This refers to prenatally induced effects that result in fetal malformations and dysfunctions.

***E. coli* WP2uvrA:** This refers to a strain of the bacterium *Escherichia coli* that is used to score mutations induced by chemicals. This assay is used to identify potential carcinogens.

Electrophiles: These represent molecules that contain electron-deficient atoms. The biotransformation of nonelectrophilic chemicals to electrophiles may render them toxic. Electrophiles react with nucleophiles (e.g., nucleic acids, proteins).

Error-prone DNA repair: When cellular DNA is damaged, the cell attempts to repair that damage. One of the mechanisms of DNA repair acts by a process that lacks fidelity. In fact, this introduces mutations. The assay is also known as "SOS repair." Tests designed to measure this error-prone DNA repair, also known as "SOS chromotest," are used to identify potential carcinogens.

Frameshift mutation: This refers to a mechanism of mutation that results from the gain or loss of one or two DNA base pairs.

G2/M phase: This describes a distinct stage in the cell cycle between the completion of DNA synthesis and cell division.

Genotoxicant: This is a chemical that induces modifications, usually damaging ones, in the cellular DNA.

$\alpha_2\mu$-Globulin nephropathy: This refers to a nephropathy (i.e., renal injury) induced by certain chemicals in male (but not female) rats that results in the accumulation of protein droplets in a segment of the proximal tubule. This may lead to a chronic nephropathy which may result in renal adenomas and carcinomas. There does not appear to be a corresponding pathology in humans.

Inhibition of intercellular gap junctional communication: This refers to the interruption of the exchange of small regulatory molecules between cells through specific channels (gap junctions). The ability to block this process has been associated with tumor promoters.

Maximum tolerated dose (MTD): The MTD is generally defined as the dose of a chemical that decreases the growth rate of a young animal by 10%.

Micronuclei in vivo, induction of: This refers to an assay performed on rodent bone marrow erythrocytes to determine the increase of chromosomal fragments induced by chemicals. This test is used to identify potential carcinogens.

Mitogenesis: This refers to induced cell proliferation. This process increases the chance of the induction of events (i.e., mutation) that increase the frequency of cancers. Toxicants can induce mitogenesis by causing cell death and subsequently this results in a process of regeneration.

Modulator: This refers to a SAR-derived substructure that affects the predicted activity associated with a specific biophore or toxicophore. A modulator may increase, decrease, or abolish the predicted activity.

NFκB: This refers to a cell transcription factor that can be activated by oxidants.

Non-congeneric chemicals: This refers a group of chemicals belonging to different chemical classes.

Oncogenes: These are genetic elements involved in pivotal events in the regulation of cell growth. When these genes are mutated (i.e., activated), this may result in the neoplatic process.

Phase II enzymes: This involves the biotransformation of a chemical to a product of greatly increased water solubility. This results in its more rapid excretion. This enzymic modification may involve conjugation with glucuronic acid, sulfate, glutathione, and others.

Postmitochondrial activation mixture: This is a cell-free extract (also known as "S9"), derived usually from rodent livers, fortified with coenzymes. It is used to "activate" (i.e., convert) promutagens and progenotoxicants to a chemical form (i.e., electrophile) that has the potential to react with cellular DNA and thereby induce DNA damage and consequently mutagenic/genotoxic effects.

***Salmonella* mutagenicity:** This refers to a widely used method (also known as the Ames test) using strains of the bacterium *Salmonella typhimurium*. The assay is used to identify chemicals that induce mutation and therefore have the potential to cause cancers.

***Salmonella typhimurium* tester strains:** There are strains of bacteria used to detect chemically induced mutations. The strains have been genetically engineered to be hypersensitive to chemical mutagens. A *Salmonella* mutagenicity assay is carried out using up to five strains, each responding to different types of mutations (e.g., base substitution, frameshift).

Sensitivity: This refers to the number of positive responses/total number of positives tested.

Sister chromatid exchanges, induction of: This is a test performed on cultured cells or whole animals. The assay determines the increase in symmetrical switches between sister chromatids on the cell's chromosomal complement as a result of exposure to chemicals. The exchanges are visualized as dye-stained sections along the length of the chromosome. This assay is used to identify potential carcinogens.

Specificity: This refers to the number of negative responses/total number of negatives tested.

Suppressor genes (tumor suppressor genes): These represent genetic elements involved in the regulation of the cell cycle. When these genes are mutated, as a consequence of exposure to chemicals, malignancy may be initiated.

TD_{50}: This refers to the dose of a chemical carcinogen that allows 50% of the chemicals to remain tumor-free.

tk^+/tk^-: This refers to the target in culture mouse lymphoma cells that is used to score mutations induced by chemicals. That assay is used to identify potential carcinogens. tk^+/tk^- indicates the thymidine locus.

Toxicophore: This refers to a SAR-derived substructure that is significantly associated with a specific toxicological activity.

Tubulin: This refers to the principal chromosomal protein of spindle fibers. Inhibition of tubulin polymerization may block cell division. Hence, this inhibition is a target of cancer therapeutic drugs.

Tumor promotion: This is considered the second stage of cancer progression. It follows tumor initiation (often due to a genotoxic event). The chemicals that induce promotion are not thought to interact with the cellular DNA.

Unscheduled DNA synthesis: This refers to an assay that detects the repair of damage to the cellular DNA. The cells of choice for this assay are rat hepatocytes. This assay is used to detect potential carcinogens.

10

OncoLogic: A Mechanism-Based Expert System for Predicting the Carcinogenic Potential of Chemicals

YIN-TAK WOO and DAVID Y. LAI

Risk Assessment Division, Office of Pollution Prevention and Toxics, U.S. Environmental Protection Agency, Washington, D.C., U.S.A.

1. INTRODUCTION

The ability to predict the potential toxicity of an untested chemical based on structure–activity relationships (SAR) analysis is dependent on the knowledge of toxicological information on structurally or functionally related compounds and of the possible mechanism(s) of action that contribute to the specific toxicity endpoint of interest. Among all the toxicity endpoints, carcinogenicity is undoubtedly one of the most difficult endpoints to predict because of the complexity of its mechanisms of action (see Sec. 2 below) and the difficulty of obtaining a robust, well-balanced database needed for effective SAR analysis. Beyond that, cancer data are often difficult to interpret or model because of variability associated with long-term studies, differences in the species/strains of testing animals used, the route of administration, and the specific testing protocol. Prospective validation and hypothesis testing are also difficult because of the high cost and the long duration of time needed for carcinogenesis bioassays.

Despite the difficulties, many different SAR, QSAR, and data mining methods have been developed to predict carcinogenic potential of untested new and existing chemicals. Several review articles (1–8) have evaluated, compared, and contrasted many of these methods, which include: (a) qualitative expert judgment; (b) classical QSAR studies using regression analysis, principal component and factor analysis, discriminant and pattern recognition analysis, similarity analysis, neuronal nets, etc.; (c) data mining methods involving machine learning to discover SAR features, classify active and inactive compounds, and/or induce knowledge rules; (d) biologically based models such as receptor modeling, docking, and ligand SAR; and (e) integrative models incorporating both chemical and biological information. To evaluate and compare the predictive capabilities of various methods, the U.S. National Toxicology Program (NTP)/National Institute of Environmental Health (NIEHS) sponsored two predictive challenges to the international scientific communities to make prospective prediction of the outcome of two series of cancer bioassays involving 44 and 30 chemical compounds, respectively (see Refs. 9,10 for review). Another international workshop, The Predictive Toxicology Challenge 2000–2001, was devoted to the simulation of prediction of rodent carcinogenicity of noncongeneric compounds using NTP bioassay database as the training set (11,12). The totality of these past experiences showed that most models, even those with great internal cross-validation, may not fare very well in prospective predictions, particularly with noncongeneric compounds. Although opinions may vary, most reviewers and modelers tend to agree that: (a) Expert judgment is crucial for the development and interpretation of predictive systems. (b) Predictive systems that incorporate both chemical and biological information tend to perform better. (c) The predictive capability of any predictive system is dependent on the quality, appropriateness, and coverage of its training database/knowledge base. Predictions based on closely related or congeneric series of compounds tend to be more accurate than those derived from noncongeneric database. (d) Knowledge of potential mechanism of action of the untested and/or related compounds may facilitate prediction by focusing on the most relevant descriptors, and by providing mechanistic support for the prediction.

In this chapter, we describe the development of a mechanism-based expert system for predicting the carcinogenic potential of chemicals based on organized, formalized, and codified knowledge rules developed by a team of chemical carcinogenesis SAR experts through decades of experience in analyzing the structural and biological bases of various classes of chemical carcinogens and in applying predictive strategy to assessing carcinogenic potential of new and untested chemicals at the U.S. Environmental Protection Agency (EPA).

2. MECHANISM-BASED STRUCTURE–ACTIVITY RELATIONSHIPS ANALYSIS

2.1. Multistage, Multifactorial Process of Chemical Carcinogenesis

To understand the basis of mechanism-based SAR analysis, a brief overview of the mechanism of chemical carcinogenesis is needed. It is now well documented that chemical carcinogenesis is a multistage, multifactorial process that involves numerous exogenous and endogenous factors intertwined in a network of pathways and feedback loops (13). Conceptually, the complete carcinogenesis process consists of three operational stages—initiation, promotion, and progression. Initiation involves a mutational event that may involve gene, mutation, chromosome aberration, translocation, and instability. Promotion involves clonal expansion of initiated cells to reach a critical mass by a variety of means such as cell proliferation, inhibition of programmed cell death, inhibition of terminal differentiation, and loss of growth control. Progression may involve a second mutational event, the loss of tumor suppressor gene, impairment of immune surveillance, and acquisition of ability to metastisize.

From the mechanistic point of view, there are basically two types of carcinogens—genotoxic and epigenetic/nongenotoxic. Genotoxic carcinogens, also known as DNA-reactive carcinogens, are chemicals that directly interact with DNA either as parent chemicals or as reactive metabolites. Epigenetic carcinogens are agents that act through a secondary mechanism that does not involve direct DNA damage. In reality, the demarcation is seldom absolute. Most potent genotoxic carcinogens are also endowed with epigenetic activities that may act synergistically to carry out the complete carcinogenic process. Most epigenetic carcinogens may indirectly cause DNA damage, facilitate mutagenesis by fixation of promutagenic DNA damage prior to repair, or increase error-prone DNA repair. It would therefore be more accurate to consider carcinogens as *predominantly* genotoxic or epigenetic. In general, genotoxic carcinogens are more likely to induce tumors in multiple targets and species, whereas epigenetic carcinogens are more likely to be single species-/target-carcinogens.

2.2. SAR of Genotoxic and Epigenetic Carcinogens

There are numerous examples of chemical carcinogens that act predominantly by genotoxic mechanisms. The classical major structural classes of genotoxic carcinogens are: direct-acting carcinogens (including epoxides, aziridines, nitrogen and sulfur mustards, α-haloethers, and lactones); aromatic amines and nitroaromatics; nitrosamines and nitrosamides; hydroazo and azoxy compounds; carbamates; organophosphates; aflatoxin-type furocoumarins; and homocyclic and heterocyclic polycyclic aromatic hydrocarbons. The SAR features of most of these major classes have been reviewed (14–18). The key common features for potent carcinogens are: (a) propensity to be or to generate electrophilic intermediate, (b) availability of a stabilizing mechanism to allow transport of reactive intermediates from the site of activation to the site of interaction for DNA covalent binding, (c) characteristics of persistent DNA adducts, and (d) ability to act on various stages of carcinogenesis.

The scientific literature on epigenetic mechanisms of chemical carcinogens has been growing at an explosive pace in the past several years because of the importance of mechanistic understanding in elucidating the molecular basis of carcinogenesis, considering human relevance of animal data, and modeling quantitative risk assessment. Epigenetic carcinogens include agents that: (a) act via receptor-mediated mechanisms, (b) generate reactive oxygen species to cause indirect DNA damage, (c) cause aberrant gene expression, (d) disturb the homeostatic status of cells, and (e) elicit cytotoxicity with subsequent compensatory regenerative hyperplasia. In general, cytotoxic agents tend to operate at relatively high dose levels, whereas receptor-mediated agents may operate at low doses. For agents that involve reversible binding to receptors, the two key common elements are favorable molecular size/shape, and long biological half life to allow sustained binding/activation of the receptor. The principal mechanisms include: (a) peroxisome proliferation, (b) arylhydrocarbon (Ah) receptor-mediated and other enzyme induction, (c) inhibition of gap junctional intercellular communication, (d) oxidative stress, (e) perturbation of DNA methylation and gene expression, (f) hormonal imbalance, (i) cytotoxicity-induced regenerative cell proliferation, (j) inhibition of microtululin polymerization, and (k) impairment of immune surveillance. The SAR features of the agents that act these mechanisms have been reviewed (19).

It is important to point out that the SAR features contributing to genotoxic mechanisms often differ from those contributing to epigenetic/nongenotoxic mechanisms. An examination (20) of the extents of overlaps between the toxicophores associated with: (a) inhibition of gap junctional intercellular communication (an epigenetic mechanism), (b) mutagenicity in Ames *Salmonella* tests (a genotoxic indicator), and (c) rodent carcinogenicity showed substantial overlap (59%) between (b) and (c), some overlap (23%) between (a) and (c), but minimal overlap (9%) between (a) and (b) suggesting that (a) and (b) lack commonality and may contribute to carcinogenicity independently. Thus, it would be paramount to consider both the genotoxic and epigenetic SAR features for integrative assessment of the carcinogenic potential of chemicals. Some potent carcinogens may use different portions of the same molecule to carry out genotoxic and epigenetic mechanisms (19).

2.3. Basic Principles of Mechanism-Based SAR Analysis

The basic principles and concepts of mechanism-based SAR analysis have been discussed (18,21,22). Essentially, mechanism-based SAR analysis involves comparison with structurally related compounds with known carcinogenic activity, identification of structural moiety(ies) or fragment(s) that may contribute to carcinogenic activity through a perceived or postulated mechanism, and evaluating the modifying role of the rest of the molecule to which the structural moiety/fragment is attached to. Chemicals with features or properties (e.g., high molecular weight, strong metal chelation) indicative of lack of bioavailability by the anticipated route(s) of exposure may also be excluded from consideration and assigned a low concern with stipulation that the concern may change if a different route of exposure may occur. For chemicals with no structural features, integration of biochemical/pathobiological properties and short-term predictive tests may provide a basis for prediction.

3. THE ONCOLOGIC EXPERT SYSTEM

3.1. Overview of the OncoLogic System

The OncoLogic Cancer Expert System is a rule-based expert system developed jointly by te Office of Pollution Prevention and Toxics of the U.S. EPA and an expert system developer, LogiChem, Inc. They used organized, formalized, and codified knowledge rules developed by a team of carcinogenesis SAR experts through decades of experience in analyzing the structural and biological bases of various classes of chemical carcinogens and in applying predictive strategy to assessing carcinogenic potential of new and untested chemicals at the U.S. EPA. It has a flexible infrastructure that allows assessment of virtually any type/class of chemical substances as long as there is sufficient knowledge, test data, and mechanistic information to develop predictive method that is specific for that chemical type/class. Depending on the type of chemical, the input to the system includes the chemical structure as well as all available chemical, biological, and mechanistic information (e.g., physicochemical properties, chemical stability, route of exposure, bioactivation and detoxification, genotoxicity, and other supportive data) critical to the evaluation of carcinogenic potential. The output from the system includes not only the prediction of the carcinogenic potential but also the underlying scientific rationale for the prediction and the existing carcinogenicity knowledge of the type/class of compounds to which the chemical belongs. The carcinogenic potential is expressed semiquantitatively in a scale of six concern levels ranging from: (a) low (unlikely to be carcinogenic), (b) marginal (equivocal or marginal carcinogen or carcinogenic only at doses at or exceeding maximum tolerated dose), (c) low-moderate (likely to be a weak carcinogen toward a single species/target, or carcinogenic at relatively high dose), (d) moderate (likely to be moderately active carcinogen toward one or more species/target), (e) high-moderate (highly likely to be an active carcinogen toward one or more species/target, or a potent carcinogen at high doses), to (f) high (highly likely to be a potent carcinogen even at relatively low doses, or carcinogenic toward multiple species/targets). These concern levels can be considered composite scoring of relative potential carcinogenic potency, as well as confidence of prediction.

3.2. Sources of Knowledge and Information and Methods for Rule Development

The major sources of information and data from which the knowledge rules were derived include: (a) a six-volume series of monographs entitled *Chemical Induction of Cancer* in which virtually all the structural classes of chemical carcinogens were systematically collected, reviewed, and analyzed with emphasis on SAR features and mechanism of action (14–18), (b) International Agency for Research on Cancer monograph series, (c) U.S. National Cancer Institute/National Toxicology Program technical report series, (d) U.S. Public Health Service publication series 149 entitled *Survey of Compounds Which Have Been Tested for Carcinogenic Activity*, (e) the Carcinogenic Potency Database (see Ref. 23 for the latest version), and (f) nonclassified chemical industry and EPA research data from various program offices such as Office of Pollution Prevention and Toxics, Office of Pesticide Program. In addition, relevant current literature on carcinogenicity, SAR, and mechanistic studies have been continuously monitored and reviewed. External domain experts were consulted as needed. The process for developing SAR rules for each class/type of compounds involves: (a) gathering all available information and data, (b) determining the need for subclassification to optimize prediction, (c) brainstorming to determine all important structural features and secondary modifying factors, and (d) assigning importance to each factor and narrowing down to critical factors that are amenable to system development. The rules were incorporated into decision trees and used along with user input for SAR analysis.

3.3. Overall Structure of the System

The overall structure of the OncoLogic system (24) consists of: (a) a structural arm that evaluates the carcinogenic potential of a chemical by mechanism-based SAR analysis, and (b) a functional arm that evaluates the carcinogenic potential of a chemical by integrating the available biological short/medium-term predictive data or tests on the chemical. The structural arm, in turn, consists of four subsystems for the prediction of carcinogenic potential of organic chemicals, polymers, metals/metalloids, and fibrous substances, respectively. The functional arm can be used to provide support for the structural arm as well as an independent method of analysis.

3.4. Structural Arm: Predicting by Structure–Activity Relationships Analysis

The Fiber Subsystem is designed for evaluation of carcinogenic potential of fibrous substances. It is well documented (18) that the most critical factors that contribute to carcinogenic potential of natural mineral and synthetic vitreous fibers are physical dimension (diameter and aspect ratio) and physicochemical properties (e.g., flexibility, durability, surface characteristics, and splitting potential). The in vivo biodegradability/biopersistence is also an important factor. The actual chemical composition appears to play a relatively minor role. Using the data of Stanton et al. (25) as a guide, fibers are given an initial carcinogenic concern based on their physical dimension with the highest concern associated with long and thin fibers. The initial concern level is then modified according to user input on the physiochemical properties or manufacturing process (e.g., thermal extrusion vs. slow crystallization), and exposure scenario to arrive at the final concern. Since fiber dimensions are usually given as a range rather than a discrete number, two methods of evaluation—the standard approach using average size and the conservative approach using the smallest diameter and longest length—are used to give the average concern and worst case scenario concern.

The Polymer Subsystem is designed for evaluation of polymeric substances with number average molecular weight (MW) exceeding 1000. Owing to the lack of or poor bioavailability, most high MW polymers are considered "safe." However, under special situations such as presence of low MW reactive monomers or prepolymers, oral exposure to polysulfated polymers, inhalational exposure to swellable water soluble polymers or lung overloading, polymeric substances could be carcinogenic. In addition, some polymers are designed to be readily chemically or biologically degradable. The ensuing breakdown products may be of concern. The knowledge rules in this Subsystem capture these situations along with graphic menus for identifying reactive functional groups of concern as well as polymeric linkages that are known to be susceptible to breakdown. In addition, some of the guiding principles that the Agency uses to evaluate polymers have been incorporated.

The Metal/Metalloid Subsystem is designed to evaluate inorganic and organic compounds containing metals or metalloids using organized information distilled out of our monograph (18). It is divided into: (a) a rule-based submodule to evaluate major carcinogenic metals/metalloids such as arsenic, chromium, or cadmium, and (b) a database submodule to cover other metals and metalloids with relatively limited carcinogenicity information. For the rule-based submodule, the critical factors of concern are: (a) the nature of metal/metalloid, (b) the type of chemical bonding, (c) dissociability and solubility of the compound, (d) oxidation state of the metal/metalloid, (e) physical state (for solids, whether it is crystalline or amorphous), (f) exposure scenario, and (g) possible breakdown products and organic ligands. By responding to the questions posed by the Subsystem, for instance, the user will get vastly different concern level for inhalation exposure to a sparingly soluble crystalline Cr(VI) compound vs. dermal exposure to a soluble Cr(III) compound. For using the database submodule, the user will need to make some judgment in comparing the untested compound to the database on related metals/metalloids. In addition to the rule-based submodule and database, the Subsystem contains information on the chelating effect of various dyes and the ensuing impact on bioavailability of the metal/metalloid.

The Organic Subsystem contains over 50 chemical classes of organic compounds (Table 1). For some major classes such as aromatic amines, further subclassifications (e.g., monocyclic, annealated bicyclic, bicyclic with intercyclic linkage, and polycyclic) may be needed to maximize predictive accuracy. For each class/subclass, specific rules unique to the class/subclass have been developed (see Ref. 24 for details)

Table 1 List of Major Structural Classes of Chemicals Covered in OncoLogic

Acylating agents (including subclasses: acyl and benzoyl halides; anhydrides; carbamyl halides; phosgene-type compounds)
Aflatoxins and microbial toxins
Aromatic amines and arylazo compounds
Carbamates and thiocarbamates
Coumarins and furocoumarins
Direct-acting alkylating agents (including subclasses: acrylamides; acrylates and related; aldehydes; alkanesulfonoxy esters; alkyl and alkenyl halides; alkyl phosphates; alkyl sulfates and alkanesulfonates; epoxides; ethyleneimines; alpha-haloethers/thioethers; lactones and sultones; nitrogen and sulfur mustards; reactive ketones and sulfones)
Direct-acting arylating agents (including subclasses: aryldiazonium salts; alpha-halogenated heterocycles; halogenated nitroaromatics; nitroarenes)
Halogenated alkanes and alkenes
Halogenated aromatic hydrocarbons (including halogenated benzenes, naphthalenes, biphenyls, terphenyls, diphenyl ethers, dibenzo-*p*-dioxins, dibenzofurans, dibenzothiphenes)
Hydrazo compounds, aliphatic azo, and azoxy compounds, triazenes
Nitroalkanes and nitroalkenes
Nitroarenes
N-Nitrosamine and nitrosamides
C-Nitroso compounds and oximes
Organophosphorous compounds and pesticides
Peroxides and peroxy compounds
Phenols and phenolics
Polycyclic aromatic hydrocarbons (including homocyclic, heterocyclic, and substituted)
Siloxanes, siloxenes, silanols, and silyl halides
Thiocarbonyls

and used in combination with general exclusion rules such as molecular weight or molecular size cutoff. For example, for polynuclear aromatic hydrocarbons (PAHs), the key factors that are deemed crucial for contributing to carcinogenicity include: (a) a favorable molecular size and shape, (b) propensity to generate reactive intermediates (e.g., carbonium ion, free radical) after metabolic activation, and (c) capability of the ring system to provide resonance stabilization of the reactive intermediates. An effective method to measure the molecular size and shape is to calculate the incumbrance area (the smallest rectangular envelope that can contain the entire planar molecule of a PAH drawn proportionally to molecular dimension) and the aspect ratio (length/width). Virtually, all PAHs with incumbrance area exceeding 185 $\mathring{A}^2$ or aspect ratio greater than 2.1 are not carcinogenic (14). Special mechanistically sensitive areas/positions include the bay/fjord, K-, and L-regions, and the "+B" and "−P" positions (Fig. 1). Software programs have been developed to calculate the incumbrance area and locate these sensitive regions/positions as an input for determining the carcinogenic potential. For ring-substituted PAHs, the impact of the substituent on carcinogenic potential is determined by assessing its ring position, hydrophilicity/hydrophibicity, bulkiness, and effect on resonance stabilization. All these considerations are integrated for assessing the overall carcinogenic concern of the compound.

For chemicals with direct-acting alkylating/arylating/acylating functional groups, the position of the functional group (terminal vs. internal), the number of functional groups (monofunctional vs. polyfunctional) and their molecular flexibility, and intergroup distance are all important consideration. In addition, information on the expected route of exposure is an important factor. In general, exposure to direct-acting chemicals by inhalational or parenteral administration is expected to be more hazardous than exposure via drinking water or diet because they have a better chance of reaching target tissue/organ before being detoxified by protective nucleophiles in the body or foodstuffs.

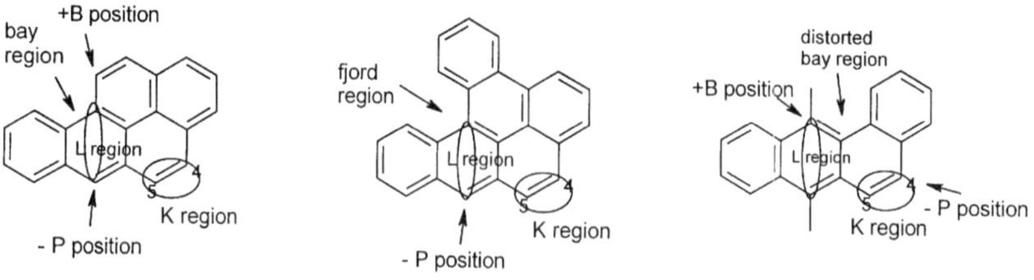

Figure 1 Important regions and positions of polynuclear aromatic hydrocarbons.

A similar information gathering and brainstorming approach has been carried out on each of the 50 chemical classes to identify structural features critical for determining the carcinogenic potential of chemicals within that class. For a number of miscellaneous chemicals for which no appropriate chemical class is available, a separate database of miscellaneous chemicals has been developed.

A sample run of an aromatic amine compound using the structural arm is shown in Appendix A along with abbreviated justification report to illustrate the scientific rationale used in the prediction.

3.5. Functional Arm: Predicting by Activity–Activity Relationships Analysis

The functional arm of the OncoLogic system uses a mechanism-based approach to organize and integrate all available noncancer short-/medium-term predictive test data of a chemical as a basis for predicting carcinogenic potential. It can be used to complement the structural arm as well as an independent method for prediction particularly for cases with little basis for SAR analysis. During the past two decades, numerous new short-term predictive tests have been developed to predict carcinogenic potential of chemicals. Most of these tests tend to perform well for some types of chemicals but not all types. The main reason is that chemical carcinogenesis is a multistage, multifactorial process, whereas most of the short-term predictive tests can only serve as an indicator of activity in one or two of these stage(s)/factor(s). Using expert judgment built into the system, the OncoLogic system either uses a structural class-specific decision tree to utilize a combination of predictive tests to cover all three mechanistic stages to complement the structural arm or an integrative approach of using all available mechanistically complementary predictive tests.

In the structural class-specific decision tree approach, a battery of predictive tests needed to provide a reasonable mechanism-based, data-supported prediction is designed for the class. For example, for organic peroxides, a decision tree—involving data input for tests for: (a) free radical formation, (b) alkylating activity, (c) genotoxicity or clastogenicity, and (d) 28-day assay in Sencar mouse skin measuring sustained inflammatory/hyperplastic response, modified DNA base (such as 8-hydroxy-deoxyguanosine), and *ras* oncogene activation—has been used to assess the carcinogenic concern level. The details of the approach has been published (26).

In the integrative approach, a variety of widely used short-term predictive tests and pathological biomarker endpoints known to be correlated to carcinogenicity have been selected and mechanistically assessed in terms of their potential contribution to tumorigenesis initiating, promoting, and/or progressing activity (Table 2). Depending on the type of the test/data, the dose/concentration eliciting positive/negative effect, the severity of the effect, a scoring system has been developed by expert judgment. Both positive and negative data are taken into consideration. If multiple tests/data are available in the same mechanistic stage, then the weight given to redundant tests/data get progressively lower. The scores for all the available tests/data are integrated to give a final score that, in turn, can be converted to concern level. The details of the scoring method have been described (27). A sample printout is shown in Appendix B. Since the method requires data to cover all three stages, it is likely that some chemicals may not have adequate data. For such cases, a partial score is shown along with explanation of what other data are needed.

3.6. Technical Specification and User Information

All the knowledge rules and information developed in the above process are organized and incorporated into decision trees and modules along with system and user interphase by the software engineer initially using rulemaster and subsequently C programming. Each of the components has been meticulously checked to ensure that the system closely mimics the human expert judgment and is free of inconsistency. The current version (5.1) of the OncoLogic system

Table 2 A Partial List of Short-/Medium-Term Predictive Tests and Their Putative Contribution to Various Stages of Caxcinogenesis

Short-/medium-term predictive tests or toxicological/pharmacological endpoints	Initiation	Promotion	Progression
(A) Oncogene/tumor suppressor genes			
(1) Activation of *ras*-like oncogene	x		
(2) Activation of *myc-like* oncogene		×	
(3) Inactivation of *p53-like* suppressor		×	x
(B) Transgenic rodent models.			
(1) *p53* deficient	×		
(2) Big Blue, MutaMouse or equivalent	×		
(3) *ras* oncogene expressed		×	×
(C) Genotoxicity and related tests			
(1) Ames	×		
(2) Chromosomc aberration	×		×
(3) Aneuploidy/polyploidy		×	x
(4) Cell transformation (e.g., SHE type)	×	×	×
(5) Cell transformation (e.g., 3T3 type)		×	×
(6) DNA binding or electrophilicity	×		
(D) Epigenetic tests			
(1) Peroxisome proliferation		×	
(2) Cell proliferation		×	
(3) DNA hypomethylation		×	×
(4) Intercell. comm. inhibition		×	
(5) Apoptosis in immune systems			×
(6) Cytochrome P450 induction		×	
(E) Subchronic toxicity endpoints			
(1) Serum increase in LH, TSH		×	
(2) Tissue hyperplasia		×	
(3) Atrophy of thymus/spleen			×
(4) Crystalluria/bladder stone		×	

was completed in 1998–1999. It is a PC-based, stand alone software with its own chemical drawing package. The hardware requirements are: (a) high density 3.5 in. drive, (b) color, VGA monitor, (c) MS-DOS version 5.0 or later, (d) at least 110 MB of free disk space, (e) 575 KB of free conventional memory, and (f) a mouse with mouse driver loaded in memory. In addition, a disk-caching program such as MS-DOS's "smartdrv" program is preferable. Currently, the OncoLogic system may be available from LogiChem, Inc. (P.O. Box 357, Boyertown, PA 19512) at distribution cost for governmental and nonprofit organizations or for purchase by for-profit organizations. At the time of this writing, the U.S. EPA is at the final stage of negotiation to purchase the distribution and copyrights to the system in order to upgrade to Window® version and make the system freely available to qualified users.

The OncoLogic system is primarily designed for users with basic knowledge of organic chemistry. In order to use the system, the user is required to place the chemical in the correct chemical class. Help screens are available to provide guidance to ensure correct class placement. Once in the correct class, the user may enter the chemical by selecting chemical name, CAS number or by drawing chemical structure using the drawing package provided. For new chemicals with no chemical names or CAS number, drawing will be the only means. In some cases, the users may be required to provide additional input such as expected route of exposure. In such cases, help screens will also be provided to explain the background and reasoning of the questions. Once all the information has been entered, the system will search for most appropriate decision tree and knowledge rules and give evaluation of the carcinogenic potential along rationale.

3.7. Validation, Performance, Strengths, and Limitations

The OncoLogic system was developed using all the available knowledge and mechanistic information that could be distilled out of available data on various classes of chemical carcinogens. Consequently, it was not possible to perform traditional internal cross-validation using leave-one-out or set aside training set and testing set types of techniques. Instead, the most appropriate techniques used were peer review process and prospective validation. During the developmental stage of individual classes of chemicals, the knowledge rules were peer reviewed by external domain experts as needed. In addition, after the completion of the versions 2.0 and 4.0, two international peer review panels of domain experts were invited to test the accuracy of the system and assess the scientific soundness of decision logics. The inputs were incorporated into the subsequent development.

For any predictive system, the ultimate foolproof method of testing the accuracy is prospective validation. At the time of the first prospective predictive exercise sponsored by NIEHS/NTP, the aromatic amine module of the OncoLogic system was completed. Table 3 summarizes the OncoLogic predictions of the carcinogenic potential of the eight aromatic amines and the actual bioassay test results. For five of these compounds, the OncoLogic system correctly predicted not only the carcinogenic potential but also the relative potency giving a low concern for the inactive chemical #1, a marginal concern for chemical #2 which was considered to have equivocal activity, a moderate concern for chemical #6 which has some evidence of carcinogenicity, and a high-moderate concern for chemicals #7 and #8 that have clear evidence of carcinogenicity. The predictions for the other three chemicals

Table 3 OncoLogic Prediction vs. NTP Bioassay Results: Aromatic Amines and Related Compounds

No.	Chemical name	NTP bioassay results[a] Rat	Mouse	Overall	OncoLogic prediction[b]
1	4,4x-Diamino-2,2x-stilbene disulfonic acid	NE/NE	NE/NE	–	L
2	*p*-Nitroaniline	NT	EE/NE	Eq	mar
3	*p*-Nitrobenzoic acid	NE/SE	NE/NE	(+)	mar
4	*p*-Nitrophenol	NT	NE/NE	–	LM
5	4-Hydroxyacetanilide	NE/EE	NE/NE	Eq	LM
6	2,4-Diaminophenol dihydrochloride	NE/NE	SE/NE	+	M
7	3,3x-Dimethylbenzidine	CE/CE	NT	+	HM
8	*o*-Nitroanisole	CE/CE	CE/CE	+	HM

[a]Bioassay results in male/female. Abbreviations used: CE, clear evidence; SE, some evidence; EE, equivocal evidence; NE, no evidence; NT, not tested. Overall rating: +, positive; Eq, equivocal; (+), weakly positive; –, negative.
[b]OncoLogic predictions (concern level): L, low; mar, Marginal; LM, Low-Moderate; M, Moderate; HM, High-Moderate.

were also quite close with a marginal concern for chemicals #3 which has weak positive data, a low-moderate concern for chemical #4 which is negative but only tested in one species, and a low-moderate for chemical #5 which has equivocal data. In the most recent NIEHS/NTP prospective predictive exercise, we made prediction on the carcinogenic potential of all 30 chemicals along with rationale using the OncoLogic approach (21). In addition, the potential target organs and relative potency were predicted and the confidence of predictions stated according to the specifications of the exercise. The predictions were completed at the end of 1996. Thus far, the bioassay results of all 30 chemicals have just been finalized. A comparison of all the predictions at this stage using close to 20 different methods has been published. The OncoLogic approach is among one of the best performers (10).

To be able to use any predictive system/model to its full potential, it is important to understand its strengths and limitations. The major strengths of the OncoLogic system are: (a) detailed prediction of carcinogenic potential with semiquantitative ranking of relative hazard, underlying scientific rationale, and information on related class of chemical, (b) its wealth of organized and carefully synthesized knowledge bases incorporating the essence of decades of research and practical experience of a team of domain experts in predicting carcinogenic potential of chemicals by SAR analysis, (c) a flexible system capable of incorporating both chemical and biological information to predict virtually any type of compounds, (d) ability to synthesize predictive information from multidisciplinary sources, and (e) predictions backed by mechanistic understanding and capable of generating testable hypothesis. The major limitations of the OncoLogic system include: (a) requirement of user to have a basic knowledge of organic chemistry and ability to place chemicals in their appropriate chemical class, (b) evaluation of chemicals by the structural arm is limited to the chemical structures covered by the available SAR knowledge of the related compounds, (c) system is mainly geared for industrial and environmental chemicals with limited coverage of pharmaceutical compounds, and

(d) system currently not amenable to large-scale, batch mode uses.

3.8. Current and Potential Uses

OncoLogic is a powerful, efficient and, economical carcinogenicity screening tool that can be used for many different purposes. These include: (a) *Product development*: OncoLogic has been used by chemical industries in their early stages of research and development of new products. It has been included as one of the predictive tools of EPA's pollution prevention framework (28) to encourage industries to design safer chemicals, search safer substitutes, and more environmentally friendly synthetic methods with fewer undesirable byproducts. A more detailed document on molecular design of safer chemicals has been published (29). (b) *Hazard identification and risk assessment*: For untested chemicals, OncoLogic has been used to identify potential hazard, to design experimental studies, and to select surrogate/analog chemicals that may be used for interim risk assessment. For tested chemicals, OncoLogic may be used to provide input to weight of evidence assessment, to elucidate potential mechanism of action, and to contribute to selection of quantitative risk assessment methodology. (c) *Health and environmental protection*: Owing to the high cost of carcinogenicity testing, hundreds of potentially hazardous chemicals in various environmental media remain to be tested. With limited resources, prioritization is the key to sensible and effective approaches. In this respect, OncoLogic has been used in many occasions. More recently, OncoLogic has been used to provide input for prioritization of disinfection by products in drinking water for testing and monitoring (22). (d) *Regulation*: OncoLogic has been used as one of the key predictive tools in regulation of new chemicals for which test data are not available or inadequate. There is also worldwide interest in expanding the use of SAR/QSAR tools for other regulatory purposes (e.g., product labeling, regulation of existing chemicals, or hazardous wastes) although the legal ramifications surrounding uncertainties associated with most of these tools remain to be resolved. With detailed descriptions of scientific rationale with most predictions, OncoLogic should be a useful tool in this respect.

3.9. Future Development

The current version of the OncoLogic system was completed in 1999. Work is underway for updating and upgrading the system. The coverage of chemicals will be expanded from the initial focus of covering virtually all the known classes of chemical carcinogens to virtually all classes of chemicals including those that are generally considered as safe so that the user will be able to run virtually any class of chemicals. The user interphase will be improved to enhance the user friendliness. Attempts will be made to incorporate significant new structural features discovered in the past several years into the OncoLogic system. The functional arm will be expanded to cover important new biological endpoints useful for predicting carcinogenicity (e.g., Ref. 30) with emphasis on human significance. In this respect, the emerging fields of "omics" technologies (genomics, proteomics, and metabonomics), high throughput screening assays, and systems biology (e.g., Refs. 31–34) may offer great promise of integrating with traditional SAR to reveal, assess, and elucidate the mechanism(s) of action of known and suspected carcinogens, to serve as a tool for biomarkers of exposure, to gain insight to in vivo metabolism, to generate hypothesis-driven research, and to improve predictive capability. The ultimate goal of the OncoLogic system will be to predict the realistic carcinogenic potential of chemicals to humans using all the available information along with reasoning and rationale that are transparent to users.

NOTE:1

The scientific views expressed are solely those of the authors and do not necessarily reflect the views or policies of the U.S. EPA. Mention of trade names, commercial products, or organizations does not imply endorsement by the U.S. Government.

ACKNOWLEDGMENTS

We thank Profs. Joseph Arcos and Mary Argus for initiating and supporting the project and Dr. Marilyn Arnott and her colleagues at LogiChem, Inc. for their contribution.

APPENDIX A

Abbreviated Justification Report from a Sample Run of an Aromatic Amine Compound Using the Structural Arm of OncoLogic

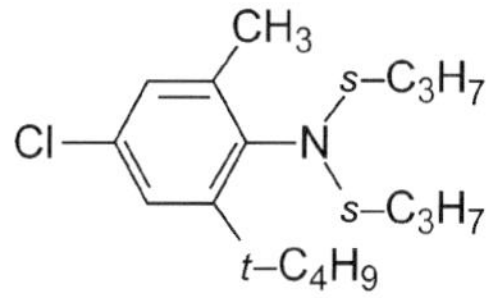

Summary

Code number: XYZ12345

Substance ID: *N,N*-Di-(*s*-propyl)-2-methyl-4-chloro-6-*t*-butylaniline

The level of carcinogenicity concern for this compound is marginal.

Justification

In general, the level of carcinogenicity concern of an aromatic amine is determined by considering the number of rings, the presence or absence of heteroatoms in the rings; the number and position of amino groups; the nature, number, and position of other nitrogen-containing "amine-generating groups"; and the type, number, and position of additional substituents. Aromatic amine compounds are expected to be metabolized to *N*-hydroxylated/*N*-acetylated deruvatuves, which are subject to further bioactivation to produce electrophilic reactive intermediates capable of interaction with DNA to initiate carcinogenesis.

Di-alkyl substituted amino groups of aryl compounds may, subject to the influence of the chain length and bulkiness of the alkyl groups, undergo *N*-dealkylation followed by *N*-acetylation and further bioactivation.

The evaluation of this compound proceeds as if the dialkyl substituted amino group, NR_1R_2 (where $R_1 = s$-propyl; $R_2 = s$-propyl) were a free amino group. The influence of the *N*-alkyl groups on the bioactivation of the compound is considered at the end of the evaluation.

The *t*-butyl and methyl groups flanking the amino group are not expected to significantly reduce the likelihood that it will be bioactivated to form electrophilic reactive intermediates. Since these groups have no effect on the bioactivation of the amino group due to their position relative to the amino group, they are considered with other ring substituents.

An aromatic compound containing one benzene ring, one amino group, and one methyl or methoxy group ortho to the amino group, has a carcinogenicity concern of high-moderate.

The additional chloro and/or bromo group(s) generally raise(s) the level of concern, but they also impose an upper limit of high-moderate on the concern level of the compound. Therefore, the level of concern remains high-moderate.

A single hydroxyalkyl, alkyl, or alkoxy ($3 < c < 6$) group lowers the level of carcinogenicity concern to moderate.

The R groups of the above listed amine–generating groups are expected to interfere with the bioactivation of this compound. The modification factor determined for this amine-generating group reduces the level of carcinogenicity concern for this aromatic amine.

The alkyl ring substituents flanking the amino/amine-generating group are not expected to exert any significantly flanking effect.

Therefore, the level of carcinogenicity concern for this compound is reduced to marginal.

APPENDIX B

Abbreviated Justification Report from a Sample Run of a Hypothetical Compound Using the Functional Arm of OncoLogic

Summary

Code number: Test007

Compound ID: Compound ABC

Test results entered:

- *p53*-deficient transgenic mouse model: Moderate positive; in vivo
- Ames test: Moderate positive; in vitro
- DNA binding: Moderate positive; in vivo, persistent and specific
- Cytochrome P450 induction: Weak positive; in vivo
- Immunosuppression: Weak negative, in vivo

Based on the reported test results, there is evidence that the compound may be active in the initiation and promotion stages of carcinogenesis but suggestive evidence that the compound may be inactive in the progression stage. Therefore, a concern level of moderate is assigned. It is suggested that additional predicted tests for progression be conducted to ascertain the negative finding.

Score: 44.58

Overall concern: moderate

Justification

In the absence of actual carcinogenesis bioassay data, a wide variety of noncancer toxicological endpoints may be used to predict carcinogenic potential of chemicals. Functional analysis of OncoLogic provides a conceptual framework to facilitate evaluations based on these endpoints.

OncoLogic considers the multistage nature of carcinogenesis in its use of functional predictive data. The assays and biological effects covered in OncoLogic have been weighted and assigned to initiation, promotion, and progression stages of carcinogenesis based on current expert judgement of their relative importance in contributing to each.

Results observed in vivo are given more weight than in vitro because it is implicit.

Stage-Specific Results

The information you have entered is organized below according to positive and negative results contributing to the assessment of activity in each stage of carcinogenesis. The strength of each response is indicated by ⟨strong⟩, ⟨moderate⟩, or ⟨weak⟩.

The compound may act as an initiator of carcinogenesis because of positive results observed in the following tests: *p53*-deficient transgenic mouse ⟨moderate +⟩ gene/point mutation: Ames ⟨moderate +⟩⟨; DNA binding: in vivo⟩, ⟨weak +⟩

The compound may act as a promoter of carcinogenesis because of the positive results observed in the following tests: cytochrome P450 induction: ⟨weak +⟩

Although the compound is negative in the following test, additional negative test results would be required to be confident that the compound is inactive in the progression stage: immunosuppression: in vivo ⟨weak −⟩

Test Interpretation

The *p53*-deficient transgenic mouse is useful for detecting genotoxic carcinogens...

REFERENCES

1. Benigni R, Richard AM. QSARs of mutagens and carcinogens: two case studies illustrating problems in the construction of models for noncongeneric chemicals. Mutat Res 1996; 371:29–46.
2. Benfenati E, Gini, G. Computational predictive programs (expert systems) in toxicology. Toxicology 1997; 119:213–225.
3. Richard AM. Structure-based methods for predicting mutagenicity and carcinogenicicty: are we there yet? Mutat Res 1998; 400:493–507.
4. Helma C, Gottmann E, Kramer S. Knowledge discovery and data mining in toxicology. Stat Methods Med Res 2000; 9:329–358.

5. McKinney JD, Richard A, Waller C, Newman MC, Gerberick F. The practice of structure–activity relationships (SAR) in toxicology. Toxicol Sci 2000; 56:8–17.

6. Hulzebos EM, Janssen PAH, Maslankiewicz L, Meijerink MCM, Muller JJA, Pelgrom SMG, Verdam L, Vermeire TG. The Application of structure–activity relationships in human hazard assessment: a first approach. RIVM Report 601516 008. Bilthoven: Rijks Instituut Voor Volksgezondheid en Milieu, 2001.

7. Greene N. Computer systems for the prediction of toxicity: an update. Adv Drug Deliv Rev 2002; 54:417–431.

8. Richard AM, Benigni R. AI and SAR approaches for predicting chemical carcinogenicity: survey and status report. SAR QSAR Environ Toxicol 2002; 13:1–19.

9. Benigni R. SARs and QSARs of mutagens and carcinogens: understanding action mechanisms and improving risk assessment. In: Benigni R, ed. Quantitative Structure–Activity Relationship (QSAR) Models of Mutagens and Carcinogens. Boca Raton, FL: CRC Press, 2003:259–282.

10. Benigni R, Zito R. The second National Toxicology Program comparative exercise on the prediction of rodent carcinogenicity: definitive results. Mutat Res. 2004; 566:49–63.

11. Helma C, Kramer S. A survey of the Predictive Toxicology Challenge 2000–2001. Bioinformatics 2003; 19:1179–1182.

12. Toivonen H, Srinivasan A, King RD, Kramer S, Helma C. Statistical evaluation of the Predictive Toxicology Challenge 2000–2001. Bioinformatics 2003; 19:1183–1193.

13. Arcos JC, Argus MF. Multifactor interaction network of carcinogenesis—a "tour guide." In: Arcos JC, Argus MF, Woo YT, eds. Chemical Induction of Cancer, Modulation and Combination Effects. Boston: Birkhauser, 1995:1–20.

14. Arcos JC, Argus MF. Chemical Induction of Cancer: Structural Bases and Biological Mechanisms. Vol. IIA. Polycyclic Aromatic Hydrocarbons. New York: Academic Press, 1974.

15. Arcos JC, Argus MF. Chemical Induction of Cancer: Structural Bases and Biological Mechanisms. Vol. IIB. Aromatic Amines and Azodyes. New York: Academic Press, 1974.

16. Arcos JC, Woo YT, Argus MF (with collaboration of Lai D). Chemical Induction of Cancer: Structural Bases and Biological Mechanisms. Vol. IIIA. Aliphatic Carcinogens. New York: Academic Press, 1982.

17. Woo YT, Lai D, Argus MF, Arcos JC. Chemical Induction of Cancer: Structural Bases and Biological Mechanisms. Vol. IIIB. Aliphatic and Poly halogenated Carcinogens. New York: Academic Press, 1985.

18. Woo YT, Lai D, Argus MF, Arcos JC. Chemical Induction of Cancer: Structural Bases and Biological Mechanisms. Vol. IIIC Natural, Metal, Fiber, and Macromolecular Carcinogens. New York: Academic Press, 1988.

19. Woo YT, Lai DY. Mechanisms of action of chemical carcinogens and their role in structure-activity relationships analysis and risk assessment. In: Benigni R, ed. Quantitative Structure–Activity Relationship (QSAR) Models of Mutagens and Carcinogens. Boca Raton, FL: CRC Press, 2003:41–80.

20. Rosenkranz HS. SAR in the assessment of carxcinogenesis: the Multi-CASE approach. In: Benigni R, ed. Quantitative Structure–Activity Relationship (QSAR) Models of Mutagens and Carcinogens. Boca Raton, FL: CRC Press, 2003: 175–206.

21. Woo Y, Lai D, Arcos J, Argus M, Cimino M, DeVito S, Keifer L. Mechanism-based structure-activity relationship (SAR) analysis of carcinogenic potential of 30 NTP test chemicals. Environ Carcinog Ecotoxicol Rev 1997; C13:139–160.

22. Woo Y, Lai D, McLain JL, Manibusan MK, Dellarco V. Use of mechanism-based structure–activity relationships analysis in carcinogenic potential ranking for drinking water disinfection byproducts. Environ Health Prespect 2002; 110(suppl 1): 75–87.

23. Gold LS, Manley NB, Slone TH, Ward JM. Compendium of chemical carcinogens by target organ: results of chronic bioassays in rats, mice, hamsters, dogs, and monkeys. (http://potency.berkely.edu). Toxicol Pathol 2001; 29:639–652.

24. Woo YT, Lai DY, Argus MF, Arcos JC. Development of structure–activity relationship rules for predicting carcinogenic potential of chemicals. Toxicol Lett 1995; 79:219–228.

25. Stanton MF, Laynard M, Tegeris A, Miller E, May M, Kent E. Carcinogenicity of fibrous glass: pleural response in the rat in relation to fiber dimension. J Nat Cancer Inst 1977; 58: 587–603.

26. Lai DY, Woo YT, Argus MF, Arcos JC. Carcinogenic potential of organic peroxides: prediction based on structure–activity relationships (SAR) and mechanism-based short-term tests. Environ Carcing Ecotoxicol Rev 1996; C14:63–80.

27. Woo YT, Lai D, Argus MF, Arcos JC. An integrative approach of combining mechanistically complementary short-term predictive tests as a basis for assessing the carcinogenic potential of chemicals. Environ Carcing Ecotoxicol Rev 1998; C16: 101–122.

28. U.S. Environmental Protection Agency: Pollution Prevention (P2) Framework. EPA Publication 748-B-00-001. Office of Pollution Prevention and Toxics, U.S. EPA, June 3, 2000.

29. Lai D, Woo YT, Argus MF, Arcos JC. Cancer risk reduction through mechanism-based molecular design of chemicals. In: DeVito SC, Garrett RL, eds. Designing Safer Chemicals, ACS Symposium Series 640. Washington, DC: American Chemical Society, 1996:62–73.

30. Woo YT, Lai DY. Aromatic amino and nitroamino compounds and their halogenated derivatives. In: Bingham E, Cohrssen B, Powell CH, eds. Patty's Toxicology. New York: Willey, 2001:969–1106.

31. Fielden MR, Zacharewski TR. Challenges and limitations of gene expression profiling in mechanistic and predictive toxicology. Toxicol Sci 2001; 60:6–10.

32. Storck T, von Brevern MC, Behrens CK, Scheel J, Bach A. Transcriptomics in predictive toxicology. Curr Opin Drug Discov Devel 2002; 5:90–97.

33. Waters M, Boorman G, Bushel P, Cunningham M, Irwin R, Merrick A, Olden K, Paules R, Selkirk J, Stasiewicz S, Weis B, Houten BV, Walker N, Tennant R. System toxicology and the chemical effects in biological system (CEBS) knowledge base. Environ Health Prespect 2003; 111: 811–824.

34. USEPA. A Framework for a Computational Toxicology Research Program in ORD, Draft. Office of Research and Development, U.S. Environmental Protection Agency, Washington, DC: June 11, 2003. (www.epa.gov/comptox/comptox frameworlhtml).

INTRODUCTORY REFERENCE

1. Woo YT, Arcos JC, Lai DY. Structural and functional criteria for suspecting chemical compounds of carcinogenic activity: state of the art predictive formalism. In: Milman HA, Weisburger EK, eds. Handbook of Carcinogen Testing. Park Ridge, NJ: Noyes Publications, 1985:1–25.

GLOSSARY

Bioavailability: The degree to which a substance becomes available to the target tissue after administration or exposure.

Carcinogenesis: The complex, multistep process of cancer induction. The complete carcinogenesis process is believed to involve at least three major steps: initiation, promotion, and progression.

Cell proliferation: Enhancement in the number of cells by DNA synthesis and mitosis.

Covalent binding: The joining together of chemicals by bond sharing electrons.

Cytotoxic: Causing irritation or death of cells.

Electrophilic intermediate: Electron-deficient reactive intermediate such as carbonium and nitrenium ions.

Epigenetic mechanism: Toxic mechanism(s) that do(es) not involve direct damage to DNA.

Expert system: An artificial intelligence application that uses the knowledge base of human expertise to aid in problem solving.

Genotoxic mechanism: Toxic mechanism that results from direct damage to DNA.

Hyperplasia: Increase in the size of a tissue through an increase in cell number.

Initiation: The first step of carcinogenesis in which normal cells are irreversibly turned into preneoplastic, initiated cells as a result of a mutational event(s). Chemicals or agents that induce initiation are called initiators.

Malignant tumor: A tumor that can invade surrounding tissues or metastasize to distant sites resulting in spreading of tumors and life-threatening consequences.

Mechanism of action: The specific event/step(s) by which a substance causes a particular toxic effect.

Metabolism: The conversion of a chemical from its original form (parent) to another (metabolite) in the body. The conversion that leads to a more or less toxic metabolite is termed bioactivation or detoxification, respectively.

Metastasis: The movement of diseased cells, in particular cancer cells, from the site of origin to another location in the body.

Mutagen: Chemical that induces change in the DNA sequence of genes and/or chromosomes.

Promotion: Clonal expansion of initiated cells to grow into a benign/premalignant tumor/neoplasm by a variety of means such as cell proliferation, inhibition of programmed cell death. Chemicals that can carry out promotion are called promoters.

Progression: Transformation of benign/premalignant tumor into invasive, metastatic, malignant tumor by overcoming suppression and barrier of surrounding healthy cells. Chemicals that can carry out progression are called progressors.

Receptor: A cell surface or intracellular biological molecule that, when occupied by a specific substance by binding, triggers a or a series of physiologic/toxic effect(s) in the cell.

SAR: The process by which the toxicity of a substance can be predicted based on its similarity in structure to that of other chemicals the toxicity is known.

Subchronic toxicity: The adverse effects of a substance resulting from repeated exposure to a toxic agent over a period of several weeks usually not exceeding three months.

11

META: An Expert System for the Prediction of Metabolic Transformations

GILLES KLOPMAN
MULTICASE Inc., Beachwood, Ohio, and Case Western Reserve University, Cleveland, Ohio, U.S.A.

ALEKSANDR SEDYKH
Department of Chemistry, Case Western Reserve University, Cleveland, Ohio, U.S.A.

1. OVERVIEW OF METABOLISM EXPERT SYSTEMS

As the number of chemicals used in industrial processes and in everyday life rapidly increases, the need to assess a priori their impact on the human organism and the environment becomes critically important. Two kinds of chemical interactions are of particular concern. Firstly, metabolic transformations in the human body and other relevant living organisms, and secondly, transformations under external factors (i.e., sunlight, hydrolysis, atmospheric oxidation, temperature, irradiation, electric and magnetic fields). If both types of interactions could be modeled, the impact of any chemical on the biosphere could be predicted and, if necessary, controlled.

Evidently, the task of modeling human metabolism towers above all the others, since the human body is the ultimate goal for most of the evaluations. Virtually, any industrially produced chemical or its derivative has a chance to end up in the human organism. Thus, the need of risk assessment (and on the other hand, demands of drug design) make the tool for predicting human metabolism indispensable.

For obvious ethical reasons, the experimental data on various chemicals in regard to human metabolism is not as abundant as desirable. Hence, animal models of human metabolism are extensively used.

Accordingly, the first application of the META expert system was to model xenobiotic metabolism of the common mammalian species (1).

2. THE META EXPERT SYSTEM

The META approach consists in compiling separate models (called "dictionaries") for each system of interest, whether it is a group of similar organisms or an environmental compartment with certain conditions. These dictionaries, essentially, are databases of relevant chemical transformations accompanied with supplemental data. The META program ("META") allows to operate any of the existing dictionaries and handles the entire work session with the user. Currently, there are available models for generic mammalian metabolism (1), biodegradation under anaerobic (2) and aerobic (3,4) conditions, and photodegradation in upper layers of lakes (5).

To automate the validation and adjustment of META dictionaries, a special program, META_TREE, was created. Based on available experimental data, it finds discrepancies in the dictionary and optimizes its transformations for best performance.

Thus, the complete META expert system can be represented by the following triad: the META program, the META dictionaries, and the META_TREE utility.

3. META DICTIONARY STRUCTURE

Aside from the structural data, each transformation record also has a priority number (X-field in Fig. 1), which reflects the speed or predominance of the underlying chemical or enzymatic reaction that the transform represents. The transform may also include information on some test compounds, which undergo this transformation. This information is stored in the D-field (Fig. 1) and used in the optimization of the dictionary priorities by META_TREE.

The R-field of the transformation record contains references in a coded form that points to the publications, which describe or support the specified transformation. The list of

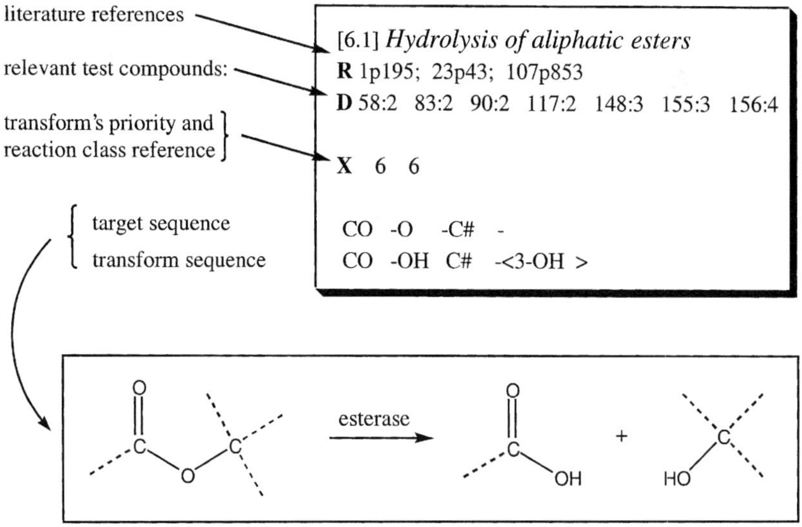

Figure 1 Example of a transformation rule stored in the dictionary.

all referenced publications is collected in an explicit form at the end of the dictionary.

4. META METHODOLOGY

A detailed description of the META program algorithm has been given in a series of publications by the Klopman laboratory (1,6,7).

As was mentioned before, META operates from a dictionary of transformations. Each transformation rule ("transform") consists of two structural fragments: a "target" sequence and a "transformation" sequence (Fig. 1).

META scans a test molecule for the presence of "target" fragments and replaces them one at a time by the corresponding "transformation" fragments, thus producing a set of primary transformation products. At the same time, the program monitors and evaluates the thermodynamic stability of all the molecules generated and consults a spontaneous reactions module that manages unstable structural moieties. Whenever a molecule is found to be unstable, it is transformed into a stable product via an appropriate spontaneous reaction transform. Upon demand, the first level products will be processed further so that a complete tree of transformation products can be obtained.

In addition, the predicted products are evaluated for excretion and toxicity. The former is based on the estimation of the Log *P* (*n*-octanol/water partition coefficient) (8), the latter employs a separate module of signal transforms, whose "target" sequences are known to occur mainly in toxic compounds (9).

A typical META work session consists of the following steps:

- Choosing a model of interest (mammalian metabolism, aerobic degradation, anaerobic degradation, and photodegradation).
- Entering a test compound as a SMILES code (10), MOL-file (11), or graphic input.
- Browsing through the tree of primary products and advancing to the next level of products when necessary.

- Retrieving results (chemical structures, calculated properties, literature references, etc.)

META can also be operated in batch-mode, which allows to screen libraries of chemicals against a particular model. In this mode, META generates and evaluates first level products for each test compound and generates a list of them. Filtered and formatted results are reported in a log-file.

5. META_TREE

According to its description, META reads a dictionary of transforms and applies them to the test molecules in order to identify possible products. Normally, a dictionary contains several hundreds of transformation rules. Most of them lack direct kinetic data, and thus, to assign properly their priorities is a difficult manual work. The META_TREE program resolves this complication. It automatically adjusts priorities of transformations in order to follow experimental metabolic and/or degradation pathways as precisely as possible. For that purpose a basic "genetic algorithm" was implemented (6,12).

5.1. Basic Genetic Algorithm Methodology

Genetic algorithms (12) imitate the evolutionary process in nature: 1. generation of diversity, 2. survival of the fittest, and 3. reproduction. Its typical task consists of finding in a multiparameter system the best configuration of parameters that satisfies certain requirements. Thus, each parameter is called a "gene" and each configuration of parameters is called an "individual." A group of individuals forms a "population" representing a pool of approximate solutions to the problem. The so-called "D-function" performs the role of the driving factor of the evolution. The D-function takes an individual and evaluates its "fitness."

The initial pool of individuals is randomly generated. Then the fitness of all individuals is evaluated. At this step, the next generation is produced by means of "crossover" and "mutation" procedures, which mimic natural reproduction patterns. Two individuals at a time are selected from the population (either randomly or depending on the fitness of the individual) and their genes are combined into a pair of "children." If crossover does not happen, the children will be genetically identical to their respective parents; otherwise each child will bear a partial set of genes. Mutation is then applied with some probability, thus randomly modifying some genes of the children. The process is repeated until the number of newly generated individuals reaches the size of the initial population.

This new generation replaces the previous one and functions now as a current population, after which the whole cycle is repeated.

The successive populations diverge from each other less and less until the best individuals in each new population are practically the same. At this point, population stability is achieved and the optimization is deemed complete.

5.2. META_TREE Methodology

When a dictionary is being processed, META_TREE identifies the transforms to be optimized. Some transforms do not have enough experimental data; others might be intentionally kept duplicates (for example, to demonstrate that the same transformation can be carried out by a different mechanism). Each optimizable transform is treated as one "gene." Thus, each "individual" represents one possible configuration of the transformation priorities in the dictionary.

To calculate the fitness of each individual, the META_TREE D-function uses compounds located at the end of the dictionary file and the D-fields of dictionary transforms (Fig. 1). For each such compound, META_TREE generates a list of the transformations it is assigned to and a list of transformations that the test compound actually undergoes. The latter should include the former, and the priorities of the unaccounted extra transformations ("side-transformations") should be inferior to the priorities of the assigned transformations.

The flexibility of the META_TREE genetic algorithm rests on a group of adjustable parameters, which enables different optimization schemes with a small population size. These trial optimizations are compared, and then the most promising direction is used for an in-depth optimization.

META_TREE was successfully employed in optimizing priorities of the photodegradation (5) and mammal metabolism (1) dictionaries, which are of particularly large size (over 1300 transforms each).

The META_TREE utility also carries out the auxiliary function of the general dictionary maintenance. This facilitates the development and further extension of a transformation database by a researcher or programmer, without the need of a comprehensive knowledge of the dictionary design (6).

REFERENCES

1. Talafous J, Sayre LM, Mieyal JJ, Klopman G. META. 2. A dictionary model of mammalian xenobiotic metabolism. J Chem Inf Comput Sci 1994; 34:1326–1333.
2. Klopman G, Saiakhov R, Tu M, Pusca F, Rorije E. Computer-assisted evaluation of anaerobic biodegradation products. Pure Appl Chem 1998; 70:1385–1394.
3. Klopman G, Tu M. Structure-biodegradability study and computer-automated prediction of aerobic biodegradation of chemicals. Environ Toxicol Chem 1997; 16:1829–1835.
4. Klopman G, Zhang Z, Balthasar DM, Rosenkranz HS. Computer-automated predictions of aerobic biodegradation of chemicals. Environ Toxicol Chem 1995; 14:395–403.
5. Sedykh A, Saiakhov R, Klopman G. META V. A model of photodegradation for the prediction of photoproducts of chemicals under natural-like conditions. Chemosphere 2001; 45:971–981.
6. Klopman G, Tu M, Talafous J. META. 3. A genetic algorithm for metabolic transform priorities optimization. J Chem Inf Comput Sci 1997; 37:329–334.

7. Klopman G, Dimayuga M, Talafous J. META. 1. A program for the evaluation of metabolic transformation of chemicals. J Chem Inf Comput Sci 1994; 34:1320–1325.

8. Klopman G, Wang S. A computer Automated Structure Evaluation (CASE) approach to calculation of partition coefficient. J Comput Chem 1991; 12:1025–1032.

9. Klopman G. The META-CASETOX system for the prediction of the toxic hazard of chemicals deposited in the environment. NATO ASI Ser, Ser 2 1996; 23:27–40.

10. Weininger D. SMILES, a chemical language and information system. 1. Introduction to methodology and encoding rules. J Chem Inf Comput Sci 1988; 28:31–36.

11. Yao Jh. Computer treatment of chemical structures. (II). Internal storage data structures and file formats of chemical structure representation in computers. Jisuanji Yu Yingyong Huaxue 1998; 15:65–69.

12. Wehrens R, Buydens LMC. Evolutionary optimization: a tutorial. Trends Anal Chem 1998; 17:193–203.

GLOSSARY

Biodegradation: Decomposition of a chemical by living systems (usually microorganisms).

D-function: Driving factor of simulated evolutionary process in genetic algorithms.

Genetic algorithms: Optimization algorithms that imitate the evolutionary process in nature.

Hydrolysis: Decomposition of a chemical compound by reaction with water.

Log*P*: Decimal logarithm of the chemicals partition ratio between *n*-octanol and water.

MOL-file: Textual file type standard (developed by MDL) to store molecular structure.

Photodegradation: Decomposition of a chemical compound under light.

SMILES code: Symbol-based representation of a chemical structure.

12

MC4PC—An Artificial Intelligence Approach to the Discovery of Quantitative Structure–Toxic Activity Relationships

GILLES KLOPMAN
MULTICASE Inc., Beachwood, Ohio, and Case Western Reserve University, Cleveland, Ohio, U.S.A.

SUMAN CHAKRAVARTI
Case Western Reserve University, Cleveland, Ohio, U.S.A.

JULIAN IVANOV and ROUSTEM SAIAKHOV
MULTICASE Inc., Beachwood, Ohio, U.S.A.

1. INTRODUCTION

The relation between the structure of chemicals and their properties has fascinated chemists ever since it was discov-

MC4PC, CASE, MultiCASE and MCASE are copyrighted programs owned by KCI LLC ered that chemicals have properties defined by their atomic composition and the spatial relation between them. In this context, it is important, we believe, to distinguish between global properties of chemicals and local properties.

Global properties depend on the complete make-up of the molecule and change drastically when any part of the molecule is altered. Examples of global properties are molecular weight, solubility, boiling and melting points, and most other physical and thermodynamic properties. The relation between the structure of molecules and their global properties can often be related to every part of the molecule. For example, the molecular weight is calculated as the sum of the atomic weights of all the atoms of the molecule. Similarly, solubility in water, for example, has been shown to be related to a contribution of every group of atoms of the molecule, sometimes corrected by terms representing interactions between these groups.

Local properties, on the other hand, are often defined by the presence of certain group of atoms, spatially organized around each other, that produce the property. While the rest of the molecule is still important, its importance is only seen in that it modulates the activity produced by the key fragment. Local properties mostly encompass chemical and possibly biochemical properties, although biochemical properties definitely depend also on global properties such as solubility properties. In chemistry, we call these key fragments functionalities because they endow the molecules that contain them with a certain chemical property. For example, COOH is a functionality that generates acidity when present in a molecule. Similarly, NH_2 generates basicity, NO_2 and –N=N– (azo group) are chromophores (produce color), a thiol group (SH) when present in a molecule gives it a foul odor, and so on.

The relationships between global properties and the structure of chemicals are often straightforward as the property is calculated by adding up the contribution of each part of the molecule (atoms, group of atoms and/or spatial relationship between them). This is done using techniques such as quantitative structure–activity relationships (QSARS) obtained from multilinear regression analysis and other similar techniques. Likewise, relationship between activity observed for molecules containing a certain functionality can also be obtained from QSARs by compiling the effect that atoms or group of atoms in the molecules can have on the activity produced by its functionality.

Over the years, chemists have identified the nature of the functionalities responsible for the majority of important chemical behavior of molecules. Therefore, many QSARs were developed to allow the practitioners to predict activity of many chemicals that contain the corresponding functionality. These chemicals are called congeneric.

The situation is much more complex in the case of biological events, both toxic and pharmacologic. Indeed, the number of biological properties is enormous, and one cannot expect that relationships will be general enough to carry from one endpoint to another. Therefore, in order to be able to move forward, it is necessary to accept the major hypothesis that the biological activities, as the local properties of chemicals, also relate to specific atomic groups characteristic of each endpoint. The actual potency, however, remains dependant on the remaining part of the molecule whose effect may play an important role not only by modulating activity of the specific functionality but also by affecting the molecule's transport and metabolic properties.

However, even when this hypothesis is accepted, a major task remains to be undertaken—that of identifying the functionalities, which we call "biophores" and others, "structural alerts" (1), responsible for each of the biological endpoints that one wishes to study. A further complication in this endeavor is that there may be several different biophores that can produce the same biological response and it is therefore difficult to sort out what is responsible for what. Attempts to identify the biophores manually have sometimes been successful, particularly for highly studied situations and when the biophores are similar to common functionalities. A case in point is the identification of the *N*-nitroso group as being responsible for the carcinogenicity of many of the molecules that contain it. This was possible because most molecules containing this group were found experimentally to be rodent carcinogens. However, while there are a few cases where such discoveries were made, there is not much known about the majority of other toxic and pharmacological endpoints.

The MCASE program and its Windows® counterpart MC4PC were developed to fill this gap and were made to discover the biophores responsible for biological endpoints when some experimental information about the activity of diverse chemicals had been measured. In addition, the program will evaluate the influence of the overall structure of the molecules in the form of a more traditional QSAR. Once these evaluations are made, the program can be queried as to the potential activity of molecules that have not been tested or even synthesized as yet.

The program can be used to create models for any number of biological endpoints, thus creating a library of intelligent modules containing information that can be used to evaluate many potential toxic or beneficial properties of molecules very rapidly and at very low cost. We find that the system can make predictions for up to 3000 chemicals in one hour thus making it particularly appropriate for testing large libraries of chemicals.

One of the major issues confronting the user of such models is the accuracy of the predictions. We find, not surprisingly, that this is a function of the chemical and activity diversity of the chemicals that were used to produce the models. Generally, the more diverse the training set, the better the predictions. Our validation standard is to withhold 10% of the chemicals that were slated to be used to create a model. A reduced model is then created and is used to predict the activity of the molecules that were withheld. We typically do this ten times and generate the predicted sensitivity, selectivity, and coverage of the model. Generally speaking, the specificity is higher than the sensitivity. Indeed, when the program has information about a group of molecules that had previously been found to be conducive to activity, the prediction of activity is probably correct (few false positives). On the other hand, when a molecule is predicted inactive, generally by default, there exist the possibility that this molecule is still active because it contains a group that was never evaluated in regard to activity for this model and is the biophore of an as yet unknown mechanism (more false negatives). Thus, predictions of activity are to be trusted more than predictions of inactivity. The problem can generally be resolved by adding more diverse structures, preferably actives, to the training set.

It is to be noted that while the bulk of this paper deals with structure–toxic activity relationships, the program is equally or even more useful to discover structure–pharmacologic activity.

The CASE and its successor, MCASE method, have been successfully applied to the study of mutagens and carcinogens (2–4), activity of antimicrobial agents (5), hallucinogens (6), opiate alkaloids (7), antileukemic agents (8), beta adrenergic agents (9), inhibition of thermolysin enzyme (10) and of sparteine monooxygenase activity (11), mutagenicity (12,13), aquatic toxicity (14), toxicity in hamsters (15,16), anti-HIV activity (17), carcinogenic potentials (18,19), genotoxic carcinogenicity of fragrances (20), rodent carcinogenicity (21), DNA reactivity (22), and many other pharmacological endpoints.

2. THE MCASE METHODOLOGY

The discovery of the specific functionalities (biophores) possibly responsible for a given biological effect is handled by the computer automated structure evaluation (CASE) program. The algorithm starts by tabulating, for each molecule of a training set, the fragments that are formed by breaking up the molecule into linear or branched subunits containing between 2 and 10 interconnected groups consisting of heavy atoms (i.e., non-hydrogen atoms), together with several labels indicating the characteristics, nature, and environment of the subunits in the molecule. These fragments are the potential functionalities or descriptors of the ensuing correlations, and their selection is entirely driven by the nature of the compounds of the database. The procedure is open-ended and completely automatic. For example, Dulcin would generate the fragments indicated in Table 1.

In the case of fused rings, though, special care is taken to include only fragments formed by the atoms located on the external surface of the molecules. Thus, in naphthalene (Fig. 1), no subunit contains the sequence of atoms C5–C6 since the bond they form is common to two rings. This type of bond is chemically inert except in very rare cases. Larger bridges are unaffected by this procedure and do appear in the corresponding fragments.

All fragments originating from active molecules are labeled active, while those from inactive molecules are labeled inactive. Once all molecules that constitute the training set have been entered, a statistical analysis of the fragments distribution is made. A binomial distribution (23) is assumed, and a fragment is considered irrelevant to activity if its distribution among active and inactive chemicals is similar to that of the total sample of molecules. Any significant discrepancy from a random distribution of subunits between the active and inactive pool is then taken as an indication that the subunit is relevant to the property being examined. Thus, a fragment is considered relevant if it is found in a binomial distribution that would have had at most 5% chance of being observed if its occurrence was random. It is labeled as activating if its distribution is skewed toward active molecules and inactivating otherwise.

The algorithm described above has been successfully implemented in the MCASE program (multiple computer automated structure evaluation) (24–26). MCASE is a knowledge-based system designed to discover the relationship between the structure of the chemicals and their activity in a specific biological assay. It has been designed to deal with noncongeneric databases, consisting of structurally unrelated molecules, that are not normally amenable to treatment with traditional QSAR type techniques. As such, its main goal is to find the structural entities that discriminate active molecules from the mass of inactive ones and its success is dependent on

Table 1 Subunits of Size 3–6, Generated for Dulcin, CCOclccc(NC(N)=O)ccl

Size 3	Size 4	Size 5	Size 6
cH=c –cH =	c =cH –cH =c –	cH =c –cH =cH –c =	cH =c –cH =cH –c =cH –
cH =c <–cH =	c <=cH –cH =c <–	cH =c <–cH =cH –c <=	cH =c <–cH =cH –c <=cH –
cH=cH –c =	cH=cH –c =cH –	cH =cH –c =cH –cH =	cH=cH –c =cH –cH=c –
cH =cH –c <=	cH =cH –c <=cH –	cH=cH –c <=cH –cH =	cH =cH –c <=cH –cH =c <–
cH_2–O –c =	CH_2–O –c =cH –	CH_2–O –c =cH –cH =	CH_2–O –c =cH –cH=c –
NH –c =cH –	CH_3–CH_2–O –c =	CH_3–CH_2–O –c =cH –	CH_2–O –c =cH –cH =c <–
NH_2–CO –NH –	NH –c =cH –cH =	NH –c =cH –cH =c –	CH_3–CH_2–O –c =cH –cH =
O –c =cH –	NH_2–CO –NH –c =	NH–c =cH –cH=c <–	NH –c =cH –cH=c –cH =
O –CH_2–CH_3	O –c =cH –cH =	NH_2–CO –NH–C =cH –	NH –c =cH –cH =c <–cH =
CO –NH –c =	CO –NH –c =cH –	O –c =cH –cH=c –	NH_2–CO –NH –c =cH –cH =
		O –c =cH –cH=c <–	O –c =cH –cH=c –cH =
		CO –NH –c =cH –cH =	O –c =cH –cH =c –NH –
			O –c =cH –cH =c <–cH =
			CO –NH –c =cH –cH =c –
			CO –NH –c =cH –cH =c <–

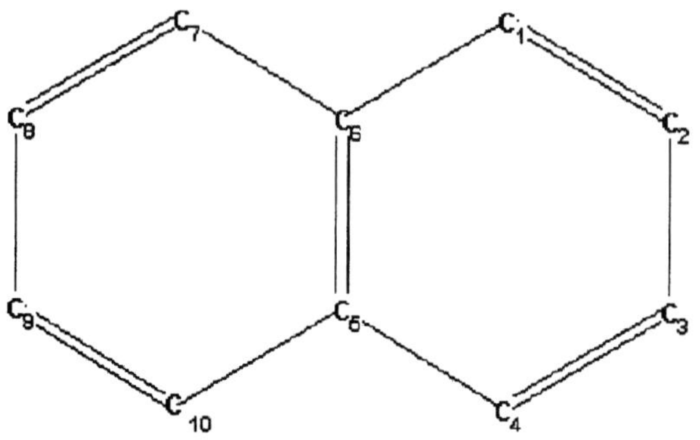

Figure 1 Structure of naphthalene.

the validity of the working hypothesis that a relationship does indeed exist between structure and activity. As indicated above, the program selects its own descriptors from a number of possible substructural units and creates a fragment dictionary without human intervention. The selected descriptors are characterized either as activating (biophore) or as inactivating (biophobe). The ability of MCASE to select biophores that are readily recognized as being part of a molecule is a major advantage of the method. Indeed, the identification by MCASE of structural components embedded in the molecule offers a foothold that human intelligence can exploit with respect to possible structural site of metabolism or receptor binding which can lead to further hypothesis testing.

The structural information input can be entered using the SMILES (27) or KLN (28) line notations as well as a MDL MOL file.

An important algorithm feature of the MCASE program is the use of a hierarchical selection of descriptors. In the first step, the program identifies only the true biophores, i.e., those fragments found to have an unquestionable relation to activity. This is done in a hierarchical way, by first identifying the substructure with the highest probability of being responsible for activity as judged by the binomial distribution among active and inactive molecules. The molecules containing this fragment (called the top biophore) are eliminated from the data set and the remaining compounds are submitted to a new analysis resulting in the identification of the next most relevant biophore. This procedure is then repeated until either the entire set is eliminated, or all statistically relevant fragments (biophores) have been identified and the remaining compounds cannot be explained by statistically significant descriptors.

The molecules in the data set are then separated into subclasses based on the presence of each of the biophores. For each subclass, a new analysis is performed to identify relevant modulators seen as responsible for the differences in activity between the molecules of the subset. Because of this algorithm, MCASE can handle many different classes of chemicals in a single database even if the observed activity results from different mechanistic paths.

The system is hierarchical in that modulators are important only in the context of molecules containing the primary biophore. These modulators may increase or decrease the intrinsic activity produced by the biophore. It is possible that the modulator is itself an essential part of the biophore. This is particularly true if most of the structures containing the biophore also contain the modulator. It is also to be noted that, unless the biophore is embedded in the modulator, the program does not recognize the relative position of the fragment and its modulators. This may be important if, for example, they each are recognized by a multidentate enzyme, where the two bonding sites are at well-defined locations relative to each other.

Once the database has been processed, test compounds can be submitted for predictions. When a new molecule is tested, the program searches its structure for the presence of a biophore. If it does not find one, the molecule will be called inactive by default. However, if it does find one, it will then search for the presence of potential modulators to arrive at a projected value for the potency.

As shown above, MCASE basically deals with several sets of congeneric systems. The main difference between these and conventional congeneric databases is that the commonality between the molecules is based on a rational evaluation of their structures rather than on the arbitrary choice of common structural features. Indeed, the biophores are essentially structural entities alerting to the strong possibility that the molecule is active, letting the modulators determine the quantitative activity potential from traditional QSARs restricted to each class of compounds.

2.1. Geometry Factor

In order to keep the geometry (cis/trans orientation) of the fragment after taking it out of the content of the molecule, a geometry index has been introduced (29). The geometry index is a seven bit number, where each bit serves to encode the geometry of each 1,4 pair in a linear fragment. Since there are seven 1,4 pairs in a linear fragment of 10 atoms, a single byte of information is sufficient to characterize the cis/trans configuration of a whole fragment.

2.2. Expanded Fragments

Expanded fragment are defined as substructural entities that are similar to a bona fide biophore, but do not exist in sufficient number in the learning set to become biophores by themselves. The following rules have been drafted for identifying expanded fragments:

- They must differ from an established biophore at only one site and only by a minimal difference.
- The only acceptable minimal differences are variation in the number of attached hydrogen atoms or, the replacement of a double bonded atom by another double bonded atom.
- The expanded fragment must have been found only in active molecules.

2.3. Automatic Design

One of the unique features of MCASE is its ability to design new molecules. The program generates a list of substructures that are relevant to activity. By searching the structure under consideration for substructures that are similar to those listed in the dictionary, the program can replace them by more potent analogs. This is done in successive steps, allowing the operator to intervene and control the design path.

3. RECENT DEVELOPMENTS: THE MC4PC PROGRAM

In order to take advantage of the flexible Windows environment and the availability of personal computers, during the last few years the MCASE technique has been successfully implemented into a new program named MC4PC. Maintaining the core features of the CASE/MULTICASE approaches, the new program offers many new opportunities for treating molecular structures, managing databases, testing chemicals of unknown activity, and for displaying the results in a user-friendly interface.

3.1. Molecular Editor Drawing Tools

The main purpose of the molecular editor is to provide a fast and easy way for entering molecular structures that can be passed for prediction or stored in a database. The drawing palette (Fig. 2) includes drop down lists with all available atom and bond types and a number of buttons associated with templates for the most common ring systems.

The MC4PC is organized into three basic modes and each of them is responsible for handling a specific problem. The three main modes of the program are Create/Modify, Reprocess, and Test modes. Below is a short description of the main features and algorithms used in the new program.

3.2. Create/Modify Mode

The general purpose of this mode is to prepare and store the input data which will be later passed to MCASE. Input structures can be entered manually by using the molecular editor drawing tools or by typing the line notation (SMILES or KLN)

Figure 2 Drawing palette.

of the structure along with their CAS Number (if available), Name and Activity. When the structures are entered and the user stores them, the program automatically calculates a number of molecular properties of the structure, including molecular weight, human intestinal absorption (30), log$P$ (31), water solubility (32), and so on.

Entering a structure manually is a somewhat slow and tedious process and it is generally more practical to enter the required information from a properly formatted text file. Each line of this text file corresponds to a single structure.

The handling of the activity of the chemicals in the program is of special interest. MULTICASE internally uses so-called CASE units, a linear scale from 10 to 99, to characterize the activity of the compounds entered for its learning process. For reference, the CASE units are interpreted as follow:

10–19, inactive
20–29, marginal
30–39, active
40–49, very active
50–99, extremely active

The CASE units are irrelevant to the user as they are only used internally by the program to scale activities into inactive, active, and active ranges. The user's choice of a break point between what is considered significant activity and insignificant activity will determine how the program will select the biophores. Helpful charts are available showing the distribution of the compounds and the curve of the relationship between experimental activity and CASE units as a function of the selected break point (Fig. 3).

Once the choice of the break point is made, the program creates the corresponding input file and MC4PC is ready for MULTICASE analysis (building the data base). The process of creating a database is completely automatic and does not need the user's involvement at all.

3.3. Reprocess Mode

Reprocessing is needed when molecules are added, deleted, or changed in the data file. The database can be partially

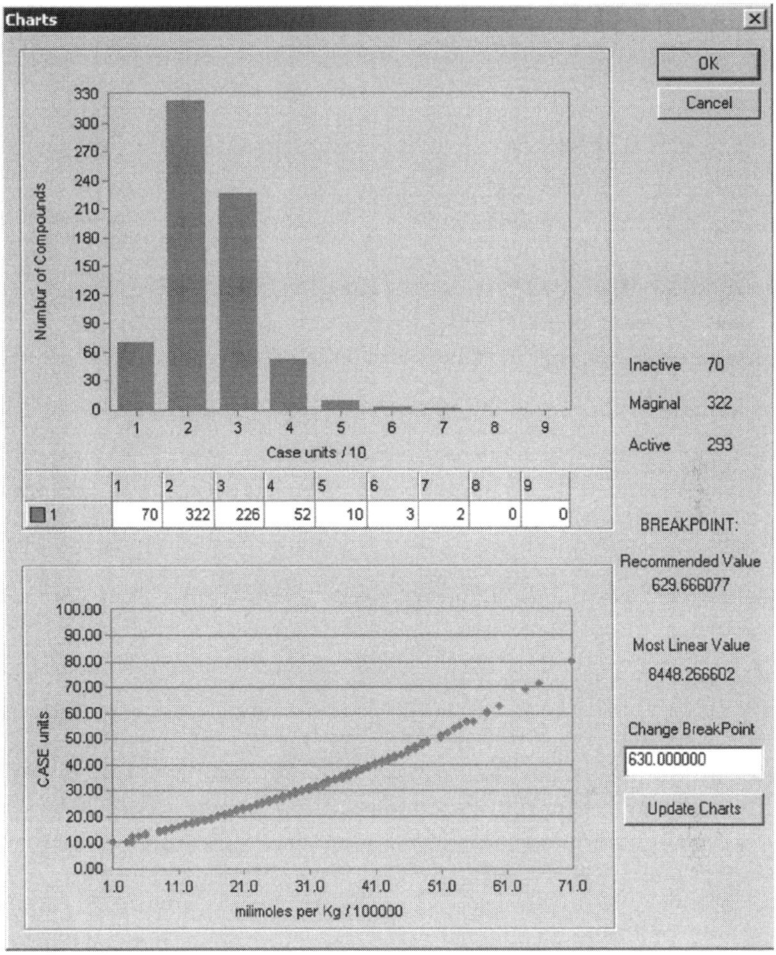

Figure 3 Distribution of the compounds as a function of the break point.

reprocessed by redoing only the statistics or the selection of biophores, or it can be completely rebuilt starting from the input file.

3.4. Test Mode

One of the most important uses of MC4PC is for predicting the biological activity of new, untested chemicals based on

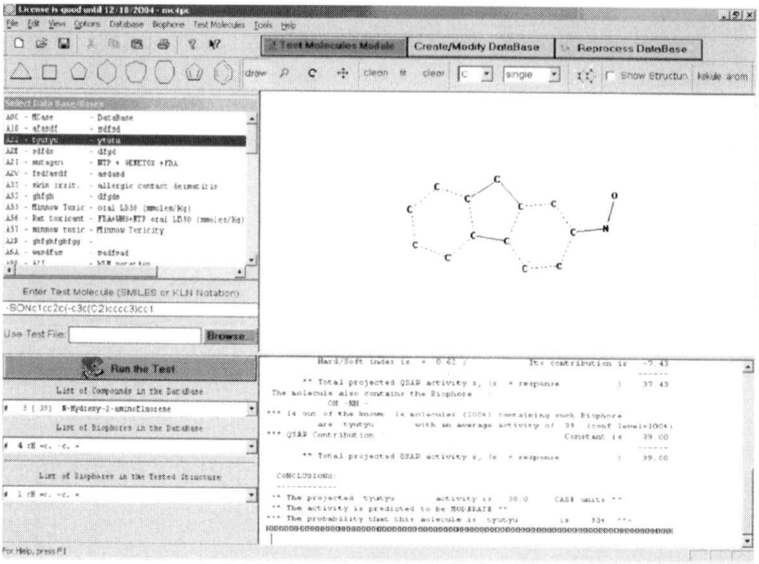

Figure 4 The Test mode of MC4PC.

structural similarities (biophores) present in the molecules used in the training data set. A typical use is the screening of candidate pharmaceuticals for toxicity such as mutagenicity in bacteria, carcinogenicity in rodents, reproductive toxicity in mammals as well as many other endpoints.

A list of all the databases available to the users together with their descriptions appears on the left side of the screen in a simple scrolling list box. The user can select one or more databases (up to 20) to run the test against (Fig. 4)

The process of testing is also fully automatic. The program reads a structure from the Molecular Editor or from an input file and runs the prediction for this compound against the selected databases consecutively. All output information is stored in text files. Alternatively, all the predictions can be seen on the bottom text output window of the program (Fig. 4).

Let us discuss a real example, the prediction of 2-(5-nitro-2-furyl)-5-*N*-methyl (SMILES: O=N(C1=CC=C(C2=NN=C(NC)S2)O1)=O) against A2I mutagenicity database (Fig. 5)

```
********************************************************************************
 Now Processing... 2-(5-nitro-2-furyl)-5-N-methyl
********************************************************************************
 A2I- mutagen      - NTP + GENETOX +FDA                          #2236 1.54
 -----------------------------------------------------------------
 Molecule satisfies the rule of 5 (bioavailable)
 MC Calculated Human Intestinal Absorption is: 91.6%
MULTICASE-3 Prediction
 ----------------------------------
 The molecule contains the Biophore    (nr.occ.= 1):
           NO2-c  =
***204 out of the known 238 molecules ( 86%) containing such Biophore
         are  mutagen       with an average activity of  34. (conf.level=100%)
*** QSAR Contribution :                                  Constant is    39.00
                                       .                               -------
       ** Total projected QSAR activity x, (x  = response           )    39.00
 The molecule also contains the expanded Biophore    :
          O  -c >=cH -
*** 35 out of the known  35 molecules (100%) containing such Biophore
         are  mutagen       with an average activity of  39. (conf.level=100%)
*** QSAR Contribution :                                  Constant is    39.00
                                       .                               -------
       ** Total projected QSAR activity x, (x  = response           )    39.00

  CONCLUSIONS:
  ------------
 ** The projected  mutagen       activity is   39.0     CASE units **
 ** The activity is predicted to be MODERATE **
*** The probability that this molecule is  mutagen       is    97%  **-
```

Figure 5 Prediction of 2-(5-nitro-2-furyl)-5-*N*-methyl against A2I mutagenicity database.

The molecule obeys Lipinski's "rule of five," which states that a chemical is unlikely to be bio-available unless there are fewer than five H-bond donors and 10 H-bond acceptors, the molecular weight is smaller than 500, and $\log P$ is smaller than 5. It is also predicted by MCASE (30) to be 91.6% absorbed. The program finds that the molecule contains a biophore. The biophore is found in 238 molecules of the database of which 204 are mutagen. Therefore, there is a good chance that the test chemical is also mutagenic. Additionally, the program finds that the test structure contains an expanded biophore present in 35 other molecules of the database. All of these are mutagen, a fact that increases the probability the test structure is a mutagen as well. The final prediction is that the test chemical is biologically active (mutagen in this example), with an activity of 39.0 CASE units and the probability that it is mutagenic is 97%.

```
*** Fragment   1 of xF3 is IDENTICAL to fragment   1 of xF4
 xF3 Cl -CH2-CH2-N  -              18  1  0  17 0.000   act
 xF4 Cl -CH2-CH2-N  -              19  0  0  19 0.000   act
*** Fragment  21 of xF3 is SIMILAR to fragment  34 of xF4
 xF3 Cl -C" -Cl                     6  0  0   6 0.016   act
 xF4 Cl -C  =CH2-                   3  0  0   3 0.125   act
*** Fragment   8 of xF4 is IMBEDDED  in fragment   8 of xF3
 xF3 Br -CH2-CH -                   5  0  0   5 0.031   act
 xF4 Br -CH2-                       9  1  0   8 0.011   act
```

Figure 6 Compare databases.

3.5. Compare Databases

The MCASE program enables comparison of related fragments in two different databases. Only fragments that are statistically active in both databases are displayed. In the example shown above (Fig. 6), two databases (AF3 and AF4) are compared. Fragment number 1 of AF3 (C1–CH_2CH_2–N) is the same as fragment number 1 in AF4. In AF3, the fragment is found in 18 molecules, of which 17 are active and 1 is inactive. In AF4, the same fragment is found in 19 molecules, of which all are active. The conclusion is that the 2-chloroethylamine fragment is strongly associated with biological activity in both databases. Fragments that are labeled as "similar" (instead of "identical") are found in slightly different structural environments in the two databases.

4. BAIA PLUS

As a part of the preprocessing routine, MCASE evaluates the possibility that the property under study is related to lipophilicity, expressed by log K_{ow}. Such correlations have been pioneered by Hansch (33) and they are still widely used in drug design and, to a lesser degree, in toxicology.

4.1. Baseline and BAIA

While classical QSAR is the most common tool for identifying and correlating linear relationships with the octanol–water partition coefficient (log K_{ow}) in congeneric series, in situations where a number of significant outliers is present, the classical regression technique usually fails (Fig. 7).

For example, aquatic toxicity data often presents a complicated pattern due to the fact that several mechanisms are operating simultaneously. In these cases, the underlying cause of toxicity has been identified as a "narcosis effect," usually represented by lipophilicity. More severe toxicity, on the other hand, has been linked to other, more specific causes. Thus, the global aquatic toxicity of chemicals is seen as resulting from the combination of some specific toxicity linked to the structure of the chemical and a baseline toxicity, determined by its lipophilicity. However, finding a baseline QSAR with log K_{ow} in a diverse noncongeneric dataset where other mechanisms are operating as well can become very challenging for classical statistical tools.

To solve this problem, the MCASE program has been outfitted with a feature called Baseline Activity Identification Algorithm, or BAIA (34), which allows it to identify and quantify the existence of a baseline activity of molecules. In some cases, the observed toxicity of the set of compounds can be well correlated with their lipophilicity. In these cases, there

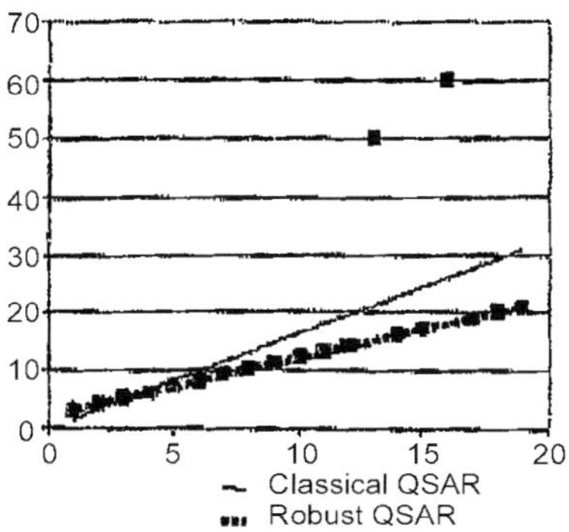

Figure 7 Classical and robust QSAR.

is no residual activity and BAIA returns to a simple linear correlation. In other cases, a baseline correlation is still obtained with log K_{ow} but many outliers exist, most if not all being more toxic than anticipated. BAIA ignores these molecules as their presence in large numbers indicates the existence of some more specific mechanism of toxicity. In such a case, the relatively low activity due to lipophilicity is only a minor component of the observed toxicity. All chemicals whose toxicity is not entirely accounted for by BAIA are thus assumed to derive their additional activity from others factors, which can then be analyzed by MCASE. These compounds therefore remain classified as actives even after BAIA treatment. Thus, for the purpose of the MCASE analysis, activity is defined as the residual activity obtained by subtracting the relevant baseline activity from the observed one.

4.2. New Algorithm

While the BAIA routine performed quite well (14,35), it had difficulties producing satisfactory results for some aquatic toxicity data (Fig. 8).

We found that "robust QSAR" seem to be the most suitable for data mining of linear sets with a large number of outliers. The procedure we used was a slightly modified version of the standard algorithm (36), where several iterative steps were added to refine the baseline parameters.

The procedure, called BAIAplus, starts off with the creation of a classical QSAR (Fig. 9).

In the next step, the procedure calculates a first approximation to the baseline by the classical robust technique (Fig. 9). Then, the routine enters into an iterative algorithm, aiming at identifying and correcting significant errors. It starts by estimating the quality of the baseline, the major criteria being the distribution of errors around the baseline. A special subroutine builds a histogram of differences between the experimental and calculated response values and looks for peaks in the positive part of the histogram (Fig. 10), identifying possible clusters of data, missed by the preliminary baseline identification.

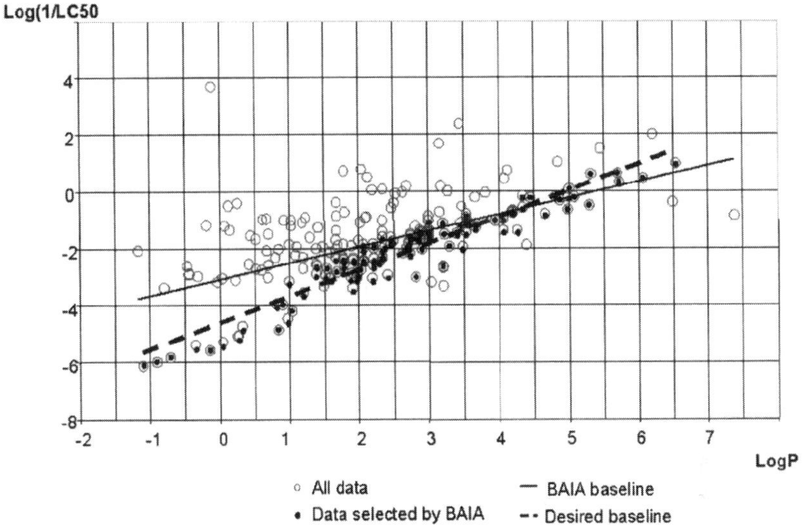

Figure 8 Base line determination by BAIA utility.

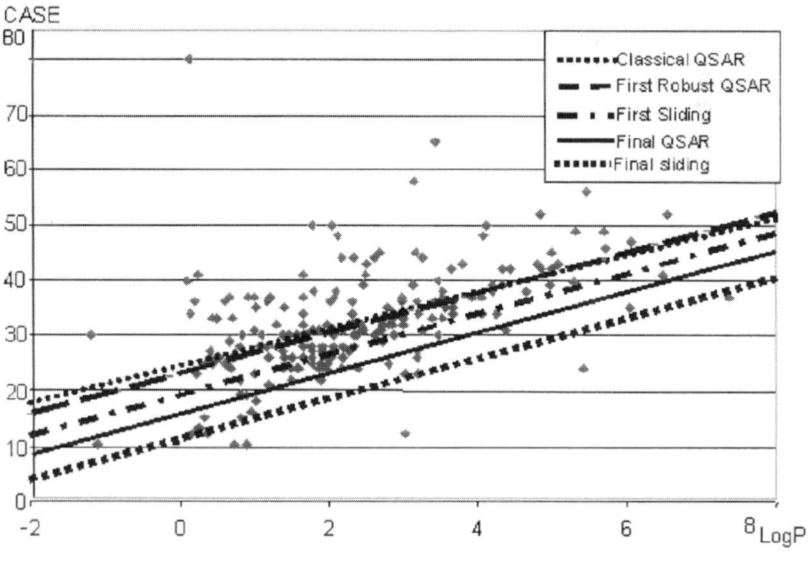

Figure 9 Consecutive steps of the BAIAplus procedure.

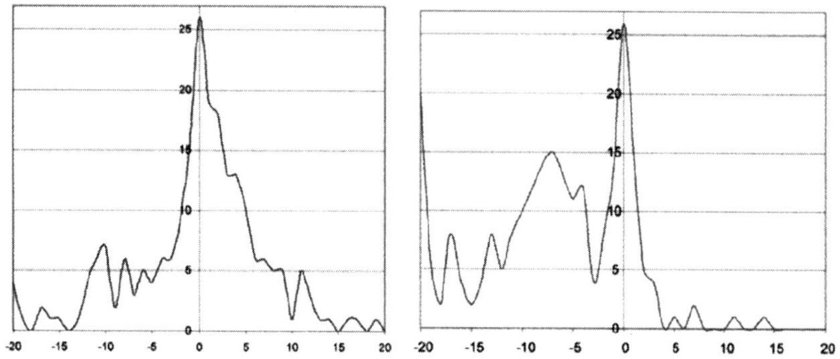

Figure 10 Histogram of baseline error before and after using BAIAplus.

For a peak to be representing a significant error and the baseline to be marked as nonperfect, three conditions need to be met:

The peak needs to be positioned in the positive part of the histogram.

The peak should contain at least 2.5% of the total number of points.

The peak's position (*X* coordinate), numerically equal to the error value should exceed the maximum allowed error value.

If such a peak is found, the baseline will slide down incrementally, based on the calculated error distributions until the significant error is decreased or even completely eliminated. As soon as these major adjustments are completed, the remaining errors are iteratively reevaluated and eliminated by a series of rotate–slide adjustments of the baseline. The final statistical parameters are then recalculated using all the points in the proximity of the newly found baseline and the results will be returned unless the number of points described by the new baseline are less than 10% of the total number of points, or the value of the slope is not within the range 0.1–0.5. In other words, if it is necessary to remove too many points to find a baseline or if the baseline is almost flat or too steep, it will be rejected.

5. DEVELOPMENT OF EXPERT SYSTEM PREDICTORS BASED ON MCASE RESULTS

The results provided by MCASE, when challenged with a new molecule, consist of a number of useful informations such as the probability that the molecule is active, its projected potency, the probability that the biophore is indeed related to activity, the eventual presence of unknown grouping of atoms, and so on. It is, however, sometimes difficult to translate this mass of results into a simple conclusion—is the molecule active or inactive? In the basic program, activity is assessed on the bases of the projected potency and ignores the other modulators. In an attempt to improve the predictivity of the models, we therefore developed what we call the Expert System Predictor program, or ESP, that reaches its conclusions based on an expert interpretation of all the results provided by MCASE.

5.1. Wizard

The first such expert system, named "the Wizard" (19,21) is a rule-based subroutine, where the rules are based on multiple logical operators IF, THEN... ELSE. Thus for every tested compound, the Wizard receives and analyzes a set of quantitative and qualitative parameters generated by MCASE. If the combination of parameters satisfies one of the rules, a decision about the activity of the molecule is made. If the rules are not satisfied, the results are deemed inconclusive.

A decision tree algorithm generates the rules used by the Wizard. We used Quinlann's C4.5 algorithm (37) as the supporting algorithm for the identification of original rules. However, the algorithm was modified to increase selectivity of statistical parameters (gain ratio) used for finding the best rules. Human expertise was used to extract the most meaningful rules from the decision tree.

We tested the applicability of the Wizard technique on our composite mutagenicity database consisting of Salmonella mutagenicity modules for TA97, TA98, TA100, TA1535, and TA1538 strains in the absence of metabolic activation and in the presence of hamster and rat liver S9 (Table 2)

Table 2 External Validation of the 15 *Salmonella* Database with and without Wizards

Database	Validation without Wizard (%)				Validation with Wizard (%)				
		Sens.	Spec.	Cone.	Cov.	Sens.	Spec.	Cone.	Cov.
TA100(-S9)	35.70	95.40	83.21	97.86	40.00	92.80	83.09	98.55	
TA1535(-S9)	31.30	98.10	89.08	99.17	43.80	96.00	88.89	95.90	
TA1537(-S9)	40.00	97.60	94.25	96.67	40.0	97.5	94.05	97.67	
TA97(-S9)	0.00	97.70	89.36	95.92	0.00	95.10	86.67	95.74	
TA98(-S9)	55.60	95.80	90.51	97.16	64.70	95.30	91.13	91.18	
TA100(HS9)	33.30	94.10	73.08	93.69	80.60	82.40	81.71	75.23	
TA1535(HS9)	30.80	97.90	90.00	98.21	87.50	95.70	95.05	90.99	
TA1537(HS9)	30.00	95.70	87.34	95.18	75.00	91.70	90.79	98.70	
TA97(HS9)	16.70	91.40	80.49	91.11	33.30	85.70	81.58	97.44	
TA98(HS9)	45.00	95.70	86.84	98.28	55.60	93.40	87.16	93.16	
TA100(RS9)	36.00	90.50	79.17	93.75	64.70	82.70	79.59	76.56	
TA1535(RS9)	54.50	100.00	95.58	97.41	72.70	100.00	97.30	94.87	
TA1537(RS9)	12.50	97.40	89.41	96.59	25.00	97.80	92.00	58.14	
TA97(RS9)	20.00	100.0	82.61	95.83	40.00	100.00	88.00	54.35	
TA98(RS9)	61.30	99.00	89.92	96.99	74.10	97.80	92.44	89.47	

For each module, the set of rules was generated as a combined result of 10 trials. To develop the rules, 16 global properties of every tested molecule, such as molecular weight, number of atoms, number of biophores, warnings, biophobes, etc., and 13 biophore dependent properties, such as predicted activity, confidence level of prediction, number of positive and negative modulators, etc., were used. Validation without Wizard: the results were taken "as is", no rules were applied to correct the predictions.

Sens.: sensitivity (percentage of correctly predicted actives); Spec: specificity (percentage of correctly predicted inactives); Cone: concordance; Cov.: coverage; -S9: no S9 activization; HS9: with hamster liver S9 activization; RS9: with rat liver S9 activization.

The Wizard generally leads to a positive outcome for all modules (see Table 2). In a few cases, the sensitivity was dramatically improved (i.e., by 50% for TA100 and A2R TA1535 with hamster S9). Generally, the Wizard improves specificity by 10–15% at the cost of reduced coverage. There are, however, a few cases (TA97 and TA1537 without activation) where the Wizard failed to improve the statistical parameters. The reported results are our first attempt to deploy the Wizards on a massive scale and we consider it to be somewhat successful. However, some of the modules require more careful refining which is going to be the subject of some future work.

In the studies described above, a different Wizard was generated for each module. We also did study the possibility to develop a "general" Wizard, a set of rules applicable to all databases. Unfortunately, this does not seem to be a reliable solution. Indeed, we found that the Wizard is very much module-dependent. In other words, a new Wizard must be developed for any new database that is built, which is not a very desirable situation. However, in cases where the quality of the prediction cannot be increased by any other means, a local Wizard could provide a reasonable solution.

5.2. WinESP: Prediction of Chemical Toxicity/Biological Activity Using Multiple MC4PC Models

Toxic properties of chemicals, like mutagenicity and carcinogenicity, are often derived from multiple biological models according to regulatory guidelines. Therefore, a discussion of MC4PC is incomplete without discussing the programs that can predict biological activity/toxicity of chemicals based on multiple MC4PC models. In this section, we describe a particular program called WinESP used in the context of multiple database activity predictions.

The most popular mutagenicity tests such as the Ames test measure the ability of a compound to cause reverse mutations in a series of bacterial strains. The assay is conducted in the absence and/or presence of liver microsomal preparations. Similarly, human carcinogenicity of a chemical is generally assessed from measured carcinogenicity experiments in different species and different genders of rodents (e.g., male and female mice and rats). Screening against multiple models increases the probability of correct detection of active and inactive compounds because they cover a wider range of biological and mechanistic domain. However, the interpretation of the results and their relevance to humans is somewhat problematic. Indeed, the observation of activity in a single cell may be relevant to human activity but may also be indicative of carcinogenic paths specific to that particular animal species or bacterial strain only. Prediction of activity in several cells may be problematic as well, particularly, if the program identifies different functionalities as responsible for activity in the different cells. This problem, however, could be alleviated if a method capable of identifying relationships between different SAR models and which can combine the results of screening against different models can be developed. Such methods should provide consistent and good results and should withstand weaknesses present in the individual models. This line of thoughts had been pioneered by our colleagues of the U.S. FDA when they relied heavily on the results of multiple MCASE analysis (19). We developed WinESP to reproduce such results and provide reliable and highly specific predictions for overall toxicity assessment of a chemical using several different MC4PC QSAR-ES models.

WinESP is a Microsoft Windows based program with a user-friendly graphical user interface (GUI). It offers several new features to facilitate a novel way of interpreting results provided by MC4PC and to obtain toxicity prediction of chemicals based on several models with high concordance.

5.2.1. Analysis of MC4PC Biophores

WinESP offers a new way to analyze the details of biophores obtained from different databases. The program maintains a comprehensive database of all biophores present in each module created by the program. Each record contains several details associated with the biophore (e.g., the structure of the biophore, its distribution in the inactive, active, and marginally active chemicals, statistical significance, average activity of the chemicals containing the biophore, various QSAR parameters obtained from the group of chemicals containing the biophore, and so on). The major advantage of such a database is that the user can retrieve relevant information for a particular biophore very easily. For example, it is possible to retrieve fragments which are similar to a particular biophore throughout different MC4PC models. The program presents this biophore database to the user in the form of GUI elements like scrollable lists (Fig. 11) on which the user can

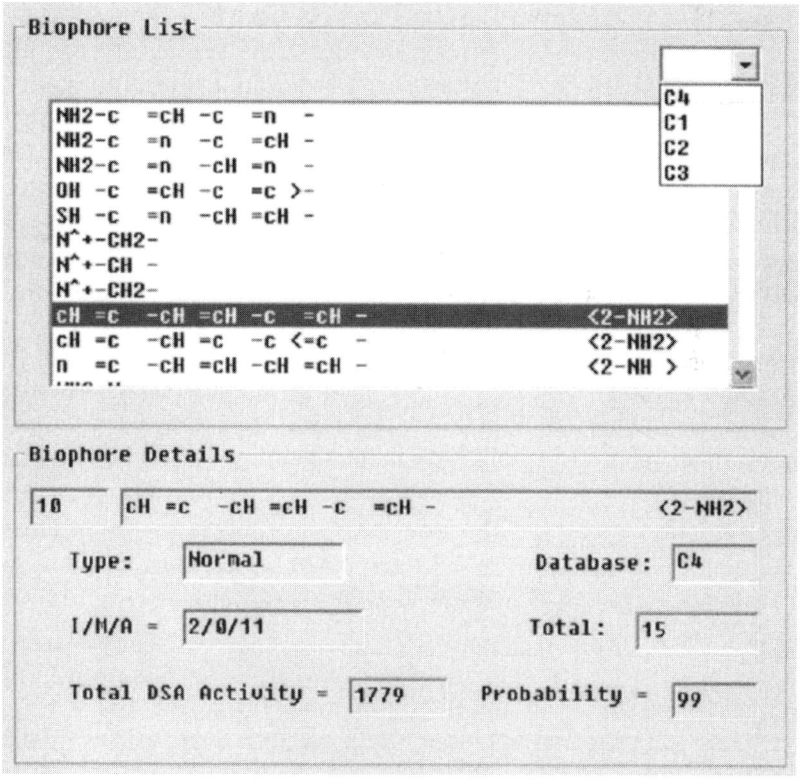

Figure 11 A biophore list embedded in the graphical user interface of WinESP.

point and click using a mouse and obtain relevant information.

To achieve a high concordance between experimentally observed and predicted activity of test chemicals, WinESP uses a unique method (19) to assign a score to each of the biophores. This score is calculated by multiplying the average CASE unit activity of the biophore by the number of chemicals containing the biophore, and then adding the calculated value from all the MC4PC models containing the biophore. Thus, rather than relying on the individual biophores identified in individual MC4PC database modules, WinESP requires confirmatory experimental findings across correlated toxicological endpoints, a high degree of representation of the biophores in the training data set, and a high biological potency for toxicological activity.

After WinESP has built a database of biophores from all the MC4PC models which are toxicologically related, it is capable of predicting the toxicity of new chemicals presented to it. It uses a variety of information obtained from the structure of the test chemicals to predict its overall toxicity, including WinESP assigned scores of the biophores, presence of any fragment which is detrimental to the activity, presence of activating and inactivating modulators, presence of substructures unknown to the training set, etc. WinESP presents the results in interface elements which can be easily browsed by the user (Fig. 12). The prediction contains a list of biophores found in the tested chemicals against different MC4PC models and other details.

5.2.2. Description of the WinESP Method

A complete run of WinESP consists of the following basic steps.

5.2.2.1. Building MC4PC Models

The first step in any WinESP run is to build the MC4PC modules which are intended to be used for multiple database prediction of new chemicals. The modules are normally toxicologically correlated. Examples include mutagenicity models

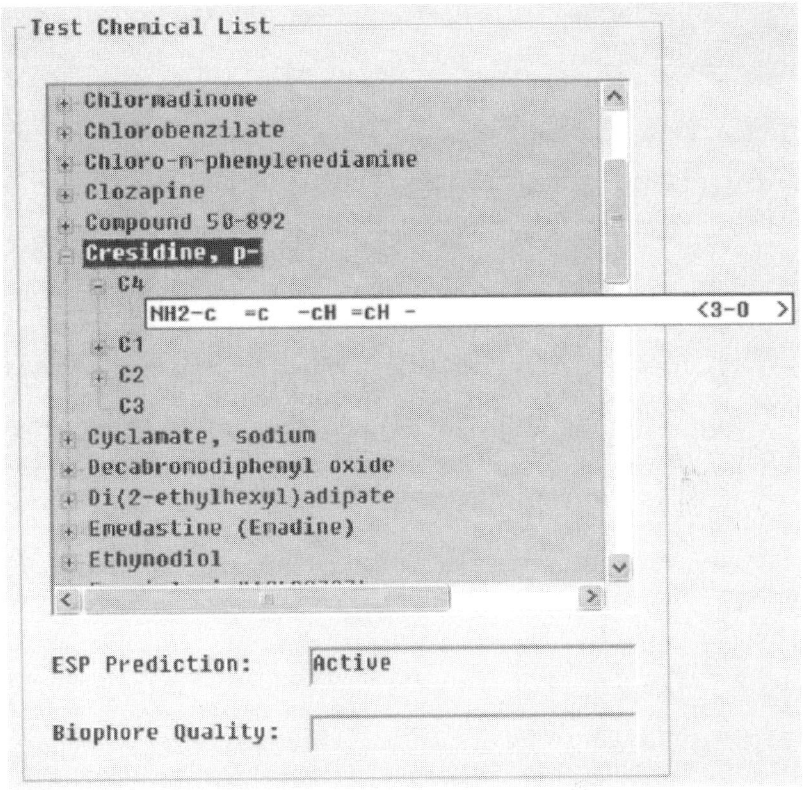

Figure 12 A list of test chemicals in the GUI of WinESP showing the results of activity prediction by WinESP.

from different *Salmonella typhimurium* strains or male, female mouse and rat models for rodent carcinogenicity.

5.2.2.2. Building a Biophore Database

In the second step, WinESP builds a master database of biophores from all the models built in the previous step. This database contains various details of each biophore and is used during the prediction phase.

5.2.2.3. Activity Prediction of New Chemicals

Once the biophore database is built, WinESP can predict the overall activity/toxicity of new chemicals presented to it using the information it has compiled from the contributing MC4PC modules.

5.2.3. Example of Application of WinESP

Based on the National Toxicology Program (NTP) report (38,39) containing results of mutagenicity studies using the Ames test, four MC4PC databases were developed for four different strains of *S. typhimurium* (TA100, TA1535, TA1537, and TA98) in the presence of standard homogenized rat liver metabolic activation. The activity distribution of chemicals in the four databases is shown in Table 3.

After building of the MC4PC databases were completed, 124 additional chemicals were collected from various sources with their relevant experimental results of Ames *Salmonella* tests in the presence and absence of homogenized rat liver metabolic activation. Among these chemicals, 100 chemicals are considered to be mutagens (showing positive result in at least one test strain). The other 24 are considered to be nonmutagenic (showing inactive results in all four test strains). These 124 chemicals were not included in any of the four previous databases. WinESP was then used to evaluate the general mutagenic potential of these "unknown" chemicals based on the information from the MC4PC models. The results of WinESP were compared with those of the straight MC4PC prediction results and are listed in Table 4.

As can be seen, when compared with the MC4PC prediction results for individual databases, WinESP program provides better results for active (mutagenic) chemicals (higher sensitivity), although the specificity of WinESP prediction is not as good as that of the sum of the MC4PC predictions. The prediction rate (concordance) and the coverage of

Table 3 Activity Distribution of Chemicals in the MC4PC *Salmonella* Mutagenicity Modules

Database	TA100	TA1535	TA1537	TA98
Chemicals (I/M/A)	1055/65/290	1049/28/124	789/25/71	1087/31/259

I/M/A: the number of inactive, marginally active and active compounds.

Table 4 Comparison of WinESP and MC4PC Prediction of Mutagenic Potential of 124 "Unknown" Chemicals

	Sens. (%)	Spec. (%)	Conc. (%)	Cov. (%)
WinESP	93	83	92	87
MCASE[a]	80	100	85	80

Sens.: sensitivity (percentage of correctly predicted actives); Spec.: specificity (percentage of correctly predicted inactives); Conc.: concordance; Cov.: coverage.
[a] The prediction was obtained by manually analyzing the prediction results of the four individual MC4PC database. Basically, a chemical was predicted to be a mutagen if it showed active prediction in any of these four models, to be nonmutagenic if it is inactive in all four models.

WinESP prediction are also higher than that of the isolated MC4PC results.

6. CONCLUSION

Computer driven models of toxic and pharmacological endpoints continue to be upgraded and continue to improve as more suitable data are assembled and processed. We see no end to this process as the models learn from previous attempts and methodologies meant to respond to new needs continue to be developed. We are not at the stage where the models can completely replace animals, and it may well be that we will never reach that point. Indeed, for the models to work, the coverage of the universe of chemicals must be complete otherwise the programs will continue to encounter cases outside of their expertise. Animals do not suffer from this weakness as they are analogic and will respond to the effect produced by a new chemical even if they have never encounter that chemical before. This is not the case with computers.

On the other hand, as the databases increase in scope, there will be a lesser need to experiment on animals, unless the programs issue a warning that the chemical's structure is outside of its domain of competence or if the endpoint is so complex as to resist to attempts to encapsulate the reasons for activity into current methodologies.

ACKNOWLEDGMENTS

The authors thank the National Institute of Health for their generous support of part of this work through SBIR grant no. 2R44CA090178–02 awarded to MULTICASE Inc.

REFERENCES

1. Ashby J, Tennant RW. Definitive relationships among chemical structure, carcinogenicity, and mutagenicity for 301 chemicals tested by the U.S. NTP. Mutat Res 1991; 257:229–306.
2. Frierson M, Klopman G, Rosenkranz HS. Structure–activity relationships (SARs) among mutagens and carcinogens: a review. Environ Mutag 1986; 8(2):283–327.
3. Klopman G, Namboodiri K, Kalos A. In: Rein R, ed. The Molecular Basis of Cancer. A. R. Liss, Inc, 1985:287–298.
4. Rosenkranz HS, Klopman G. Mutagens, carcinogens and computers. Proc Clin Biol Res 1986; 209A:71–104.
5. Klopman G, Macina OT, Levinson ME, Rosenkranz HS. Computer automated structure evaluation of quinolone antibacterial agents. Antimicro Agents Chemother 1987; 31:1831–1840.
6. Klopman G, Macina OT. Use of the Computer Automated Structure Evaluation program in determining quantitative structure–activity relationships within hallucinogenic phenylalkylamines. J Theor Biol 1985; 113(4):637–648.
7. Klopman G, Macina OT, Simon EJ, Hiller JM. Computer automated structure evaluation of opiate alkaloids. Theochemistry 1986; 134:299–308.
8. Klopman G, Macina, OT. Computer-automated structure evaluation of antileukemic 9-anilinoacridines. Mol Pharmacol 1987; 31:457–476.
9. Klopman G, Kalos AN. Quantitative structure–activity relationships of beta-adrenergic agents. Application of the computer automated structure evaluation (CASE) technique of molecular fragment recognition. J Theor Biol 1986; 118: 199–214.

10. Klopman G, Bendale RD. Computer automated structure evaluation (CASE): a study of inhibitors of the thermolysin enzyme. J Theor Biol 1989; 136(1):67–77.

11. Klopman G, Venegas RE. CASE study of in vitro inhibition of sparteine monooxygenase. Acta Pharma Jugosl 1986; 36: 189–208.

12. Rosenkranz HS, Mersch-Sundermann V, Klopman G. SOS chromotest and mutagenicity in *Salmonella*: evidence for mechanistic differences. Mutat Res-Fund Mol M 1999; 431(1):31–38.

13. Rosenkranz HS, Karol MH. Chemical carcinogenicity: can it be predicted from knowledge of mutagenicity and allergic contact dermatitis?. Mutat Res-Fund Mol M 1999; 431(1):81–91.

14. Klopman G, Saiakhov R, Rosenkranz HS, Hermens JLM. Multiple computer-automated structure evaluation program study of aquatic toxicity 1: Guppy. Environ Toxicor Chem 1999; 18(11):2497–2505.

15. Gomez J, Macina OT, Mattison DR. Structural determinants of developmental toxicity in hamsters. Teratology 1999; 60(4):190–205.

16. Rosenkranz HS, Mersch-Sundermann V, Klopman G. Structural determinants of developmental toxicity in hamsters. Teratology 1999; 60(4):190–205.

17. Klopman G, Tu M. Diversity analysis of 14156 molecules tested by the National Cancer Institute for anti-HIV activity using the quantitative structure-activity relational expert system MCASE. J Med Chem 1999; 42(6):992–998.

18. Zhu XY, Zhang YP, Klopman G, et al. Thalidomide and metabolites: indications of the absence of'genotoxic' carcinogenic potentials. Mutat Res-Fund Mol M 1999; 425(1):153–167.

19. Matthews EJ, Contrera JF. A new highly specific method for predicting the carcinogenic potential of pharmaceuticals in rodents using enhanced MCASE QSAR-ES software. Regul Toxicol Pharmacol 1998; 28(3):242–264.

20. Rosenkranz HS, Zhang YP, Klopman G. Studies on the potential for genotoxic carcinogenicity of fragrances other chemicals. Food Chem Toxicol 1998; 36(8):687–696.

21. Lee Y, Buchanan BG, Aronis JM. Knowledge-based learning in exploratory science: learning rules to predict rodent carcinogenicity. Mach Learn 1998; 30(2–3):217–240.

22. Labbauf A, Klopman G, Rosenkranz HS. Dichotomous relationship between DNA reactivity and the induction of sister chromatid exchanges in vivo and in vitro. Mutat Res-Fund Mol M 1997; 377(1):37–52.

23. Bevington PR. Data Reduction and Error Analysis for the Physical Sciences. New York: McGraw-Hill, 1969.

24. Klopman GJ. Artificial intelligence approach to structure-activity studies. Computer automated structure evaluation of biological activity of organic molecules. Am Chem Soc 1984; 106:7315–7321.

25. Klopman G, Rozenkranz HS. Structural requirements for the mutagenicity of environmental nitroarenes. Mutat Res 1984; 126(3):227–238.

26. Klopman G, Contereras R. Use of artificial intelligence in structure-activity correlations of anticonvulsant drugs. J Mol Pharmacol 1985; 27:86–93.

27. Weininger D. SMILES, a chemical language and information system. 1. Introduction to methodology and encoding rules. J Chem Inf Comput Sci 1988; 28:31–36.

28. Klopman G, McGonigal M. Computer Simulation of physical chemical properties of organic molecules. J Chem Inf Comp Sci 1981; 21:48–52.

29. Klopman G, Dimayaga M. Computer Automated Structure Evaluation (CASE) of the teratogenicity of retinoids with the aid of a novel geometry index. J Comput Aided Mol Des 1990; 4(2):117–130.

30. Klopman G, Stefan L, Saiakhov R. ADME evaluation 2. A computer model for the prediction of intestinal absorption in humans. Eur J Pharm Sci 2002; 17:253–263.

31. Klopman G, Namboodiri K, Schochet M. Simple Method of Computing the Partition coefficient. J Comput Chem 1985; 6:28–38.

32. Klopman G, Zhu H. Estimation of the aqueous solubility of organic molecules by the group contribution approach. J Chem Inf Comput Sci 2001; 41:439–445.

33. Hansch C. A quantitative approach to biochemical structure–activity relationships. Acct Chem Res 1969; 2:232–239.

34. Klopman G. The MultiCASE Program II. Baseline Activity Identification Algorithm (BAIA). J Chem Inf Comput Sci 1998; 38:78–81.

35. Klopman G, Saiakhov R, Rosenkranz HS. Multiple computer-automated structure evaluation study of aquatic toxicity II. Fathead minnow. Environ Toxicol Chem 2000; 19:441–447.

36. Press WH, Flannery BP, Teukolsky SA, Vetterling WT. . Numerical Recipes in Fortran. 2nd. ed. Cambridge: University Press1992.

37. Quinlan JR. C4.5: Programs for Machine Learning. San Mateo: Morgan Kauffman, 1992.

38. Zeiger E, Anderson B, Haworth S, Lawlor T, Mortelmans K. *Salmonella* mutagenicity tests: V. Results from the testing of 311 chemicals. Environ Mol Mutagen 1992; 19(suppl 21):2–141.

39. Haworth S, Lawlor T, Mortelmans K, Speck W, Zeiger E. *Salmonella* mutagenicity test results for 250 chemicals. Environ Mutagen 1983; 5(suppl 1):3–142.

RECOMMENDED READING

1. Hansch C, Leo A. Exploring QSAR: Fundamentals and Applications in Chemistry and Biology. American Chemical Society 1995.

2. Klopman G. Artificial intelligence approach to structure–activity studies. Computer automated structure evaluation of biological activity of organic molecules. J Am Chem Soc 1984; 106:7315–7320.

3. Klopman G. MULTICASE: a hierarchical computer automated structure evaluation program. *Quant Struct–Act Relat* 1992; 11:176–184.

4. Klopman G Rosenkranz HS. International commission for protection against environmental mutagens and carcinogens. Approaches to SAR in carcinogenesis and mutagenesis. Prediction of carcinogenicity/mutagenicity using MULTI-CASE. *Mutat Res* 1994; 305(1):33–46.

GLOSSARY

BAIA: Baseline Activity Identification Algorithm.

BAIAplus: Modified BAIA utilizing robust QSAR technique.

Binomial distribution: The binomial distribution gives the discrete probability distribution $P_p(n|N)$ of obtaining exactly n successes out of N trials (where the result of each trial is true with probability p and false with probability $q = 1 - p$).

Biophobe: Group of atoms (molecular subfragments) responsible for decrease of activity observed for molecules that contain it.

Biophore: Group of atoms (molecular subfragments) believed to be responsible for a given biological effect.

Concordance: The ratio of correct predictions divided by the number of conclusive predictions that were made.

Congeneric molecules: Belonging to the same genus; in this case containing common functionality.

Decision tree algorithm: A type of expert system comprised of a branching structure of questions and possible responses designed to lead a system to an appropriate solution to a supplied problem or provide needed information.

ESP: Expert System Predictor.

Gain: A parameter representing the difference between the information needed to identify an element of decision tree and the information needed to identify an element of the decision tree after the value of attribute X has been obtained, that is, this is the gain in information due to attribute X.

Sensitivity: Percentage of correctly predicted positives.

Histogram: Method calculating individual and cumulative frequencies for a range of data and data bins. This method generates data for the number of occurrences of a value in a data set.

Lipophilicity: This represents the affinity of a molecule or a moiety for a lipophilic environment. It is commonly measured by its distribution behavior in a biphasic system, either liquid–liquid or solid–liquid systems.

Log K_{ow}: Logarithmic octanol–water partition coefficients, used to assess a lipophilicity.

Modulators: Group of atoms (molecular subfragments) believed to modulate activity of molecules containing the primary biophore (biophobe).

Narcosis effect: The narcosis effect is a nonspecific mode of toxic action, which is recognized as resulting from the disruption of a cell's cytoplasmic membrane by the sheer physical presence of lipophilic chemicals. It was postulated that for aquatic organisms, the mechanism of narcosis is not a "specific" process. This means, that compounds with identical octanol-water partition coefficients should have the same intrinsic toxicity, irrespective of their structures.

QSAR: Quantitative structure–activity relationships (QSAR) are mathematical relationships linking chemical structure and pharmacological activity in a quantitative manner for a series of compounds. Methods which can be used in QSAR include various regression and pattern recognition techniques.

Robust QSAR: A regression model which includes the possibility of measurement outliers. It also refers to parameter estimation that can handle outliers.

Specificity: Percentage of correctly predicted negatives.

13

PASS: Prediction of Biological Activity Spectra for Substances

VLADIMIR POROIKOV and
DMITRI FILIMONOV
Institute of Biomedical Chemistry of Russian
Academy of Medical Sciences, Moscow, Russia

1. INTRODUCTION

Each pharmaceutical research and development project is aimed at discovering new drugs for the treatment of certain diseases. The investigation of new pharmaceuticals is carried out in a stepwise manner. This is because drug discovery is a time-consuming process involving enormous financial resources and manpower, and with a substantially high risk factor. On average, it requires 12 years and approximately $800 million for introducing a new medicine to the market (1) with a high risk of negative results (1 out of 10,000 substances studied is developed to a safe and potent drug). Drug research starts with identification of a "lead molecule" with required biological activity. Subsequently, the lead molecule is developed to get more potent compounds with appropriate pharmacodynamic and pharmacokinetic properties that can qualify as drug candidates (2). General biological potential of any molecule under study is also evaluated in stages. The emphasis is first laid on testing for specific activity followed by general pharmacology and toxicology study, clinical trials, postmarketing registration of adverse effects, etc. As a result, adverse/toxic actions are often discovered at a stage when a lot of time and money are already expended (3). At the same time, it is practically impossible to test experimentally all compounds against each known kind of biological activity and possible toxic effects. So, a computer-aided prediction is the "method of choice" at the early stage of drug research. Relying on predicted results, one may establish the priorities for testing a particular compound and the basis for selecting the most prospective hits/leads/candidates from the set of compounds available for screening. Application of computational methods has significantly decreased the time required for obtaining a compound with the required properties with reduction in financial expenditure. In addition, it helps to obtain more effective and safety medicines.

Both computer-aided analysis of quantitative structure–activity/structure–property relationships (QSAR/QSPR) and molecular modeling are widely used for finding and optimizing lead compounds. However, the majority of such methods are constrained by studying a single targeted biological activity within the particular chemical series (4–6). Typically, they are applied step-by-step to analyze different activities/properties in correspondence with the sequential study of biologically active compounds mentioned above. On the other hand, most of the known biologically active compounds demonstrate several or even many kinds of biological activity, which constitute the so-called "biological activity spectrum" of the compound (3). Some components of the biological activity spectrum may serve as a basis for the treatment of certain pathologies, while others may be a source for adverse/toxic effects. For instance, thalidomide was prescribed worldwide (1950s to early 1960s) to pregnant women as treatment for morning sickness. Subsequently, it was discovered that thalidomide was teratogenic (~12,000 babies were born with tiny or no limbs, flipper-like arms and legs, with serious facial deformities and defective organs). Because of this, the drug was withdrawn from the market in 1962 (7). However, now thalidomide is again considered as a prospective pharmaceutical agent because of some newly discovered activities, e.g., angiogenesis inhibitor, tumor necrosis factor antagonist, and others (8). If, at the early stage of study, researchers could predict the most probable biological activities in drugs like thalidomide, they might avoid the dramatic consequences of their adverse/toxic action and could suggest wider pharmacotherapeutic applications.

2. BRIEF DESCRIPTION OF THE METHOD FOR PREDICTING BIOLOGICAL ACTIVITY SPECTRA

The computer program PASS (Prediction of Activity Spectra for Substances) was developed as a tool for evaluation of general biological potential in a molecule under study (9). There had been several earlier attempts to develop such a kind of computer system (10–13). In particular, the feasibility for computer-aided prediction of biological activity of chemical compounds on the basis of their structural formulae was studied within the State System for Registration of New Chemical Compounds Synthesized in the USSR in 1972–1990 (14). For some objective and subjective reasons, this problem was not completely solved, but the studies carried out at that time provided the background and experience necessary for development of such a computer program.

The latest version of PASS (1.911) predicts about 1000 kinds of biological activity with the mean prediction accuracy of about 85%. PASS could predict only 541 kinds of biological activity in 1998 (15) and 114 kinds in 1996 (16) (mean prediction accuracy was only 78% in 1996). The default list of predictable biological activities includes main and side pharmacological effects (e.g., antihypertensive, hepatoprotective, sedative, etc.), mechanisms of action (5-hydroxytryptamine agonist, acetylcholinesterase inhibitor, adenosine uptake inhibitor, etc.), and specific toxicities (mutagenicity, carcinogenicity, teratogenicity, etc.).

Information about novel activities and new compounds can be straightforwardly included into PASS, and used for further prediction of biological activity spectra for new chemical compounds. A complete list of biological activities predicted by PASS along with a detailed description of the algorithm, applications, and efficiency of PASS is available on the web site (17). Besides, it is also possible to get predictions of biological activity spectra or estimate the accuracy of prediction of the biological activity by submitting substances with known activities and obtaining results of prediction via the internet (18).

2.1. Biological Activity Presentation

In PASS, biological activities are described qualitatively (active or inactive). Reflecting the result of chemical compound's interaction with a biological object, the biological activity depends on both the compound's molecular structure and the terms and conditions of the experiment. Therefore, structure–activity relationship analysis based on qualitative presentation of biological activity describes general "biological potential" of the molecule being studied. On the other hand, qualitative presentation allows integrating information concerning compounds tested under different terms and conditions and collected from many different sources as in the PASS training set.

Any property of chemical compounds determined by their structural peculiarities can be used for prediction by PASS. It is clear that the applicability of PASS is broader than the prediction of biological activity spectra. For example, we use this approach to predict drug-likeness (19) and biotransformation of drug-like compounds (20).

2.2. Chemical Structure Description

The 2D structural formulae of compounds were chosen as the basis for description of chemical structure, because this is the only information available in the early stage of research (compounds may only be designed but not synthesized yet). Plenty of characteristics of chemical compounds can be calculated on the basis of structural formulae (21). Earlier (22), we applied the substructure superposition fragment notation (SSFN) codes (23). But SSFN, like many other structural descriptors, reflects the abstraction of chemical structure by the human mind rather than the nature of the biological activity revealed by chemicals. The multilevel neighborhoods of atoms (MNA) descriptors (24–26) have certain advantages in comparison with SSFN. These descriptors are based on the molecular structure representation, which includes the hydrogens according to the valences and partial charges of other atoms and does not specify the types of bonds. MNA descriptors are generated as recursively defined sequence:

- zero-level MNA descriptor for each atom is the mark A of the atom itself, and
- any next-level MNA descriptor for the atom is the sub-structure notation $A(D_1D_2 \cdots D_i \cdots)$,

where D_i is the previous-level MNA descriptor for ith immediate neighbor of the atom A.

The mark of the atom may include not only the atomic type but also any additional information about the atom. In particular, if the atom is not included into the ring, it is marked by "–". The neighbor descriptors $D_1D_2 \cdots D_i \cdots$ are arranged in a unique manner, e.g., in lexicographic order. Thus iterative process of MNA descriptors generation can be continued covering first, second, etc., neighborhoods of each atom.

For instance, starting from N atom in the piperidine-2,6-dione part of thalidomide molecule, the following MNA descriptors of the zero to the third level can be generated:

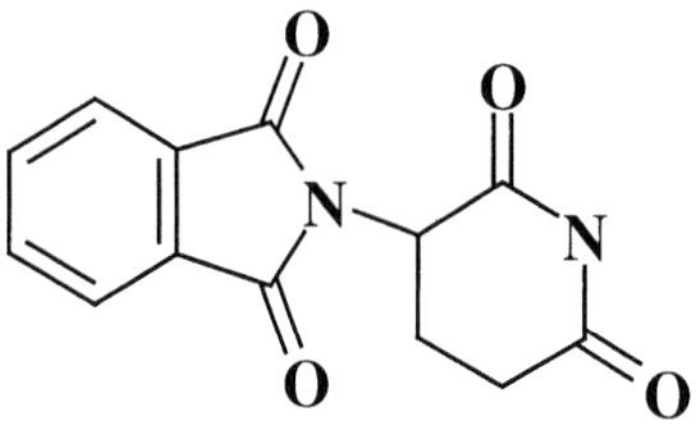

MNA/0: N
MNA/1: N(CCC)
MNA/2: N(C(CCN–H)C(CN–O) C(CN–O))
MNA/3: N(C(C(CCC)N(CCC)–O(C))C(C(CCC)N(CCC)–O(C)) C(C(CC–H–H) C(CN–O) N(CCC)–H(C)))

In the latest version of PASS (1.911), which is discussed in this paper, molecular structure is represented by the set of unique MNA descriptors of the third level (MNA/3). The list of thalidomide's MNA/3 descriptors is given below:

1. C(C(C(CCC)C(CC–H)C(CN–O))C(C(CCC)C(CC–H)–H(C))C(C(CCC)N(CCC)–O(C)))
2. C(C(C(CCC)C(CC–H)C(CN–O))C(C(CC–H)C(CC–H)–H(C))–H(C(CC–H)))
3. C(C(C(CCC)C(CC–H)C(CN–O))N(C(CCN–H)C(CN–O)C(CN–O))–O(C(CN-O)))
4. C(C(C(CCC)C(CC–H)–H(C))C(C(CC–H)C(CC–H)–H(C))–H(C(CC–H)))
5. C(C(C(CCN–H)C(CC–H–H)–H(C)–H(C))C(C(CCN–H)N(CC–H)–O(C))N(C(CCN–H)C(CN–O)C(CN–O))–H(C(CCN–H)))
6. C(C(C(CCN–H)C(CC–H–H)–H(C)–H(C))C(C(CC–H–H)N(CC–H)–O(C))–H(C(CC–H–H))–H(C(CC–H–H)))
7. C(C(C(CC–H–H)C(CN–O)N(CCC)–H(C))C(C(CC–H–H)C(CN–O)–H(C)–H(C))–H(C(CC–H–H))–H(C (CC–H–H)))
8. C(C(C(CC–H–H)C(CN–O)N(CCC)–H(C))N(C(CN–O)C(CN–O)–H(N))–O(C(CN-O)))
9. C(C(C(CC–H–H)C(CN–O)–H(C)–H(C))N(C(CN–O) C(CN–O)–H(N))–O(C(CN–O)))

11. N(C(C(CCC)N(CCC)–O(C))C(C(CCC)N(CCC)–O(C)) C(C(CC–H–H)C(CN–O)N(CCC)–H(C)))
12. N(C(C(CCN–H)N(CC–H)–O(C))C(C(CC–H–H)N (CC–H)–O(C))–H(N(CC–H)))
13. –H(C(C(CCC)C(CC–H)–H(C)))
14. –H(C(C(CCN–H)C(CC–H–H)–H(C)–H(C)))
15. –H(C(C(CC–H–H)C(CN–O)N(CCC)–H(C)))
16. –H(C(C(CC–H–H)C(CN–O)–H(C)–H(C)))
17. –H(C(C(CC–H)C(CC–H)–H(C)))
18. –H(N(C(CN–O)C(CN–O)–H(N)))
19. –O(C(C(CCC)N(CCC)–O(C)))
20. –O(C(C(CCN–H)N(CC–H)–O(C)))
21. –O(C(C(CC–H–H)N(CC–H)–O(C)))

The substances are considered to be equivalent in PASS if they have the same set of MNA descriptors. Since MNA descriptors do not represent the stereochemical peculiarities of a molecule, the substances, whose structures differ only stereochemically, are formally considered as equivalent.

2.3. Training Set

The PASS estimations of biological activity spectra of new compounds are based on the structure–activity relationships knowledgebase (SARBase), which accumulates the results of the training set analysis. The in-house–developed PASS training set includes about 50,000 known biologically active substances (drugs, drug candidates, leads, and toxic compounds). Since new information about biologically active compounds is discovered regularly, we perform the special informational search and analyse the new information, which is further used for updating and correcting the PASS training set.

2.4. Algorithm of Activity Spectra Estimation

The algorithm of prediction was chosen from a large number of options examined in the past several years. It is based on the specially designed *B*-statistics, in which the well-known

Fisher's arcsine transformation is used. On the basis of a molecule's structure represented by the set of m MNA descriptors $\{D1, \ldots, D_m\}$ for each kind of activity A_k, the following B_k values are calculated:

$$B_k = (S_k - S_{0k})/(1 - S_k \cdot S_{0k})$$

$$S_k = Sin[\Sigma_i \mathrm{ArcSin}(2P(A_k|D_i) - 1)/m]$$

$$S_{ok} = 2P(A_k) - 1$$

where $P(A_k|D_i)$ is a conditional probability of activity of kind A_k if the descriptor D_i is present in a set of molecule's descriptors; $P(A_k)$ is a priori probability to find a compound with activity of kind A_k. For any kind of activity A_k, if $P(A_k|D_i)$ is equal to 1 for all descriptors of a molecule, then $B_k = 1$; if $P(A_k|D_i)$ is equal to 0 for all descriptors of a molecule, then $B_k = -1$; if there is no relationship between the molecule's descriptors and activity of kind A_k, and, so, $P(A_k|D_i) \approx P(A_k)$, then $B_k \approx 0$.

Up to the PASS version 1.703, the algorithm of prediction was based on the following data:

- n is the total number of compounds in the SARBase;
- n_i is the number of compounds containing descriptor D_i in the structure description;
- n_k is the number of compounds containing the kind of activity A_k in the activity spectrum;
- n_{ik} is the number of compounds containing both the kind of activity A_k and the descriptor D_i.

And the estimations of probabilities $P(A_k)$, $P(A_k|D_i)$ are given by

$$P(A_k) = n_k/n, \quad P(A_k|D_i) = n_{ik}/n_i$$

In PASS version 1.703 and later, instead of integers n_i and n_{ik}, the sums g_i and g_{ik} of descriptors weights w are used, where $w = 1/m$, and m is the number of MNA descriptors of individual molecule. This modification increases the accuracy of prediction significantly. So, right now the estimations of probabilities $P(A_k|D_i)$ are given by

$$P(A_k|D_i) = g_{ik}/g_i$$

The main purpose of PASS application is to predict the activity spectra for new substances. To provide more accurate predictions, if the compound under prediction has the equivalent structure in the SARBase, this structure is "excluded" from the SARBase during the prediction with all associated information about its biological activities. The calculations are done by using $n-1$, $g_i - w$, and, when the kind of activity A_k is contained in its activity spectrum in the SARBase, by using $n_k - 1$ and $g_{ik} - w$. Here $w = 1/m$, and m is a number of MNA descriptors in molecule under prediction and its equivalent in the SARBase. The B_k values are calculated using MNA descriptors, which are found in SARBase, i.e., for descriptors of a molecule under prediction with $g_i > 0$ or $g_i - w > 0$, in the case of structure "exclusion."

To take the "yes/no" qualitative prediction, it is necessary to determine B-statistics threshold values for each kind of activity A_k. Using theory of statistical decision, this can be done on the basis of risk function's minimization. But nobody can a priori specify the risk functions for all activity kinds and all possible practical tasks. Therefore, the predicted activity spectrum in PASS is presented by the rank-order list of activities with probabilities "to be active" Pa and "to be inactive" Pi, which are the functions of B-statistics for a molecule under prediction. The B-statistics functions Pa and Pi are the results of the training procedure described below. The list is arranged in descending order of $Pa - Pi$; thus, the more probable activity kinds are at the top of the list. The list can be shortened at any desirable cutoff value, but $Pa > Pi$ is used by default. If the user chooses a rather higher value of Pa as a cutoff for selection of probable activities, the chance to confirm the predicted activities by the experiment is also high, but many existing activities will be lost. For instance, if $Pa > 80\%$ is used as a threshold, about 80% of real activities will be lost; for $Pa > 70\%$, the portion of lost activities is 70%, etc.

2.5. Training Procedure

For each compound from the training set, MNA descriptors are generated and its known activity spectrum and set of descriptors are stored in the SARBase. If this compound has the equivalent structure in SARBase, only new activities are added to activity spectrum. After inclusion of all information from the training set(s) into SARBase, the values n, g_i, n_k, g_{ik} are calculated. For each compound in the SARBase and for each activity kind A_k, values B_k of B-statistics are calculated. Calculations are done taking into account the described above "exclusion" of processed compound. For each activity kind A_k, the calculated values B_k are subdivided into two samples: for active and inactive compounds. These obtained samples are used for calculation of the smooth estimations of B-statisties distribution functions on the following basis.

Suppose we have the sample $x_1, \ldots, x_n$ of n values of random variable X, which has an unknown distribution function $F(x)$. Using an empirical step-function for approximation of F often faults because of small n. To provide the smooth estimation of $F(x)$, the inverse function $x(F)$ is calculated as the conditional expectation of random variable X:

$$x(F) = \Sigma_i\ (n-1)! \cdot F^{i-1}/(i-1)! \cdot (1-F)^{n-i}/(n-i)! \cdot x'_i$$

where $(n-1)! \cdot F^{i-1}/(i-1)! \cdot (1-F)^{n-i}/(n-i)!$ is the binomial distribution, and $x'_1,\cdots,x'_n$ $(x'_1 < x'_2 < \cdots < x'n)$ is the ranked sample $x_1, \ldots, x_n$. The distribution function $F(x)$ is given reciprocal function of quantiles $x(F)$.

Each sample of B values for active compounds is arranged in the ascending order; each sample of B values for inactive compounds is arranged in descending order. The above described quantiles $b(F)$ are calculated. As a result, for each appropriate kind of activity, the probabilities Pa and Pi are given by

$$b_{active}(Pa) = B, \quad b_{inactive}(Pi) = B$$

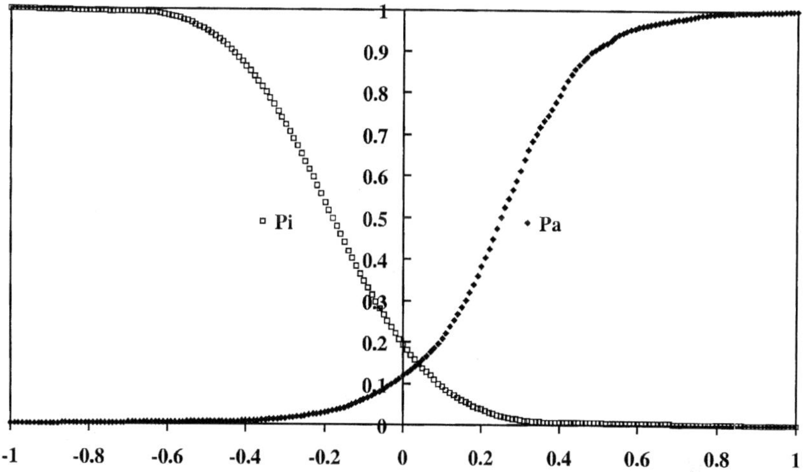

Figure 1 Initial estimation of functions Pa(B) and Pi(B) for alpha-adrenergic antagonists.

By definition, the probabilities *Pa* and *Pi* are also the probabilities of the first and second kinds of prediction error at the threshold *B*, respectively. They can be also interpreted as the measures of belonging to fuzzy subsets of "active" and "inactive" compounds. Both interpretations of probabilities *Pa* and *Pi* are equivalent and can be used for interpreting the result of prediction. They can also be used for construction of different criteria for prediction result analysis corresponded to specific practical problems.

The example of the probabilities *Pa*(*B*) and *Pi*(*B*) for activity "alpha-adrenoreceptor antagonist" as functions of *B*-statistics value is shown in Fig. 1.

The point of the curves' intersection corresponds to the parity of the first and second kinds of prediction error probabilities. The value $\text{MEP} = Pa(B^*) = Pi(B^*)$ is the estimation of maximal error of prediction (MEP), which is used in PASS as a measure of prediction accuracy. General PASS prediction accuracy is estimated using average MEP for all predictable activity kinds.

2.6. PASS Validation

Leave one out cross-validation for all ~1000 kinds of biological activity and ~50,000 substances provides the estimate of PASS prediction accuracy at the training procedure. Average accuracy of prediction is about 86% according to the LOO CV estimation, while that for particular kinds of activity varies from 63% (antacid, multiple sclerosis treatment) to 99% (urokinase-type plasminogen activator receptor antagonist). Accuracy of prediction data for all kinds of biological activity predicted by PASS is presented at the web site (17).

The accuracy of PASS predictions depends on several factors, from which the quality of the training set seems to be the most important one. A perfect training set should include the comprehensive information about biological activities known or possible for each compound. In other words, the whole *biological activity spectrum* should be thoroughly investigated for each compound included into the PASS training set. Actually, no database exists with information about biologically active compounds tested against each kind of biological activity. Therefore, the information concerning known biological activities for any compound is always incomplete. We investigated the influence of the information's incompleteness on the prediction accuracy for new compounds. About 20,000 principal compounds from MDDR database (27) were used to create the heterogeneous training and evaluation sets. At random 20%, 40%, 60%, 80% of information were excluded from the training set. Either structural data or biological activity data were removed in two separate computer experiments. In both cases, it was shown that even if up to 60% of information is excluded, the results of prediction are still satisfactory (26). Thus, despite the incompleteness of information in the training set, the method used in PASS is robust enough to provide the reasonable results of prediction.

Incompleteness of data on biologically active compounds significantly restricts the possibilities for evaluation of PASS prediction abilities, because many biological activities, which are probable according to the predictions, were never tested. Fortunately, there exists the NCI database with 42,689 heterogeneous compounds each being tested in anti-HIV assays (28). We used this database to estimate to what extent the application of PASS predictions enriches the population of active compounds in the subsets selected from the sample. It was shown in this experiment that, depending on the cutoff value of *Pa*, the enrichment of "actives" varies from 2.3 to 16.7 (29). Thus, PASS predictions significantly increase the probability of finding particular activity in compounds under study.

3. APPLICATION OF PREDICTED BIOLOGICAL ACTIVITY SPECTRA IN PHARMACEUTICAL RESEARCH AND DEVELOPMENT

Using thalidomide as an example, let us consider how the predicted biological activity spectra can be applied to study the efficacy and safety of new pharmaceuticals. Thalidomide's biological activity spectrum predicted by PASS 1.911 is given below (only activities with $Pa > 30\%$ are shown). Here, *Pa* and *Pi* are the probabilities to be active and inactive respectively.

Pa	*Pi*	*Biological activity*
0.988	0.011	4-Aminobutyrate transaminase inhibitor
0.964	**0.002**	**Inflammatory bowel disease treatment**
0.883	0.013	Lysase inhibitor
0.742	**0.007**	**Antiarthritic**
0.751	0.023	Ovulation inhibitor
0.731	0.008	Steroid synthesis inhibitor
0.776	0.085	Ligase inhibitor
0.626	**0.003**	**Tumor necrosis factor antagonist**
0.623	**0.006**	**Arachidonic acid antagonist**
0.619	**0.022**	**Teratogen**
0.603	**0.016**	**Sedative**
0.581	0.034	Antiviral (picornavirus)
0.576	0.036	Tocolytic

(*Continued*)

Pa	*Pi*	*Biological activity*
0.509	0.006	GABA receptor agonist
0.560	0.071	Cardiovascular analeptic
0.536	0.056	Antidyskinetic
0.569	0.115	Oxidoreductase inhibitor
0.465	0.015	Lipocortins synthesis antagonist
0.646	0.201	Superoxide dismutase stimulant
0.450	**0.005**	**Hypnotic**
0.521	0.105	Platelet adhesion inhibitor
0.426	**0.063**	**Immunomodulator**
0.374	0.086	Antiinflammatory
0.458	0.176	Alzheimer's disease treatment
0.342	0.062	L-lactate dehydrogenase stimulant
0.332	0.062	Cardioprotectant
0.405	**0.154**	**Tumor necrosis factor alpha release inhibitor**
0.431	**0.183**	**Autoimmune disorders treatment**
0.429	0.186	Acetylcholine Ml receptor antagonist
0.451	0.211	Antischistosomal
0.304	0.098	Uricosuric
0.337	0.132	Neurotransmitter antagonist
0.338	0.135	Complement inhibitor
0.365	**0.163**	**Cytokine modulator**
0.333	0.144	Interferon agonist
0.301	0.134	Antisecretoric
0.338	0.172	Membrane integrity antagonist
0.325	0.160	Interleukin agonist
0.307	**0.146**	**Immunosuppressant**
0.448	0.296	Antiviral (herpes)
0.480	0.335	Antihypoxic
0.322	**0.190**	**Angiogenesis inhibitor**
0.338	**0.295**	**Nootropic**
0.301	0.268	Calcium channel agonist

Forty-four kinds of biological activity are predicted as probable with $Pa > 30\%$. Most known pharmacotherapeutic and adverse/toxic effects of thalidomide from the literature appeared in the PASS predicted biological activity spectrum (marked in bold). Some new activities are predicted too, such as 4-aminobutyrate transaminase inhibitor, lysase inhibitor, ovulation inhibitor, etc. After confirming experimentally, these biological activities may become a reason for new applications of this drug. However, one should also note that PASS could predict the teratogenic effect for thalidomide with a *Pa* of 0.619. Keeping this in mind, the researcher either has to choose the thalidomide analogs without predicted teratogenicity (if any) or has to stipulate the strict conditions for thalidomide's use in order to prevent teratogenic adverse effects.

Similar strategy may serve as the basis for selecting compounds with required pharmacotherapeutic action but without unwanted adverse/toxic effects among the compounds available for screening. PASS prediction for 10,000 compounds can be performed in a few minutes, thus the program can be applied to the databases with hundreds of thousands or even millions of drug-like substances. The list of unwanted adverse/toxic effects predicted by the current version of PASS is given below (*N* is the number of compounds with particular activity in the training set; MEP is the maximum error of prediction in LOO CV):

N	*MEP* (%)	*Biological activity*
21	5.345	Abortion inducer
551	18.067	Carcinogenic
74	36.116	Cardiotoxic
141	35.234	Convulsant
658	23.221	Embryotoxic
70	25.359	Membrane integrity antagonist
700	16.949	Mutagenic
79	19.144	Narcotic
17	19.954	Skin irritative effect
261	21.459	Spasmogenic
473	24.348	Teratogen
929	26.616	Toxic
26	19.654	Ulcerogenic

All these activities are certainly unwanted. By application of PASS to the top 200 drugs, it was shown that about 83% of adverse and toxic effects were predicted (30).

In particular situations, some other activities predicted by PASS can also be considered as unwanted, e.g., sedative effect in anxiolytics used by drivers; application of irreversible inhibitors of acetylcholinesterase may lead to neurotoxic effect, etc. Moreover, PASS construction allows an addition of new pharmacotherapeutic, adverse, and toxic activities if appropriate training set can be created.

4. FUTURE TRENDS IN BIOLOGICAL ACTIVITY SPECTRA PREDICTION

Using current version of PASS, one may predict about 1000 kinds of biological activity with mean accuracy of about 85%. However, PASS training set already covers more than 1500 kinds of biological activity (some of them cannot be predicted yet because of the small number of compounds with the appropriate activity in the training set). Due to the achievements of genomics and proteomics, increase in the number of molecular targets is expected to rise from 500 to 5000 or even 10,000 in the coming years (31). New activities and new biologically active compounds can be added to the training set for extension of PASS prediction abilities. However, such mechanistic increase of the training set is not the only way of improving the prediction. The relationships between different kinds of biological activity (32) should be taken into account for investigation of general chemical–biological interactions, thus providing the basis for integrative computational toxicology.

REFERENCES

1. http://www.ifpma.org.
2. Wermuth C. The Practice of Medicinal Chemistry. New York: Academic Press, 2003.

3. Poroikov VV, Filimonov DA. How to acquire new biological activities in old compounds by computer prediction. J Comput Aided Mol Des 2002; 16:819–824.

4. Holtje HD, Sippl W. Rational Approaches to Drug Design. Barcelona: Prous Science, 2001.

5. van de Waterbeemd H. Structure–Property Correlations in Drug Research. Austin: Landes, 1996.

6. Livingstone D. Data Analysis for Chemists. Applications to QSAR and Chemical Product Design. Oxford: Oxford University Press, 1995.

7. Chu YH, Cheng CC. Affinity capillary electrophoresis in biomolecular recognition. Cell Mol Life Sci 1998; 54: 663–683.

8. Deplanque G, Harris AL. Anti-angiogenic agents: clinical trial design and therapies in development. Eur J Cancer 2000; 36:1713–1724.

9. Poroikov V, Filimonov D. Computer-aided prediction of biological activity spectra. Application for finding and optimization of new leads. In: Holtje HD, Sippl W, eds. Rational Approaches to Drug Design. Barcelona: Prous Science, 2001.

10. Avidon VV. The criteria of chemical structures similarity and the principles for design of description language for chemical information processing of biologically active compounds. Chem Pharm J (Rus) 1974; 8:22–25.

11. Piruzyan LA, Avidon VV, Rozenblit AB, et al. Statistical analysis of the information file on biologically active compounds. I. Data base on the structure and activity of biologically active compounds. Chem Pharm J. (Rus) 1977; 11:35–40.

12. Golender VE, Rozenblit AB. Logical structural approach to computer assisted drug design. Drug Design. Vol. 9. New York: Academic Press1980.

13. Rozenblit AB, Golender VE. Logical Combinatorial Methods in Drug Design. Riga: Zinatne, 1983.

14. Burov YuV, Poroikov VV, Korolchenko LV. National system for registration and biological testing of chemical compounds:

facilities for new drugs search. Bull Natl Cent for Biol Active Comp (Rus) 1990; 1:4–25.

15. Gloriozova TA, Filimonov DA, Lagunin AA, Poroikov VV. Chem Pharm J (Rus) 1998; 32:32–39.
16. Poroikov VV, Filimonov DA. QSAR and Molecular Modelling Concepts, Computational Tools and Biological Applications. Barcelona: Prous Science Publishers, 1996:49–50.
17. See http://www.ibmh.msk.su/PASS.
18. Sadym A, Lagunin A, Filimonov D, Poroikov V. Prediction of biological activity spectra via Internet. SAR QSAR Environ Res 2003; 14:339–347.
19. Anzali S, Barnickel G, Cezanne B, Krug M, Filimonov D, Poroikov V. Discriminating between drugs and nondrugs by prediction of activity spectra for substances (PASS). J Med Chem 2001; 44:2432–2437.
20. Borodina Yu, Sadym A, Filimonov D, Blinova V, Dmitriev A, Poroikov V. Predicting biotransformation potential from molecular structure. J Chem Inform Comput Sci 2003; 43:1636–1646.
21. Guba W. Representation of chemicals. In: Helma C., ed. Predictive Toxicology. Marcel Dekker, 2003; 11–35.
22. Filimonov DA, Poroikov VV, Karaicheva EI, et al. Computer-aided prediction of biological activity spectra of chemical substances on the basis of their structural formulae: computerized system PASS. Exp Clin Pharmacol (Rus) 1995; 58:56–62.
23. Avidon VV, Pomerantsev IA, Rozenblit AB, Golender VE. Structure–activity relationship oriented language for chemical structure representation. J Chem Inf Comput Sci 1982; 22:207–214.
24. Filimonov D, Poroikov V, Borodina Yu, Gloriozova T. Chemical similarity assessment through multilevel neighborhoods of atoms: definition and comparison with the other descriptors. J Chem Inf Comput Sci 1999; 39:666–670.
25. Lagunin A, Stepanchikova A, Filimonov D, Poroikov V. PASS: prediction of activity spectra for biologically active substances. Bioinformatics 2000; 16:747–748.

26. Poroikov VV, Filimonov DA, Borodina YuV, Lagunin AA, Kos A. Robustness of biological activity spectra predicting by computer program PASS for non-congeneric sets of chemical compounds. J Chem Inf Comput Sci 2000; 40:1349–1355.

27. See http://www.mdli.com.

28. Voigt JH, Bienfait B, Wang S, Nicklaus MC. Comparison of the NCI open database with seven large chemical structural databases. J Chem Inf Comput Sci 2001; 41:702–712.

29. Poroikov W, Filimonov DA, Ihlenfeldt WD, Gloriozova TA, Lagunin AA, Borodina YuV, Stepanchikova AV, Nicklaus MC. PASS biological activity spectrum predictions in the enhanced open NCI database browser. J Chem Inf Comput Sci 2003; 43:228–236.

30. Poroikov V, Akimov D, Shabelnikova E, Filimonov D. Top 200 medicines: can new actions be discovered through computer-aided prediction?. SAR QSAR Environ Res 2001; 12: 327–344.

31. Drews J. Drug discovery: a historical perspective. Science 2000; 287:1960–1964.

32. Lagunin AA, Filimonov DA, Poroikov VV. clustering of chemical compounds with required biological activity spectra in large databases. Abstract of Sixth International Conference on Chemical Structures 2002:25–26.

GLOSSARY

Biological activity spectrum: The biological activity spectrum is the "intrinsic" property of the compound that reflects all kinds of its biological activity, which can be found in the compound's interaction with biological entities.

First kind error of prediction: "False-positives," when an inactive compound is predicted to be active.

LOO CV: Leave-one-out cross-validation is applied to all compounds from the training set. Each compound is excluded from the training set with the information about its activities, and prediction is obtained for this compound based on the structure–activity relationships calculated for the remaining compounds. The procedure is repeated iteratively for all compounds from the training set, and average accuracy of prediction is calculated through all compounds and all types of biological activity.

Molecular modeling: Molecular modeling is the investigation of molecular structures and properties using computational chemistry and graphical visualization techniques to provide a plausible three-dimensional representation under a given set of circumstances.

MNA: Multilevel neighborhoods of atoms are structural descriptors that can be applied for the analysis of structure–property/structure–activity relationships. These descriptors are based on the molecular structure representation, which includes the hydrogen atoms according to the valences and partial charges of other atoms and do not specify the types of bonds. MNA descriptors are generated as recursively defined sequence: zero-level MNA descriptor for each atom is the mark A of the atom itself; any next-level MNA descriptor for the atom is the substructure notation $A(D_1D_2\cdots D_i\cdots)$, where D_i is the previous-level MNA descriptor for ith immediate neighbor of the atom A. If the atom is not included into the ring it is marked by "–". For details see Ref. 24.

PASS: Prediction of activity spectra for substances. Computer program developed and updated by Vladimir Poroikov, Dmitri Filimonov, and associates (Institute of Biomedical Chemistry of Russian Academy of Medical Sciences).

QSAR: Quantitative structure–activity relationships are mathematical relationships linking chemical structure and pharmacological activity in a quantitative manner for a series of compounds. Methods, which can be used in QSAR, include various regression and pattern recognition techniques.

SAR: Structure–activity relationship is the relationship between chemical structure and pharmacological activity for a set of compounds.

Second kind error of prediction: "False-negatives," when an active compound is predicted to be inactive.

Teratogen: A teratogen is a substance that produces a malformation in a fetus.

14

lazar: Lazy Structure–Activity Relationships for Toxicity Prediction

CHRISTOPH HELMA

Institute for Computer Science, Universität Freiburg, Georges Köhler Allee, Freiburg, Germany

1. INTRODUCTION

The development of the `lazar` system was more or less a byproduct of editing this book. The intention was to demonstrate how to use some of the concepts from previous chapters as building blocks for the creation of a simple predictive toxicology system that can serve as a reference for further developments. Most of the machine learning and data mining techniques have been described in an earlier chapter (1). Some of the basic ideas (utilizing linear fragments for predictions) were developed already in the 1980s by Klopman and

---

`lazar` stands for Lazy Structure–Activity Relationships.

CAS	SMILES	*Salmonella* Mutagenicity
26148-68-5	NC1C=CC2=C(N=1)NC3=CC=CC=C23	1
75-07-0	CC=O	0
16568-02-8	CC=NN(C)C=O	0
107-29-9	CC=NO	0
60-35-5	CC(=O)N	0
103-90-2	C1(=CC=C(C=C1)O)NC(C)=O	0
968-81-0	O=S(=O)(C1=CC=C(C=C1)C(=O)C)NC(=O)NC2CCCCC2	0
75-05-8	CC#N	0
34627-78-6	CC(=O)NC1=CC=C(C=C1)C2=CC=CC=C2	1
114-83-0	C1(NNC(C)=O)=CC=CC=C1	1
53-96-3	C12C3=C(C=CC=C3)CC1=CC(=CC=2)NC(C)=O	1
28322-02-3	C12C3=C(C=CC=C3)CC1=CC=CC=2NC(C)=O	1
107-02-8	C=CC=O	1
79-06-1	NC(=O)C=C	0
79-10-7	OC(=O)C=C	0
107-13-1	C=CC#N	1
628-94-4	NC(=O)CCCCC(=O)N	0
3688-53-7	NC(=O)/C(=C/C1=CC=C(O1)[N+](=O)[O-])C2=CC=CO2	1
1162-65-8	C12=C3C(C4=C(C(O3)=O)C(=O)CC4)=C(C=C1OC5C2C=CO5)OC	1
116-06-3	CC(C=NOC(=O)NC)(SC)C	0
309-00-2	Cl[C@]12C3C(C4C=CC3C4)[C@](C1(Cl)Cl)(Cl)C(=C2Cl)Cl	0
107-05-1	C=CCCl	1
106-92-3	C=CCOCC1CO1	1
57-06-7	C=CCN=C=S	1
2835-39-4	O=C(CC(C)C)OCC=C	0
81-49-2	O=C1C2=C(C(=CC(=C2C(=O)C3=C1C=CC=C3)Br)Br)N	1
17026-81-2	NC1=C(C=CC(=C1)NC(=O)C)OCC	1
132-32-1	CCN1(C2C(=CC=CC=2)C3=C1C=CC(=C3)N)	1
82-28-0	O=C1C2=C(C(=CC=C2C(=O)C3=C1C=CC=C3)C)N	1
38514-71-5	C1(N=C(SC=1)N)C2=CC=C(O2)[N+]([O-])=O	1
99-57-0	O=[N+](C1=CC(=C(C=C1)O)N)[O-]	1
121-88-0	O=[N+](C1=CC(=C(C=C1)N)O)[O-]	1
119-34-6	OC1=C(C=C(C=C1)N)[N+](=O)[O-]	1
121-66-4	O=[N+](C1=CN=C(S1)N)[O-]	1
117-79-3	O=C1C2=CC(=CC=C2C(=O)C3=C1C=CC=C3)N	1
97-56-3	CC1=C(C=CC=C1)/N=N/C2=CC(=C(C=C2)N)C	1
92-67-1	NC1=CC=C(C=C1)C2=CC=CC=C2	1
92-67-1	NC1(=CC=C(C=C1)C2=CC=CC=C2)	1
...		

Figure 1 Example learning set for `lazar`. Compound structures are written in SMILES notation (17); "1" indicates active compounds, "0" indicates inactivity.

are implemented in the MC4PC system. [a description of MC4PC is given elsewhere in this book (2)]. Please note that `lazar` is *not* a reimplementation of an existing system, because it uses its own distinct algorithms to generate descriptors and to achieve the predictions.

CAS	SMILES	*Salmonella* Mutagenicity
3544-23-8	NC1=CC=C(C=C1OC)/N=N/C2=CC=CC=C2	?
72-43-5	ClC(C(C1=CC=C(C=C1)OC)C2=CC=C(C=C2)OC)(Cl)Cl	?
150-76-5	COC1=CC=C(C=C1)O	?
298-81-7	COC1=C2C(=CC3=C1OC=C3)C=CC(=O)O2	?
74-83-9	CBr	?
1634-04-4	CC(OC)(C)C	?
598-55-0	NC(=O)OC	?
55-80-1	CN(C1=CC=C(C=C1)/N=N/C2=CC(=CC=C2)C)C	?
80-62-6	O=C(C(=C)C)OC	?
66-27-3	CS(=O)(=O)OC	?
70-25-7	N=C(N(N=O)C)N[N+](=O)[O-]	?
129-15-7	O=C1C2=C(C(=CC=C2C(=O)C3=C1C=CC=C3)C)[N+](=O)[O-]	?
21638-36-8	C1(NC(CN1/N=C/C2=CC=C(O2)[N+](=O)[O-])C)=O	?
63412-06-6	C(C1C=CC=CC=1)(=O)N(N=O)C	?
298-00-0	S=P(OC1=CC=C(C=C1)[N+](=O)[O-])(OC)OC	?
872-50-4	O=C1N(CCC1)C	?
98-85-1	C1=CC=C(C(O)C)C=C1	?
452-86-8	OC1=C(C=CC(=C1)C)O	?
56-49-5	CC1=C2C3=C(C=C4C(=C3CC2)C=CC5=CC=CC=C45)C=C1	?
555-30-6	C(C(O)=O)(CC1=CC(=C(O)C=C1)O)(C)N	?
101-14-4	ClC1=C(C=CC(=C1)CC2=CC(=C(C=C2)N)Cl)N	?
101-14-4	ClC1=C(C=CC(=C1)CC2=CC(=C(C=C2)N)Cl)N	?
838-88-0	CC1=C(C=CC(=C1)CC2=CC(=C(C=C2)N)C)N	?
75-09-2	ClCCl	?
101-61-1	CN(C1=CC=C(C=C1)CC2=CC=C(C=C2)N(C)C)C	?
101-77-9	C(C1=CC=C(C=C1)N)C2=CC=C(C=C2)N	?
60-34-4	CNN	?
90-12-0	CC1=C2C(=CC=C1)C=CC=C2	?
33868-17-6	N#CN(C)N=O	?
924-42-5	O=C(NCO)C=C	?
113-45-1	C(C1=CC=CC=C1)(C2CCCCN2)C(OC)=O	?
91-62-3	CC1=CC2=CC=CN=C2C=C1	?
611-32-5	CC1=CC=CC2=CC=CN=C12	?
56-04-2	CC1=CC(=O)NC(=S)N1	?
443-48-1	N1(C(=CN=C1C)[N+](=O)[O-])CCO	?
90-94-8	O=C(C1=CC=C(C=C1)N(C)C)C2=CC=C(C=C2)N(C)C	?

Figure 2 Example test set. Compound structures are written in SMILES notation (17); toxicological activities are unknown.

Initially in this chapter, I review some observations from the application of machine learning and data mining techniques to *non-congeneric* (compounds that do not have a common core structure) data sets that led to the basic `lazar` concept. After a description of the `lazar` algorithm, I present an example application for the prediction of *Salmonella* mutagenicity and draw some conclusions for further improvements from the analysis of misclassified compounds.

2. PROBLEM DEFINITION

First, a brief review of the problem setting: We have a data set with chemical structures and measured toxicological activities.[a] This is the *training set* (Fig. 1). We have a second data set with untested structures (the *test set*, Fig. 2) and intend to predict the toxicological activities of these untested compounds.

More specifically, we want to infer from the chemical structure (or some of its properties) of the test compound to its toxicological activity, assuming that the biological activity of a compound is determined by the chemical structure (or some of its properties).[b]

For this purpose, we need:

- A description of the chemicals features that are responsible for toxicological activity, and
- A model that makes predictions based on these features.

Most efforts in predictive toxicology have been devoted to the second task: The identification of predictive models based on a given set of chemical features (Table 1). Especially with toxicological effects, we frequently face the problem that biochemical mechanisms are diverse and unknown. It is therefore hard to guess (and select) the chemical features that are relevant for a particular effect, and we risk making one of the following mistakes:

- Omission of important features that determine activity/inactivity by selecting too few features.

---

[a] Classifications (e.g., active/inactive) or numerical values (e.g., LC_{50}s), for the sake of simplicity we will cover only the classification case in this chapter.

[b] This is of course an oversimplification, because toxic action is caused by an interaction between the compound and the biological target. In the chapter by Marchal et al.(3), you can find more information about this topic.

Table 1 Summary of Chemical Properties that Can Be Used for the Prediction of Toxicological Effects

Presence of substructures
Physico/chemical properties of the whole molecule (e.g., log P, HOMO, LUMO, ...)
Graph theoretic descriptors
Biological properties (e.g., from screening assays)
Spectra (IR, NMR, MS, ...)

- Deterioration of the performance (accuracy and speed) of statistical and machine learning systems[c] by selecting too many features.

Recently several schemes for feature selection have been devised in the machine learning community (see *Journal of Machine Learning Research* 3, 2003, that contains a Special Issue on Variable and Feature Selection).[d] In my experience *recursive feature extraction* (RFE) (8) performs very well on a variety of problems, but at the risk of overfitting the data (9). In theory, it is possible to counteract by performing another cross-validation step for feature selection, but this leads frequently to practical problems: long computation times and fragmented learning data sets due to nested cross-validation loops.[e]

Furthermore, there are a lot of other problems that are frequently encountered in predictive toxicology:

- Models that are hard to interpret in terms of toxicological mechanisms (10).

[c] Most (if not all) of these systems are sensitive towards large numbers of irrelevant features. Based on my experience, this is also true for systems that claim to be insensitive in this respect (e.g., support vector machines).

[d] In predictive toxicology, Klopman (7) used an automated feature selection process in the CASE and MULTICASE systems since the 1980s.

[e] *Recursive feature extraction* needs an internal (10-fold) cross-validation step to evaluate the results (8). If we use another 10-fold cross-validation for feature selection, we have to generate 10×10 models. On the other hand, the data for a single model shrink to $0.9 \times 0.9 = 81\%$ of the original size, which can be problematic for small data sets.

- Models that are too general (improper consideration of compounds with a specific mode of action) (11).
- Limitations of the models (e.g., substructures, that are not in the training set) are often unclear. No indication if a structure falls beyond the scope of the model.
- Sensitivity toward skewed distributions of actives/inactives in the learning set.
- Handling of missing values in the training set.
- Ambiguous parameter settings (e.g., cutoff frequencies in MolFea (12), Kernel type, gamma, epsilon, tolerance, ... parameters in support vector machines).

My intention was to address these problems with the development of `lazar`.

3. THE BASIC `lazar` CONCEPT

`lazar` is in contrast to the majority of the approaches described in this book— a *lazy learning* scheme. Lazy learning means that we do not generate a global model from the complete training set. Instead, we are creating small models on demand: one individual model for each test structure. This has the key advantage that the prediction models are more specific (13) because we can consider the properties of the test structure during model generation. If we want to predict, e.g., the mutagenicity of nitroaromatic compounds, information from chlorinated aliphatic structures will be of little value. As we will see below, the selection of relevant features and relevant examples from the training set is done automatically by the system and does not require any input of chemical concepts. On a practical side, we have integrated model creation and prediction. Therefore, we need no computation time for model generation (and validation), but predictions may require more computation time than predictions from a global model (this is of course very implementation dependent).

`lazar` uses presently linear fragments to describe chemical structures (but it is easy to use more complex fragments, e.g., subgraphs or to include other features like molecular properties, e.g., log*P*, HOMO, LUMO, etc.). Linear fragments are defined as chains of heavy atoms with connecting bonds. Branches or cycles are not considered explicitly in linear fragments.[f] 1,1,1-Trichloroethane ($CCCl_3$), e.g., will contain the linear fragments {C, Cl, C–C, C–Cl, C–C–Cl, Cl–C–Cl}. Formally linear fragments are valid SMARTS expressions (see http://www.daylight.com for further references), which can be handled by various computational chemistry libraries (e.g., OpenBabel http://openbabel.sourceforge.net).

For the sake of clarity, we will separate by discuss the following three steps that are needed to obtain a prediction for an untested structure:

1. Identification of the linear fragments in the test structure.
2. Identification of the fragments that are relevant for the prediction.
3. Prediction of the activity of the test structure.

In the following section, we will give a more detailed description of the algorithms for each of these steps.

4. DETAILED DESCRIPTION

4.1. Fragment Generation

In `lazar`, we are presently using a very simplified variant of the MolFea (14) algorithm to determine the fragments of a given structure. The procedure is as follows.

As a starting point, we use all elements from the periodic table of elements (including aromatic atoms). These are the candidate fragments for the first level. First, we examine which of them occur in (or match) the test structure and eliminate those that do not match. Then we check which of the remaining fragments occur in the training structures. If a candidate fragment does not occur in the training structures, we remove it from the current level

[f] But they are frequently implicitly considered: chains with more than six aromatic carbons, for example, indicate condensated aromatic rings.

initialize *level* with table of elements
while *level* not empty
    for all *fragments* from *level*
        if *fragment* does not occur in the test structure
            remove *fragment* from *level*
            evaluate next *fragment*
        if *fragment* does not occur in the training set
            remove *fragment* from *level* and store it in the set of *unknownfragments*
    refine *level* (i.e. create *fragments* with an additional bond and atom)

Figure 3 Procedure for fragment generation.

and store it in the set of unknown fragments, because we cannot determine if it contributes to activity or inactivity. From the remaining candidates (i.e., those that occur in the test structure and the training structures), we generate the candidates for the next level (i.e., candidates with an additional bond and atom), this step is called *refinement* and will be described below. The whole procedure is repeated until the candidate pool has been depleted [i.e., all fragments of the test structure that occur also in the training set have been identified (Fig. 3)].

4.2. Fragment Refinement

A naive way to refine level n to level $n+1$ would be to attach a bond and an atom to each fragment of level n. This is, of course, very inefficient because we generate (and match) way too many fragments. We want to avoid to generate unnecessary fragments because fragment matching is the time critical step of our algorithm.[g]

Fortunately, we can determine in many cases if a fragment cannot match before generating the fragment. For this purpose, we can use an important property of the language of molecular fragments—the generality relationship:

[g] lazar delegates this to the OpenBabel libraries http://openbabel.sourceforge.net.

We define that one fragment g is *more general* than a fragment s (notation: $g \preceq s$) if g is a substructure of s (e.g., C–O is more general than N–C–C–O). This has the consequence that g matches whenever s does.

Linear fragments are symmetric, which means that two syntactically different fragments are equivalent when they are a reversal of one another (e.g., C–C–O and O–C–C denote the same fragment).

Therefore we can conclude that g is more general than s ($g \preceq s$) if and only if g is a subsequence of s or g is a subsequence of the reversal of s (e.g., C–O $\preceq$ C–C–O and O–C $\preceq$ C–C–O).

We can use this generality relationship to refine fragments efficiently: As we know, all subfragments of a new candidate fragment have to match; therefore, we need to combine only the fragments of the present level (i.e., those that match on the test and training compounds) to reach the next level. The two fragments C–C and C–O, e.g., can be refined to C–C–C, C–C–O, and O–C–O. This reduces the number of candidates considerably in comparison to attaching naively a bond and an atom.

Another method to reduce the search space is to utilize the known matches of the current level to determine the potential matches of the new fragment. If we know, e.g., that C–C occurs in compounds {A, B, C} and C–O occurs in {B,D} we can conclude that:

- C–C–C can occur in compounds {A, B, C}.
- C–C–O can occur in compounds {B}.
- O–C–O can occur in compounds {B, D}.

Knowing the potential matches of a new fragment allows us: (i) to remove candidates if they have no potential matches, and (ii) to perform the time consuming matching step only on the potential matches and not the complete data set.

As predictions are usually performed for a set of test compounds, fragments (especially those from the lower levels, like C, C–C, etc.) are frequently to be reevaluated on the training set. Storing the matches of the fragments in a database (that can be saved permanently) helps to prevent this reevaluation.

4.3. Identification of Relevant Fragments

After the successful identification of fragments of the test structure, we have the following information:

- The set of linear fragments that occurs in the test structure and in the training structures.
- The set of the most general fragments that occur in the test structure but not in the training structures (i.e., the shortest unknown fragments).
- For each fragment, the set of training structures, where the fragment matches.
- The activity classifications for the training structures.

Let us consider now each fragment *f* as a hypothesis that indicates if a compound C with this fragment is active or inactive. First we have to evaluate if a fragment indicates activity or inactivity, then we have to distinguish between more or less predictive hypotheses (i.e., fragments) and select the most predictive ones.

Fortunately, it is rather straightforward to evaluate the fragments on the training set because we know which compounds the fragment matches as well as the activity classifications for these compounds. If a fragment matches only active or inactive compounds, the decision is obvious: We will call a fragment activating if it matches only active compounds or inactivating if it matches only inactive compounds. In real life, however, most fragments will occur in active as well as inactive compounds. It is tempting to call a fragment activating if it matches more active compounds than inactives. This is certainly true for training sets that contain an equal number of active and inactive compounds. But let us assume that only 10% (e.g., 10 of 100) of the training structures are active and we have identified a fragment that matches five active and five inactive compounds (Fig. 4). We realize that this fragment matches 5 from 10 (50%) active compounds and 5 from 90 (5.6%) inactive compounds and that it is justified to call

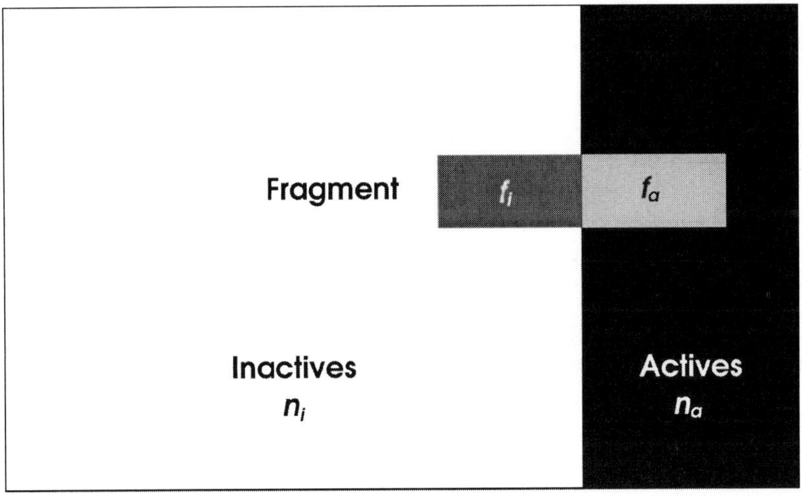

Figure 4 An activating fragment. f_a: number of active compounds with fragment f, f_i: number of inactive compound with fragment f, f_{all}: total number of compounds with fragment f ($f_{all} = f_a + f_i$), n_a number of active compounds in the training set, n_i number of inactive compounds in the training set, n_{all} total number of compounds in the training set ($n_{all} = n_a + n_i$).

it an activating fragment. This consideration leads to the following definition[h] with abbreviations from Fig. 4:

- A fragment indicates *activity*, if compounds with this fragment are *more frequently active* than compounds in the training set, i.e., $p_a = f_a/f_{all} - n_a/n_{all} > 0$
- A fragment indicates *inactivity*, if compounds with this fragment are *more frequently inactive* than compounds in the training set, i.e., $p_a = f_a/f_{all} - n_a/n_{all} < 0$.

Furthermore, we would like to know if the observed deviation from the default distribution is due to chance or not. For this purpose, we can use the Chi-square test (with corrections for small sample sizes) to calculate the probability $p(\chi^2)$ that the deviation from the default distribution is not

[h] Which is equivalent to the *leverage* in association rule mining (15).

due to chance. To obtain the total probability that a fragment indicates (in) activity, we multiply both contributions:

$$p_f = p_a * p(\chi^2)$$

Fragments with positive p_f indicate activity, fragments with a negative p_f indicate inactivity.

4.4. Redundancy of Fragments

Figure 5 lists a set of fragments that have been generated for a particular compound. It is obvious that the chemical meaning of most of the fragments is almost identical. The first six fragments point towards two aromatic rings that are connected by a single bond (probably a biphenyl). Using the same information six times would lead to an overestimation of the effect of the biphenyl structure—therefore we use only the most predictive fragment (i.e., the fragment with the highest $|p_f|$) from a set of redundant fragments and define redundancy as follows.

Fragment	f_a	f_i	p_a	p_chisq	p_f	
c-c	22	10	0.360227	0.999986	0.360227	<-
c-c:c	22	10	0.360227	0.999986	0.360227	
c-c:c:c	21	10	0.350147	0.999967	0.350135	
c:c-c:c:c	15	9	0.297727	0.99812	0.297167	
c-c:c:c:c	15	9	0.297727	0.99812	0.297167	
c:c-c:c	15	9	0.297727	0.99812	0.297167	
N-c:c	86	61	0.257761	1	0.257761	<-
N-c:c:c:c	80	57	0.256669	1	0.256669	
c-N	86	64	0.246061	1	0.246061	
N-c:c:c	80	61	0.240103	1	0.240103	
N-c:c:c:c:c:c	61	54	0.203162	0.999997	0.203161	
N-c:c:c:c:c	61	55	0.198589	0.999995	0.198588	
C-C=O	18	94	-0.166558	0.999828	-0.166529	<-
c:c-c:c:c:c:c	8	8	0.172727	0.859106	0.148390	
c-c:c:c:c:c:c	8	8	0.172727	0.859106	0.148390	
N	120	142	0.130743	0.999994	0.130742	
c-c:c:c:c:c	8	9	0.143316	0.792093	0.113519	
c:c-c:c:c:c	8	9	0.143316	0.792093	0.113519	

Figure 5 A set of redundant fragments for a particular compound. The most predictive, non-redundant fragments are marked by an arrow.

Two fragments f_1, f_2 are redundant if the matches of f_1 are a subset of the f_2 matches or vice versa.[i] if e.g., `c:c:c:n:n` matches compounds {A,B,C} with p_f=0.7 and `c:c:c:c:n:n` matches {A,C} with p_f=0.6, we will use `c:c:c:n:n` for our predictions.

4.5. Classification

With this set of most predictive, non-redundant fragments, it is possible to classify the test structure. As each fragment is a hypothesis that indicates activity or inactivity with a certain confidence (p_f), we can use the "majority vote" among all fragments to classify the test structure. For this purpose we sum up the p_f of all non-redundant fragments, add the default probabilities for activity and inactivity from the training set[j] and predict "active" if $\Sigma p_f > 0$, "inactive" if $\Sigma p_f < 0$. In addition, we can calculate the confidence in our prediction by defining

$$P_A = \frac{\Sigma p_{\mathrm{activating}}}{\Sigma p_{\mathrm{all}}}$$

(probability that the prediction "active" is true)

and similarly,

$$P_I = \frac{\Sigma p_{\mathrm{activating}}}{\Sigma p_{\mathrm{all}}}$$

(probability that the prediction "active" is true)

In addition, we will output the set of unknown fragments to warn the user for potential misclassifications. Fig. 6 demonstrates an example `lazar` prediction.

5. RESULTS

As an initial test, this algorithm was used for predicting mutagenicity in *Salmonella typhimurium.* The learning set

[i] Note that it is also possible to define redundancy with the generality relationship, but testing for matches is computationally more efficient.

[j] This can be changed if we *know* that the test compounds have a different distribution between actives and inactives than the training set.

```
Compound #13
CC(=O)NCc1nc(no1)c1ccc(o1)[N+](=O)[O-]
Removing CC(=O)NCc1nc(no1)c1ccc(o1)[N+](=O)[O-] from training set.

Activity: Salmonella Mutagenicity

p(active)       0.327273
p(inactive)     -0.672727
Fragment        f_a      f_i      p_a           p_chisq      p_f
N-C=O           15       2        0.55508       0.999999     0.55508
c-c:o           11       2        0.518881      0.999933     0.518846
N=O             69       13       0.514191      1            0.514191
c-c:n           13       3        0.485227      0.999965     0.48521
N-c:c           86       61       0.257761      1 .          0.257761
C-C=O           18       94      -0.166558      0.999828    -0.16653
C-c             48       127     -0.052987      0.864792    -0.0458227
C-C-N           35       81      -0.0255486     0.442418    -0.0113032
Unknown Fragments: n:c:o n:c-c:o

Prediction:     1        (0.747835/0.252165)
Measured Activity:       1
```

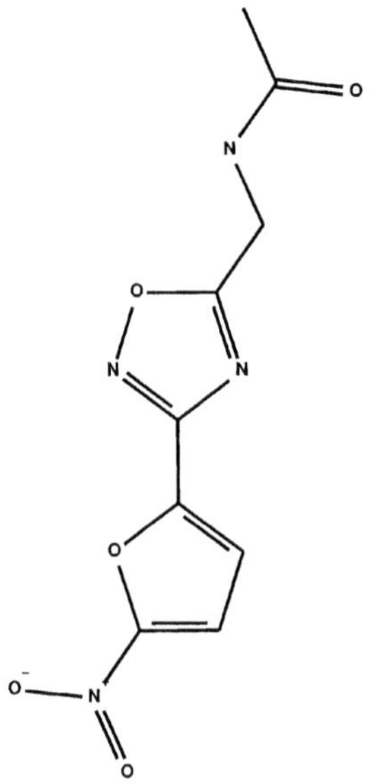

Figure 6 A typical lazar prediction. As the test structure has been detected in the training set, it was removed to obtain an unbiased prediction. Numbers in brackets after the prediction indicate the probability of being active and inactive. The first unknown fragment indicates a heteroaromatic ring with nitrogen and oxygen, the second fragment indicates two heteroaromatic rings—one with nitrogen, one with oxygen at 1,2 position, connected by a single bond.

was extracted from the *Carcinogenic Potency Database* (CPDB) (16) and contained 496 compounds. The CPDB contains only an overall classification for *Salmonella* mutagenicity, but no information about metabolic activation or tester strains. The results from leave-one-out cross-validation are summarized in Table 2; a comparison with results from the literature on similar (non-congeneric) data sets can be found in Table 3. Please keep in mind that neither the data set nor the validation methods are identical in Tables 2 and 3.

6. LEARNING FROM MISTAKES

For a model developer, the inspection of misclassified instances is the most instructive task. A case-by-case inspection of each compound can reveal systematic errors in the prediction system. It avoids, on the other hand, desperate efforts to increase the predictive accuracy and overfit the model if there are good reasons for wrong predictions (e.g., compounds that are beyond the scope of the training set, controversial measurements). The inspection of misclassified compounds from the mutagenicity data set led to the following reasons for misclassifications:

Unknown fragments: Thirty-one from 107 (29%) misclassified compounds contained unknown fragments (i.e., substructures that are not in the training set). It is likely that at least some of them are activating or deactivating and

Table 2 Accuracy of `lazar` Predictions for *Salmonella* Mutagenicity Determined by Leave-One-Out Cross-Validation

Salmonella Mutagenicity	
True positive rate $\left(\frac{t_p}{n_p}\right)$	0.66
True negative rate $\left(\frac{t_n}{n_n}\right)$	0.84
Total accuracy $\left(\frac{t_p+t_n}{n_{all}}\right)$	0.78

t_p: number of true positive predictions, n_p: total number of positive predictions, t_n: number of true negative predictions, n_n: total number of negative predictions, n_{all}: number of all predictions (compounds) $n_{\text{all}} = n_p + n_n$.

Table 3 Accuracy of *Salmonella* Mutagenicity SAR Models for Non-congeneric Compounds from the Public Literature

Author	Citation	Method	Accuracy (%)
Perotta et al.	(18)	—	74[a]
Klopman and Rosenkranz	(19)	CASE	72
Klopman and Rosenkranz	(19)	MULTICASE	80
Klopman and Rosenkranz	(19)	CASE/GI	47
Helma, Kramer and DeRaedt	(20)	MolFea + ML	78

[a]Leave-one-out cross-validation.

therefore relevant for mutagenicity. To obtain the lacking information, it is necessary to test additional compounds that contain these fragments and to add the results to the training set. Unknown fragments can give valuable guidelines for a rational selection of testing candidates and the optimization of testing procedures.

Improper balance between activating and deactivating fragments: In 47 from 107 (44%) misclassified compounds, activating as well as inactivating fragments were detected. The misclassifications were caused by an improper balance between activating and inactivating fragments. In approximately half of these cases, the predictions are associated with a low confidence (i.e., P_A and P_I close to 0.5) and should be treated with caution.

The most likely reason for an improper balance between activating and inactivating fragments is a non-linear relationship between the presence of fragments and activity. It is, e.g., conceivable that fragment F_1 alone indicates activity, and fragments F_2 and F_3 alone indicate inactivity. The occurrence of fragment F_1 together with F_2 might be, however, a strong indication of activity, whereas the occurrence of F_1 together with F_3 might indicate in contrary inactivity.

If this is the case there are a number of possible solutions: We can search, e.g., for *sets of fragments* (e.g., $\{F_1, F_2\}$, $\{F_1, F_3\}$) that discriminate better between active and inactive compounds than individual fragments, and to use these *sets of fragments* as predictors.

Another option is to search for substructures that are more specific than linear fragments. A very general solution (that we are currently investigating) is to search for subgraphs (i.e., fragments of any shape). Such a procedure enables the detection of branched fragments, ring systems as well as the consideration of stereochemistry, but it comes at a cost: increased computation time.

Missing (in)activating features: In 29 cases (27%), `lazar` failed to detect activating substructures in active compounds or inactivating substructures in inactive compounds. The main reason for this failure were inherent limitations of linear fragments (it is, e.g., impossible to describe epoxides with linear fragments). Such situations can be solved by two strategies:

Ad hoc solutions: Addition of features that describe the concepts that cannot be represented with linear fragments. In the epoxide example, we might add a flag for the presence of a three-ring or an epoxide ring, or use the smallest set of smallest rings (SSSR) as additional descriptors.

General solutions: Use, e.g., graphs instead of linear fragments.

Intuitively, the last solution seems to be preferable, because it is more general. In practice, however, we have to consider the costs: increased computation time and maybe also memory requirements. If, e.g., epoxides are the only compounds where (explicit) ring systems[k] are needed, it might be more efficient to use the three-ring or epoxide flag as an additional feature.

7. CONCLUSION

In this chapter, I have presented a brief analysis of the major problems of current predictive toxicology systems. Based on this analysis, I have applied some of the concepts from this book to develop a simple system called `lazar` and used it for

[k] Note that ring systems are in many cases implicitly indicated by linear fragments, e.g., by aromatic atoms.

the prediction of *Salmonella* mutagenicity. The initial accuracy of `lazar` predictions was comparable to the best results reported in the literature for the same effect. Finally, I suggest some possible improvements, based on the inspection of misclassified compounds. In the future, `lazar` will be applied to further data sets to explore the reasons for misclassifications systematically. The implementation of new or alternative algorithms and features in future versions will depend on the analysis of these results. A web interface to `lazar` can be found at http://www.predictive-toxicology.org/lazar/.

ACKNOWLEDGMENTS

I would like to thank Stefan Kramer and Luc DeRaedt for the development of MolFea, and Victor Horal-Gurfinkel and Peter Reutemann for programming assistance.

REFERENCES

1. Kramer S, Helma C. Machine learning and data mining. In: Helma S, ed. Predictive Toxicology. New York: Marcel Dekker, 2005.
2. Klopman G, Ivanov J, Saiakhov R, Chakravarti S. MC4PC—an artificial intelligence approach to the discovery of quantitative structure toxic activity relationships (QSTAR). Predictive Toxicology. New York: Marcel Dekker, 2005.
3. Marchal K, De Smet F, Engelen K, De Moor B. Computational biology and toxicogenomics. Predictive Toxicology. New York: Marcel Dekker, 2005.
4. Frasconi P. Neural networks and kernel machines for vector and structured data. Predictive Toxicology. New York: Marcel Dekker, 2005.
5. Eriksson L, Johansson E, Lundstedt T. Regression- and projection-based approaches in Predictive Toxicology. Predictive Toxicology. New York: Marcel Dekker, 2004.

6. Guba W. Description and representation of chemicals. Predictive Toxicology. New York: Marcel Dekker, 2004.

7. Klopman G. Artificial intelligence approach to structure–activity studies: computer automated structure evaluation of biological activity of organic molecules. J Am Chem Soc 1984; 106:7315–7321.

8. Guyon I, Weston J, Barnhill S, Vapnik V. Gene selection for cancer classification using support vector machines. Machine Learning 2002; 46(1–3):389–422.

9. Reunanen J. Overfitting in making comparisons between variable selection methods. Machine Learning Res 2003; 3:1371–1382.

10. Helma C, Kramer S. A survey of the Predictive Toxicology challenge. Bioinformatics 2003; 19:1179–1182.

11. Benigni R, Giuliani A. Putting the Predictive Toxicology challenge into prespective: reflections on the results. Bioinformatics 2003; 19:1194–1200.

12. Kramer S, Frank E, Helma C. Fragment generation and support vector machines for inducing SARs. SAR QSAR Environ Res 2002; 13:509–523.

13. Mitchell TM. Machine Learning. The McGraw-Hill Companies, Inc., 1997.

14. Helma C, Kramer S, DeRaedt L. The molecular feature miner MolFea. In: Hicks M, Kettner , eds. Proceedings of the Beilstein Workshop 2002: Molecular Informatics: Confronting Complexity. Frankfurt: Beilstein Institut, 2003:79–93.

15. Piatetsky-Shapiro G. Discovery, analysis, and presentation of strong rules. In: Piatetsky-Shapiro G, Frawley WJ, eds. Knowledge Discovery in Databases. Menlo Park, CA: AAAI Press/the MIT Press, 1991:229–248.

16. Gold LS, Zeiger E. Handbook of Carcinogenic Potency and Genotoxicity Databases. CRC Press, 1997.

17. Weininger D. SMILES, a chemical language and information system 1. Introduction and encoding rules. J Chem Inf Comput Sci 1988; 28:31–36.

18. Perotta A, Malacarne D, Taningher M, Pesenti R, Paolucci R, Parodi S. A computerized connectivity approach for analyzing the structural basis of mutagenicity in Salmonella and its relationship with rodent carcinogenicity. Environ Mol Mutagen 1996; 28:31–50.

19. Klopman G, Rosenkranz HS. Testing by artificial intelligence: computational alternatives to the determination of mutagenicity. Mutat Res 1992; 272:59–71.

20. Helma C, Kramer S, DeRaedt L. Data mining and machine learning techniques for the identfication of mutagenicity inducing substructures and structure—Activity relationships of noncongeneric compounds. J Chem Inf Comput Sci 2004; 44:1402–1411.

INTRODUCTORY LITERATURE

1. Witten IH, Frank E. Data Mining. San Francisco, CA: Morgan Kaufmann Publishers, 2000.

2. Mitchell TM. Machine Learning. The McGraw-Hill Companies, Inc., 1997.

GLOSSARY

Aromatic: A planar ring system with delocalized electrons, e.g., a benzene ring.

Congeneric: Chemical structures with a common *substructure.*

Cross-validation: A method for estimating the accuracy of a prediction model. The *training set* is divided into *n* mutually exclusive subsets (the "folds") of approximately equal size. The model is trained and tested *n* times. Each time it is trained on the *training set,* minus a fold and test on that fold. The accuracy estimate is the average accuracy for the *n* folds.

Fragment: A *substructure* of a chemical structure; hydrogens are usually neglected.

Lazy learning: Creation of individual models for each compound in the *test set.*

Linear fragment: A *fragment* without branches or cycles, i.e., a chain of heavy atoms and the connecting bonds.

MolFea: An algorithm for the identification of (in-) frequent *linear fragments* in databases with chemical structures.

Overfitting: The phenomenom that a learning algorithm adapts so well to a *training set,* that the performance on the *test set* suffers. *Cross-validation* techniques approaches have been developed to cope with this problem.

Refinement: The procedure to expand *(linear) fragments* with an additional atom and bond.

***Salmonella* mutagenicity:** Inheritable alterations in the genes of the bacteria *Salmonella typhimurium.* The *Salmonella* mutagenicity test (also known as Ames test) was introduced in the 1970s as a short-term test for carcinogenicity.

SMARTS: A language for the description of molecular patterns and properties, *fragments* can be written in SMARTS notation. See http://www.daylight.com for more information.

SMILES: A notation for chemical structures. See http://www.daylight.com for more information.

Substructure: A part of a (chemical) structure.

Test set: Chemical structures with unknown toxic activities that have to be predicted.

Training set: Data (in our case chemical structures and toxic activities) that are used to generate a prediction model.

Index